Lecture Notes in Mathematics

Edited by J.-M. Morel, F. Takens and B. Teissier

Editorial Policy
for the publication of monographs

1. Lecture Notes aim to report new developments in all areas of mathematics and their applications – quickly, informally and at a high level. Mathematical texts analysing new developments in modelling and numerical simulation are welcome.

   Monograph manuscripts should be reasonably self-contained and rounded off. Thus they may, and often will, present not only results of the author but also related work by other people. They may be based on specialised lecture courses. Furthermore, the manuscripts should provide sufficient motivation, examples and applications. This clearly distinguishes Lecture Notes from journal articles or technical reports which normally are very concise. Articles intended for a journal but too long to be accepted by most journals, usually do not have this „lecture notes" character. For similar reasons it is unusual for doctoral theses to be accepted for the Lecture Notes series, though habilitation theses may be appropriate.

2. Manuscripts should be submitted (preferably in duplicate) either to Springer's mathematics editorial in Heidelberg, or to one of the series editors (with a copy to Springer). In general, manuscripts will be sent out to 2 external referees for evaluation. If a decision cannot yet be reached on the basis of the first 2 reports, further referees may be contacted: The author will be informed of this. A final decision to publish can be made only on the basis of the complete manuscript, however a refereeing process leading to a preliminary decision can be based on a pre-final or incomplete manuscript. The strict minimum amount of material that will be considered should include a detailed outline describing the planned contents of each chapter, a bibliography and several sample chapters.

   Authors should be aware that incomplete or insufficiently close to final manuscripts almost always result in longer refereeing times and nevertheless unclear referees' recommendations, making further refereeing of a final draft necessary.

   Authors should also be aware that parallel submission of their manuscript to another publisher while under consideration for LNM will in general lead to immediate rejection.

3. Manuscripts should in general be submitted in English. Final manuscripts should contain at least 100 pages of mathematical text and should always include

   – a table of contents;
   – an informative introduction, with adequate motivation and perhaps some historical remarks: it should be accessible to a reader not intimately familiar with the topic treated;
   – a subject index: as a rule this is genuinely helpful for the reader.

   For evaluation purposes, manuscripts may be submitted in print or electronic form (print form is still preferred by most referees), in the latter case preferably as pdf- or zipped ps-files. Lecture Notes volumes are, as a rule, printed digitally from the authors' files. To ensure best results, authors are asked to use the LaTeX2e style files available from Springer's web-server at:

   ftp://ftp.springer.de/pub/tex/latex/mathegl/mono/ (for monographs) and

   ftp://ftp.springer.de/pub/tex/latex/mathegl/mult/ (for summer schools/tutorials).

   Additional technical instructions, if necessary, are available on request from lnm@springer-sbm.com.

Continued on inside back-cover

Lecture Notes in Mathematics

1893

Editors:
J.-M. Morel, Cachan
F. Takens, Groningen
B. Teissier, Paris

Heinz Hanßmann

Local and Semi-Local Bifurcations in Hamiltonian Dynamical Systems

Results and Examples

Springer

Author

Heinz Hanßmann
Mathematisch Instituut
Universiteit Utrecht
Postbus 80010
3508 TA Utrecht
The Netherlands

Library of Congress Control Number: 2006931766

Mathematics Subject Classification (2000): 37J20, 37J40, 34C30, 34D30, 37C15, 37G05, 37G10, 37J15, 37J35, 58K05, 58K70, 70E20, 70H08, 70H33, 70K30, 70K43

ISSN print edition: 0075-8434
ISSN electronic edition: 1617-9692
ISBN-10 3-540-38894-x Springer Berlin Heidelberg New York
ISBN-13 978-3-540-38894-4 Springer Berlin Heidelberg New York

DOI 10.1007/3-540-38894-x

Springer is a part of Springer Science+Business Media
springer.com
© Springer-Verlag Berlin Heidelberg 2007

Typesetting by the author and SPi using a Springer LaTeX package
Cover design: WMXDesign GmbH, Heidelberg

Printed on acid-free paper   SPIN: 11841708   VA41/3100/SPi   5 4 3 2 1 0

to my parents

Preface

The present notes are devoted to the study of bifurcations of invariant tori in Hamiltonian systems. Hamiltonian dynamical systems can be used to model frictionless mechanics, in particular celestial mechanics. We are concerned with the nearly integrable context, where Kolmogorov–Arnol'd–Moser (KAM) theory shows that most motions are quasi-periodic whence the (invariant) closure is a torus. An interesting aspect is that we may encounter torus bifurcations of high co-dimension in a single given Hamiltonian system. Historically, bifurcation theory has first been developed for dissipative dynamical systems, where bifurcations occur only under variation of external parameters.

Bifurcations of equilibria and periodic orbits

The structure of any dynamical system is organized by its invariant subsets, the equilibria, periodic orbits, invariant tori and the stable and unstable manifolds of all these. Invariant subsets form the framework of the dynamics, and one is interested in the properties that are persistent under small perturbations.

The most simple invariant subsets are equilibria, points that remain fixed so that no motion takes place at all. Equilibria are isolated in generic systems, be that within the class of Hamiltonian systems or within the class of all dynamical systems. In the latter case the dynamics is dissipative and an equilibrium may attract all motion that starts in a (sufficiently small) neighbourhood.

Such a dynamically stable equilibrium is also structurally stable in that a small perturbation of the dynamical system does not lead to qualitative changes. If we let the system depend on external parameters, then the equilibrium may lose its dynamical stability under parameter variation or cease to exist. A typical example is the $\mathbb{Z}_2$-symmetric pitchfork bifurcation where an attracting equilibrium loses its stability and gives rise to a pair of two attracting equilibria. Other examples are the saddle-node and the Hopf bifurcation. Such bifurcations have been studied extensively in the literature, cf. [129, 173] and references therein.

The dynamics around equilibria in Hamiltonian systems can be more complicated since it is not generic for a Hamiltonian system to have only hyperbolic equilibria. This also influences possible bifurcations, cf. [61, 43]. For instance, in the Hamiltonian counterpart of the above pitchfork bifurcation it is an elliptic (rather than attracting) equilibrium that loses its stability and gives rise to a pair of two elliptic equilibria. In [254, 78] dynamically stable equilibria are studied for which the nearby dynamics nevertheless changes under variation of external parameters.

Periodic orbits form 1-parameter families in Hamiltonian systems, usually parametrised by the value of the energy. In fact, where continuation with respect ot the energy fails a bifurcation[1] takes place, while other bifurcations are triggered by certain resonances between the Floquet multipliers. For more details see [3, 38] and references therein, and also Chapter 3 of the present notes.

Bifurcation from periodic orbits to invariant tori

In (generic) dissipative systems periodic orbits are isolated and one needs again external parameters μ to encounter bifurcations. One of these is the periodic Hopf [154, 155] or Neĭmark–Sacker [252, 14] bifurcation. Under parameter variation a periodic orbit loses stability as a pair of Floquet multipliers passes at $\pm \exp(i\nu T)$ through the unit circle, where T denotes the period. In the supercritical case the stability is transferred to an invariant 2-torus that bifurcates off from the periodic orbit, with two frequencies $\omega_1 \approx 1/T$ and $\omega_2 \approx 2\pi\nu$ coming from the internal and normal frequency of the periodic orbit. The subcritical case involves an unstable 2-torus with these frequencies that shrinks down to the periodic orbit and results in a "hard" loss of stability.

The frequency vector $\omega = (\omega_1, \omega_2)$ that in the above description is rather naively attached to the merging invariant tori exemplifies the problems brought by bifurcations to invariant tori. First of all we need non-resonance conditions $2\pi k/T + \ell\nu \neq 0$ for all $k \in \mathbb{Z}$ and $\ell \in \{1, 2, 3, 4\}$. Where these are violated one speaks of a strong resonance as the Floquet multipliers $\pm \exp(i\nu T) \in \{\pm 1, -\frac{1}{2} \pm \frac{i}{2}\sqrt{3}, \pm i\}$ are ℓth order roots of unity, see [272, 173] for more details. While excluding these low order resonances does lead to an invariant 2-torus bifurcating off from the periodic orbit, the motion on that torus need not be quasi-periodic.

For irrational rotation number ω_1/ω_2 the motion is indeed quasi-periodic and fills the invariant torus densely. In case the quotient ω_1/ω_2 is rational (but

---

[1] For generic Hamiltonian systems this is a periodic centre-saddle bifurcation.

now with denominator $q \geq 5$) we expect phase locking with a finite number of periodic orbits with period $\approx qT$ and all other orbits on the torus heteroclinic between two of these. The invariance (and smoothness) of the torus is guaranteed by normal hyperbolicity, an important property of dissipative systems that does not have the same consequences in the Hamiltonian context.

In the present simple situation it suffices to require that the rotation number ω_1/ω_2 on the invariant torus has non-zero derivative with respect to the bifurcation parameter μ. A more transparent approach is to consider the rotation number as an additional external parameter and it is more convenient to work with both ω_1 and ω_2 as (independent and thus two) additional parameters. In (μ, ω)-space this yields the following description. The bifurcation occurs as μ passes through the bifurcation value $\mu = 0$ and the dynamics on the torus is quasi-periodic except where $\omega = (\omega_0 q, \omega_0 p)$ is a multiple $\omega_0 \in \mathbb{R}$ of an integer vector $(q, p) \in \mathbb{Z}^2$ and thus resonant.

Torus bifurcations in dissipative systems

Bifurcations involving invariant n-tori may similarly be described using external parameters $(\mu, \omega) \in \mathbb{R}^d \times \mathbb{R}^n$. An additional complication is that the flow on an n-torus may be chaotic for $n \geq 3$ and that the torus may be destroyed altogether in the absence of normal hyperbolicity. One therefore excludes resonances $k_1\omega_1 + \ldots + k_n\omega_n = 0$ by means of Diophantine conditions[2]

$$\bigwedge_{k \in \mathbb{Z}^n \setminus \{0\}} |\, k_1\omega_1 + \ldots + k_n\omega_n \,| \geq \frac{\gamma}{|k|^\tau} \tag{0.1}$$

where $\gamma > 0$, $\tau > n - 1$ and $|k| = k_1 + \ldots + k_n$.

A first result along these lines concerns n-tori that bifurcate off from equilibria, cf. [23] and references therein. Here $d = n$ and the parameters μ are used to let n pairs $\mu_j \pm i\omega_j$ pass through the origin $\mu = 0$ in μ-space. This yields quasi-periodic n-tori for ω in the nowhere dense but measure-theoretically large subset of $\mathbb{R}^n$ defined by (0.1), and also quasi-periodic m-tori where only $m < n$ pairs $\mu_j \pm i\omega_j$ have crossed the imaginary axis.

Furthermore there are invariant tori of dimension $l > n$. In the simplest case $n = 2$ this has been proved in [32], establishing a quasi-periodic flow on the resulting 3-tori. The procedure in [24] does yield l-tori for general n, but no information on the flow on these tori.

Normal hyperbolicity yields invariant $(n + 1)$-tori bifurcating off from a family of invariant n-tori in [68, 260, 119]. At the bifurcation the invariant n-tori momentarily lose hyperbolicity and the Diophantine conditions (0.1) are needed. As shown in [33, 34] one can similarly use Diophantine conditions

---

[2] The $\bigwedge$ at the beginning signifies that the inequalities that follow have to hold true for all non-zero integer vectors.

involving the normal frequency at the bifurcation to establish a quasi-periodic flow on the $(n+1)$-tori. The "gaps" left open where the frequency vector is too well approximated by a resonance are then filled by normal hyperbolicity. On this measure-theoretically small but open and dense collection of $(n + 1)$-tori the flow remains unspecified. See also [55, 77] for more details.

Notably, these results require the bifurcating n-tori to be in Floquet form, with normal linearization independent of the position on the torus. The skew Hopf bifurcation where this condition is violated is a generic torus bifurcation that has no counterpart for periodic orbits. As shown in [282, 60, 62, 273] one has also in this case quasi-periodic $(n+1)$-tori bifurcating off from n-tori. The gaps left by the necessary Diophantine conditions are again filled by normal hyperbolicity, but to a lesser extent.

From the period doubling bifurcation [223, 173] of periodic orbits one inherits the frequency halving bifurcation of quasi-periodic tori. Under variation of the external parameter μ an invariant n-torus loses stability as a Floquet multiplier passes at -1 through the unit circle. In the supercritical case the stability is transferred to another n-torus that bifurcates off from the initial family of n-tori with the first[3] frequency divided by 2. The subcritical case involves an unstable n-torus with one frequency halved that meets the initial family and results in a "hard" loss of stability.

This situation is clarified in [34]. As μ passes through the bifurcation value $\mu = 0$ a frequency-halving bifurcation takes place for the Diophantine tori satisfying (0.1). By means of normal hyperbolicity the gaps around resonances $k_1\omega_1 + \ldots + k_n\omega_n = 0$ are filled by invariant tori on which the flow need not be conditionally periodic. This leaves small "bubbles" in the complement of Diophantine tori at and near the bifurcation value where normal hyperbolicity is too weak to enforce invariant tori. In [186, 187] this scenario has been obtained along a subordinate curve in the 2-parameter unfolding of a periodic orbit having simultaneously Floquet multipliers -1 and $\pm\exp(\mathrm{i}\nu T) \notin \{\pm 1, -\frac{1}{2} \pm \frac{\mathrm{i}}{2}\sqrt{3}, \pm\mathrm{i}\}$.

The quasi-periodic saddle-node bifurcation is studied in [65] where it appears subordinate to a periodic orbit undergoing a degenerate periodic Hopf bifurcation. The general theory is (again) given in [34], where it appears as the most difficult of the three quasi-periodic bifurcations inherited from generic bifurcations of periodic orbits. For an extension to the degenerate case see [284, 285].

Bifurcations in Hamiltonian systems

Compared to the above rich theory of torus bifurcations in dissipative dynamical systems, there are few results on conservative systems prior to [139] that I am aware of. In [41, 42, 32] invariant tori of dimension 2 and 3 are established in the universal 1-parameter unfolding of a volume-preserving vector

---

[3] Here a convenient choice of a basis on $\mathbb{T}^n$ is assumed.

field with an equilibrium having eigenvalues $0, \pm i$ or $\pm i\omega_1, \pm i\omega_2$, respectively. In the Hamiltonian case the existence of invariant tori near an elliptic equilibrium is due to the excitation of normal modes and generalizes the Lyapunov centre theorem, see [55] and references therein.

This lack of a bifurcation theory for invariant tori in Hamiltonian systems is all the more surprising as no external parameters are necessary. Indeed, every angular variable on a torus has a conjugate action variable whence n-tori form n-parameter families. The present notes aim to fill this gap in the literature.

In the "integrable" case, when there are sufficiently many symmetries, the situation can be reduced to bifurcations of (relative) equilibria. For this reason we develop the latter theory in a systematic way. From the various families of equilibria one can easily reconstruct the bifurcation scenario of invariant tori in an integrable Hamiltonian system.

While integrable systems have received a lot of attention – not to the least because their dynamics *can* be completely understood – it is highly exceptional for a Hamiltonian system to be integrable. Still, one often takes an integrable system as starting point and studies Hamiltonian perturbations away from integrability. Also if explicitly given a non-integrable Hamiltonian system, one of the few methods available is to look for an integrable approximation, e.g. given by normalization, and to consider the former as a perturbation of the latter. By a dictum of Poincaré the problem of studying the effects of small Hamiltonian perturbations of an integrable system is the fundamental problem of dynamics.

KAM theory is a powerful instrument for the investigation of this problem. It states that most[4] of the quasi-periodic motions constituting the integrable dynamics survive the perturbation, provided that this perturbation is sufficiently (and this means *very*) small. In a more geometric language these motions correspond to invariant tori. Under Kolmogorov's non-degeneracy condition one may consider the (internal) frequencies as parameters, and the Diophantine conditions (0.1) bounding the latter away from resonances lead to the Cantor families of tori one is confronted with in the perturbed system.

In its first formulation KAM theory addressed the "maximal" tori, and only later generalizations were formulated and proven for families of invariant tori that derive from hyperbolic and/or elliptic equilibria. For an overview over this still active research area see [55]. The present notes further generalize these results to families of invariant tori that lose (or gain) hyperbolicity during a bifurcation. Such bifurcations are governed by the nonlinear terms of the vector field. In this way singularity theory both governs the bifurcation scenario and helps deciding how these nonlinear terms are dealt with during the KAM-iteration procedure. As a result, the various smooth families

---

[4] The relative measure of those parametrising internal frequencies for which the torus is destroyed vanishes as the size of the perturbation tends to zero.

of invariant tori of the integrable system get replaced by Cantor families of invariant tori organizing the perturbed dynamics.

Acknowledgements

These notes derive from my habilitation thesis [142]. It is my pleasure and privilege to thank those who helped me in one way or another during the past years. First I thank Volker Enß, Hans Duistermaat and Robert MacKay for reading [142]. Furthermore I like to thank Mohamed Barakat, Larry Bates, Giancarlo Benettin, Henk Broer, Alain Chenciner, Gunther Cornelissen, Richard Cushman, Holger Dullin, Scott Dumas, Francesco Fassò, Sebastián Ferrer, Giovanni Gallavotti, Phil Holmes, Jun Hoo, Igor Hoveijn, Bert Jongen, Àngel Jorba, Wilberd van der Kallen, Jeroen Lamb, Naomi Leonard, Anna Litvak Hinenzon, Eduard Looijenga, Jan-Cees van der Meer, James Montaldi, Martijn van Noort, Jesús Palacián, Jürgen Pöschel, Tudor Raţiu, Mark Roberts, Jürgen Scheurle, Michail Sevryuk, Carles Simó, Troy Smith, Britta Sommer, Floris Takens, Ferdinand Verhulst, Jordi Villanueva, Florian Wagener, Patricia Yanguas and Jiangong You. Finally I thank the reviewers for their detailed comments.

My research was helped by the kindness of several institutions to support me. I thank both the Deutsche Forschungsgemeinschaft and the Max Kade Foundation for their grants that allowed me to stay for an extended period of time at the Universitá di Padova and Princeton University, respectively. During the past years I furthermore strongly benefitted from the activities of the European research and training network *Mechanics And Symmetry In Europe*. I also thank the Stichting Fondamenteel Onderzoek der Materie and the Alexander von Humboldt Stiftung for financial aid. Last but not least I wish to acknowledge the support of the two Aachen Graduiertenkollegs *Analyse und Konstruktion in der Mathematik* and *Hierarchie und Symmetrie in mathematischen Modellen* and their Sprecher Volker Enß and Gerhard Hiß.

Utrecht, May 2006   *Heinz Hanßmann*

Contents

The sequential order in these notes is from equilibria to invariant tori, which means that the various types of bifurcations appear and re-appear in different chapters. Therefore the following overview on the main Hamiltonian bifurcations of co-dimension one might be helpful.

bifurcation	for equilibria	periodic orbits	quasi-periodic
centre-saddle	Examples 2.6, 2.19, 2.22 and 2.23	Theorem 3.1 Example 3.3	Corollary 4.2 Theorem 4.4 Examples 4.5 and 4.8
Hamiltonian flip period-doubling frequency halving	Theorem 2.17	Theorem 3.4	Theorem 4.18
Hamiltonian Hopf	Theorem 2.20 Theorem 2.25	Theorem 3.7 Example 3.8 Example 4.5	Theorem 4.27
Hamiltonian pitchfork	Example 2.6 Example 2.22	Theorem 3.6	Theorem 4.13 Example 4.15

1

Introduction

Dynamical systems describe the time evolution of the various states $z \in \mathcal{P}$ in a given state space. When this description includes both (the complete) past and future this leads to a *group action*[1]

$$
\begin{aligned}
\varphi : \quad \mathbb{R} \times \mathcal{P} &\longrightarrow \mathcal{P} \\
(t, z) &\longmapsto \varphi_t(z)
\end{aligned}
$$

of the time axis $\mathbb{R}$ on $\mathcal{P}$, i.e. $\varphi_0 = \text{id}$ (the present) and $\varphi_s \circ \varphi_t = \varphi_{s+t}$ for all times $s, t \in \mathbb{R}$. Immediate consequences are $\varphi_s \circ \varphi_t = \varphi_t \circ \varphi_s$ and $\varphi_t^{-1} = \varphi_{-t}$. In case φ is differentiable one can define the vector field

$$
X(z) = \left. \frac{\mathrm{d}}{\mathrm{d}t} \varphi_t(z) \right|_{t=0}
$$

on $\mathcal{P}$ and if e.g. $\mathcal{P}$ is a differentiable manifold then φ can be reconstructed from X as its flow. Note that

$$
\dot{z} = X(z) \tag{1.1}
$$

defines an autonomous ordinary differential equation on $\mathcal{P}$.

Given a state $z \in \mathcal{P}$ the set $\{ \varphi_t(z) \mid t \in \mathbb{R} \}$ is called the orbit of z. Particularly simple orbits are equilibria, $\varphi_t(z) = z$ for all $t \in \mathbb{R}$, and periodic orbits which satisfy $\varphi_T(z) = z$ for some period $T > 0$ and hence $\varphi_{t+T}(z) = \varphi_t(z)$ for all $t \in \mathbb{R}$. All other orbits define injective immersions $t \mapsto \varphi_t(z)$ of $\mathbb{R}$ in $\mathcal{P}$. By definition unions of orbits form sets $M \subseteq \mathcal{P}$ that are invariant under φ, and if M is a differentiable manifold we call M an invariant manifold.

A complete understanding of a dynamical system φ is equivalent to finding (and understanding) all solutions of (1.1) whence one often concentrates on the long time behaviour as $t \to \pm\infty$. One approach is to determine all attractors[2]

---

[1] *Technical terms* are explained in a glossary preceeding the references.

[2] Since there are no attractors in Hamiltonian dynamical systems we do not give a formal definition.

in $\mathcal{P}$, compact invariant subsets A satisfying $\varphi_t(z) \stackrel{t \to +\infty}{\longrightarrow} A$ for all z near A, that are minimal with this property. Such attractors can be equilibria, periodic orbits, invariant manifolds, or even more general invariant sets. If A is an invariant manifold without equilibrium, then the Euler characteristic of A vanishes and the simplest such manifolds are the n-tori T, submanifolds of $\mathcal{P}$ that are diffeomorphic to $\mathbb{T}^n = \mathbb{R}^n/_{\mathbb{Z}^n}$. Where we speak of n-tori we always assume $n \geq 2$ in these notes.

The flow φ on a torus T is parallel or *conditionally periodic* if there is a global chart

$$\begin{array}{ccc} \mathsf{T} & \longrightarrow & \mathbb{T}^n \\ z & \mapsto & x \end{array}$$

and a frequency vector $\omega \in \mathbb{R}^n$ such that[3]

$$\bigwedge_{x \in \mathbb{T}^n} \bigwedge_{t \in \mathbb{R}} \varphi_t(x) = x + \omega t \ .$$

In case there are no resonances $\langle k, \omega \rangle = 0$, $k \in \mathbb{Z}^n$ every orbit on T is dense. If there are $n-1$ independent resonances then ω is a multiple of an integer vector and all orbits on T are periodic. For $m \leq n - 2$ independent resonances the motion is *quasi-periodic* and spins densely around invariant $(n - m)$-tori into which T decomposes. The flow on a given invariant torus may be much more complicated, this is often accompanied by a loss of differentiability. However, if the flow is equivariant with respect to the $\mathbb{T}^n$-action $x \mapsto x + \xi$ then all motions are necessarily conditionally periodic. Our starting point is therefore a family of tori carrying parallel flow, and we hope for persistence under small perturbations for the measure-theoretically large subfamily where the frequency vector satisfies a strong non-resonance condition.

Considering the long time behaviour for $t \to -\infty$ attractors are replaced by repellors and more generally one is interested in "minimal" invariant sets M. Where the dynamics on M itself is understood – for equilibria, periodic orbits and invariant tori with conditionally periodic flow – one concentrates on the dynamics nearby. Equilibria and periodic orbits are (under quite weak conditions) *structurally stable* with respect to small perturbations of the dynamical system, while invariant tori and more complicated, strange invariant sets may disintegrate. This makes it preferable to study parametrised families of such invariant sets.

In applications the equations of motion are known only to finite precision of the coefficients. Giving these coefficients the interpretation of parameters leads to a whole family of dynamical systems. Under variation of the parameters the invariant sets may then bifurcate. Bifurcations of equilibria are fairly well understood, at least for low co-dimension, cf. [129, 173] and references therein. Since these bifurcations concern a small neighbourhood of the equilibrium, we speak of *local bifurcations*. Using a *Poincaré mapping*, periodic orbits can be

---

[3] We use the same letter φ for the flow in the chart as well.

studied as fixed points of a discrete dynamical system. In addition to the analogues of bifurcations of equilibria, periodic orbits may undergo period doubling bifurcations, cf. [223, 58].

For a family of invariant n-tori with conditionally periodic flow the frequency vector ω varies in general with the parameter; let us therefore now consider $\omega \in \mathbb{R}^n$ itself as the parameter. Clearly both the resonant and the non-resonant tori are dense in the family. Under an arbitrary small perturbation (breaking the $\mathbb{T}^n$-symmetry that forces the toral flows to be conditionally periodic) the situation changes drastically. Using KAM-techniques one can formulate conditions under which most invariant tori survive the perturbation, together with their quasi-periodic flow; the families of tori are parametrised over a Cantor set of large n-dimensional (Hausdorff)-measure, see [159, 56, 55]. Within the gaps of the Cantor set completely new dynamical phenomena emerge; the dynamics on the torus may cease to be conditionally periodic[4] even in case there are circumstances like *normal hyperbolicity* that force the torus to persist. Note that the union of the gaps of a Cantor set is open and dense in $\mathbb{R}^n$. This is an exemplary instance of coexisting complementary sets, one of which is measure-theoretically large and the other topologically large, cf. [231].

It turns out that the bifurcations of equilibria and periodic orbits have quasi-periodic counterparts, see [34, 284] and references therein. In the integrable case where the perturbation respects the $\mathbb{T}^n$-action this is an immediate consequence of the behaviour of the reduced system obtained after reducing the torus symmetry. In the nearly integrable case where the torus symmetry is broken by a small perturbation one can use KAM theory to show that the bifurcation persists on Cantor sets. Notably the bifurcating torus has to be in *Floquet form*. In the same way the higher topological complexity of periodic orbits leads to period doubling bifurcations, tori that are not in Floquet form can bifurcate in a skew Hopf bifurcation, see [282, 60].

Bifurcations of invariant tori have a *semi-local* character, they concern a neighbourhood of the invariant torus which need not be confined to a small region of $\mathcal{P}$. Exceptions are bifurcations subordinate to local bifurcations and these were in fact the motivating examples for the above results. In contrast, global bifurcations lead to new interactions of different parts of $\mathcal{P}$ not present before or after the bifurcation. Examples are *connection bifurcations* involving heteroclinic orbits (these also exist subordinate to local or semi-local bifurcations).

The quasi-periodic persistence results in [159, 56, 55] are formulated and proven in terms of Lie algebras of vector fields and this allowed for a generalization to volume-preserving, Hamiltonian and reversible dynamical systems,

---

[4] For instance, if $\omega \in \omega_0 \cdot \mathbb{Z}^n$ only finitely many periodic orbits are expected to survive and the perturbed flow may consist of asymptotic motions between these. The structural stability of surviving periodic orbits is in turn the reason why a simple resonant frequency vector opens a whole gap of the Cantor set.

see also [216]. We will henceforth speak of dissipative systems when there is no such structure preserved. A dynamical system is Hamiltonian if the vector fields derives[5] from a single "Hamiltonian" function by means of a *Poisson structure*, a bilinear and alternating composition on $\mathcal{A} \subseteq C(\mathcal{P})$ that satisfies the Jacobi identity and Leibniz' rule. An important feature of integrable Hamiltonian systems is that the torus symmetry yields conjugate actions by Noether's theorem. Accordingly, invariant n-tori in integrable Hamiltonian systems with d degrees of freedom, $d \geq n$, occur as "intrinsic" n-parameter families, without the need for external parameters.

In particular, periodic orbits form 1-parameter families, or 2-dimensional cylinders (while equilibria remain in general as isolated as in the dissipative case). Thus, periodic orbits in (single) integrable Hamiltonian systems may undergo co-dimension one bifurcations, without the need of an external parameter. The ensuing possibilities were analysed in [205, 207], see also [208, 38, 232, 227, 228]. This yields transparent explanations for common phenomena like the gyroscopic stabilization of a sleeping top, cf. [13, 84, 81, 147].

Interestingly, results on bifurcations of invariant n-tori (which form n-parameter families in a Hamiltonian system) were first derived in the dissipative context (where external parameters are needed), see again [34] and references therein. Our aim is to detail the Hamiltonian part of the theory, extending the results in [139, 50] to more general bifurcations. At the same time we seize the occasion to put the well-known results on Hamiltonian bifurcations of equilibria, which are scattered throughout the literature, into a systematic framework. See also [75, 76, 45, 44] for recent progress concerning torus bifurcations in the reversible context.

1.1 Hamiltonian systems

A Hamiltonian system is defined by a Hamiltonian function on a phase space. The latter is a *symplectic manifold*, or, more generally a *Poisson space*, where the Hamiltonian H determines the vector field

$$X_H \; : \; \dot{z} \; = \; \{z, H\} \; .$$

If all solutions of X_H exist for all times, the flow φ^H is a group action

$$\varphi^H \; : \; \begin{aligned} \mathbb{R} \times \mathcal{P} &\longrightarrow & \mathcal{P} \\ (t, z) &\longmapsto & \varphi_t^H(z) \end{aligned} \tag{1.2}$$

on the phase space $\mathcal{P}$ – in case there are orbits that leave $\mathcal{P}$ in finite time (1.2) is only a *local group action*.

Despite this simple construction where a single real valued function defines a whole vector field, the study of Hamiltonian systems is a highly non-trivial

---

[5] Similar to gradient vector fields defined by means of a Riemannian structure.

task. The first systems that were successfully treated were integrable and the study of Hamiltonian systems still starts with the search for the integrals of motion. Since $\{H, H\} = 0$ the Hamiltonian is always[6] an integral of motion, whence all systems with one *degree of freedom* are integrable.

However, already in two degrees of freedom integrable systems are the exception rather than the rule, cf. [239, 117, 26]. This led to the so-called *ergodic* hypothesis that the flow of a Hamiltonian system is "in general" ergodic on the *energy shell*. That this hypothesis does not hold for *generic* Hamiltonian systems, see [191], is one of the consequences of KAM theory.

KAM theory deals with small perturbations of integrable systems and may in fact be thought of as a theory on the integrable systems themselves. Indeed, in applications the special circumstances that render a Hamiltonian system integrable may not be satisfied with absolute precision and only properties that remain valid under the ensuing small perturbations have physical relevance.

An integrable Hamiltonian system with, say, compact energy shells gives the phase space $\mathcal{P}$ the structure of a *ramified torus bundle*. The regular fibres of this bundle are the maximal invariant tori of the system. The singular fibres define a whole hierarchy of lower dimensional tori, in case of (dynamically) unstable tori together with their *(un)stable manifolds*. In this way there are two types of "least degenerate" singular fibres: the *elliptic* tori with one *normal frequency* and the *hyperbolic* tori T with stable and unstable manifolds of the form $\mathsf{T} \times \mathbb{R}$. These two types of singular fibres determine the distribution of the regular fibres. Different families of maximal tori are separated by (un)stable manifolds of hyperbolic tori and may shrink down to elliptic tori.

On the next level of the hierarchy of singular fibres of the ramified torus bundle we can distinguish four or five different types. Lowering the dimension of the torus once more we are led to elliptic tori with two normal frequencies, to *hypo-elliptic* tori and to hyperbolic tori with four *Floquet exponents*. For these latter we might want to distinguish between the focus-focus case of a quartet $\pm\mathfrak{R}\pm i\mathfrak{S}$ of complex exponents and the saddle-saddle case of two pairs of real exponents. This decision would relegate hyperbolic tori with a double pair of real exponents to the next level of the hierarchy of singular fibres. We can do the same with elliptic tori with two resonant normal frequencies. Where the two normal frequencies are in $1:-1$ resonance, the torus may undergo a quasi-periodic *Hamiltonian Hopf bifurcation* and we *always* relegate these elliptic tori to the third level of the hierarchy of singular fibres of the ramified torus bundle.

The last type of second level singular fibres consists of invariant tori (and their (un)stable manifolds) of the same dimension as the first level tori, but with *parabolic* normal behaviour. Such tori may for instance undergo a quasi-periodic *centre-saddle bifurcation*. We see that the kth level singular fibres determine the distribution of the $(k-1)$th level singular fibres (where we could abuse language and address the regular fibres as 0th level singular fibres).

---

[6] Our Hamiltonians are autonomous, there is no explicit time dependence.

Notably all invariant n-tori of the ramified torus bundle are *isotropic*, having a (commuting) set $y_1, \ldots, y_n$ of actions conjugate to the toral angles. Locally these may be used to parametrise the various families of n-tori. There is a branch of KAM theory that explores non-isotropic (in particular *co-isotropic*) invariant tori. In such a situation, the symplectic structure is necessarily non-exact and it is moreover the symplectic structure that should satisfy certain non-resonance conditions. For more information see [262] and references therein.

The aim of KAM theory is to study the fate of this ramified torus bundle under small perturbations of the integrable Hamiltonian system. Traditionally, this has been done on phase spaces that are symplectic manifolds where the perturbation of the phase space may be neglected and only the Hamiltonian gets perturbed (but see also [175]). Furthermore, a non-degeneracy condition forces the maximal tori to be *Lagrangean*, whence their dimension equals the number d of degrees of freedom. Consequently, for *superintegrable systems*[7] one uses part of the perturbation to construct from the unperturbed ramified torus bundle a non-degenerate ramified torus bundle, see [6, 196, 268, 116].

Persistence of Lagrangean tori under small perturbations was first proven in [166] under the condition that the (internal) frequencies satisfy *Diophantine conditions*, a strong form of non-resonance. This allows to solve the "homological equation" at every step of an iteration scheme, the convergence of which is ensured by the superlinear convergence of a Newton-like approximation. This set-up was modified in [5], restricting to only finitely many resonances in the homological equation by means of an *ultraviolet cut-off* (which is in turn increased at every iteration step). This allowed to successfully treat perturbations of superintegrable systems that *remove the degeneracy* in [6].

The above results were obtained for analytic Hamiltonians. In an attempt to verify the statement of [166] the validity was extended in [215] to Hamiltonians that are only finitely often differentiable. Subsequently the necessary order of differentiability could be brought down in [250]. A lower bound was provided by a counterexample in [270], sharper bounds are discussed in [109]. The machinery of the KAM iteration was condensed in [298, 299] to abstract theorems. In [121, 71, 108] convergence of the KAM iteration scheme was directly proven, without the need for a Newton-like approximation.

While (Lebesgue)-almost all frequency vectors are non-resonant, the complement of Diophantine frequency vectors is an open and dense set. Still, the relative measure of Diophantine frequency vectors is close to 1. In [72, 240] the local structure of persistent tori was shown to inherit the Cantor-like structure of Diophantine frequency vectors. The local *conjugacies* that relate the persistent tori to their unperturbed counterparts are patched together in [46] to form a global conjugacy. This should allow to recover the geometry of the bundle of maximal tori in the perturbed system.

---

[7] In the literature these are also called properly degenerate systems.

The first proof of persistence of elliptic tori in [216] only addressed the case of a single normal frequency. A more general result had already been announced in [204], but proofs appeared much later; see [55] for an extensive bibliography. In case of hyperbolic tori one can always resort to a centre manifold, cf. [211, 160], although this generally results in finite differentiability. For a direct approach see [249] and references therein. Hypo-elliptic tori can either be treated directly, cf. [159, 56, 251], or by first getting rid of the hyperbolic part by means of a centre manifold. As pointed out in [162, 55, 279, 163] the latter approach may yield additional tori that are not in Floquet form.

Parabolic tori are generically involved in quasi-periodic bifurcations and may in particular cease to exist. Correspondingly, one cannot expect persistence of the "isolated" family of parabolic tori; but the whole bifurcation scenario has a chance to persist, in this way including the bifurcating (parabolic) tori. A first such persistence result appeared in [139], which was generalized in [50] to all parabolic tori one can generically encounter in Hamiltonian systems with finitely many degrees of freedom. Additional hyperbolicity may again be dealt with by means of a centre manifold, while additional normal frequencies can be successfully carried through the KAM iteration scheme, cf. [296].

KAM theory does not predict the fate of close-to-resonant tori under perturbations. For fully resonant tori the phenomenon of frequency locking leads to the destruction of the torus under (sufficiently rich) perturbations, and other resonant tori disintegrate as well. In two degrees of freedom surviving 2-tori form barriers on the 3-dimensional energy shells, from which one can infer that all motions are bounded, cf. [222]. Where the system has three or more degrees of freedom there is no such obstruction to orbits connecting distant points of the phase space. The existence of this kind of orbits has been termed *Arnol'd diffusion*, for an up-to-date discussion see [91] and references therein.

While KAM theory concerns the fate of "most" trajectories and for all times, a complementary theorem has been obtained in [220, 221, 226]. It concerns all trajectories and states that they stay close to the unperturbed tori for *long* times that are exponential in the inverse of the perturbation strength. Here a form of smoothness exceeding the mere existence of ∞ many derivatives of the Hamiltonian is a necessary ingredient, for finitely differentiable Hamiltonians one only obtains polynomial times. Most results in this direction are formulated for analytic Hamiltonians, in [190] the neccessary regularity assumptions have been lowered to Gevrey Hamiltonians. For trajectories starting close to surviving tori the diffusion is even superexponentially slow, cf. [213, 214].

A new type of invariant sets, not present in integrable systems, is constructed for generic Hamiltonian systems in [192, 203], using a construction from [25]. Starting point is an elliptic periodic orbit around which another elliptic periodic orbit encircling the former is shown to exist. Iterating this procedure yields a whole sequence of elliptic periodic orbits which converges to a *solenoid*. The construction in [25, 192] not only yields the existence of one solenoid near a given elliptic periodic orbit, but the simultaneous existence of representatives of all homeomorphy-classes of solenoids.

Hyperbolic tori form the core of a construction proposed in [7] of trajectories that venture off to distant points of the phase space. The key ingredience are resonant tori that disintegrate under perturbation leading to lower dimensional hyperbolic tori, cf. [275, 276]. In the unperturbed system the union of a family of hyperbolic tori, parametrised by the actions conjugate to the toral angles, forms a *normally hyperbolic manifold*. The latter is persistent under perturbations, cf. [151, 211], and carries again a Hamiltonian flow, with fewer degrees of freedom.

Perturbed resonant lower dimensional tori that bifurcate according to a quasi-periodic *Hamiltonian pitchfork bifurcation* are studied in [180, 178, 181, 182]. Such parabolic resonances (PR) exhibit large dynamical instabilities. This effect can be significantly amplified by increasing the number of degrees of freedom. This is not only due to multiple resonances (m-PR), but can also be induced by an additional vanishing derivative of the unperturbed Hamiltonian at the parabolic torus for so-called tangent (or 1-flat) parabolic resonances. This latter condition makes a larger part of the energy shell accessible in the perturbed system. In high degrees of freedom, combinations like l-flat m-PR become a common phenomenon as well.

1.1.1 Symmetry reduction

To fix thoughts, let the phase space $\mathcal{P}$ be a symplectic manifold of dimension $2(n + 1)$, on which a locally free symplectic n-torus action

$$\tau : \mathbb{T}^n \times \mathcal{P} \longrightarrow \mathcal{P}$$

is given. Reduction then leads to a one-degree-of-freedom problem. If the action τ is free then the symmetry reduction is regular, cf. [206, 194, 3], and the reduced phase space is a (2-dimensional) symplectic manifold.

Singularities of the reduced phase space are related to points with nontrivial isotropy group $\mathbb{T}^n_z$, cf. [4, 265, 230]. Note that all points in the orbit

$$\mathbb{T}^n(z) = \left\{ \tau_\xi(z) \in \mathcal{P} \;\middle|\; \xi \in \mathbb{T}^n \right\}$$

have, up to conjugation, that same isotropy group, which can be given the form

$$\mathbb{T}^n_z \cong \mathbb{Z}_{k_1} \times \ldots \times \mathbb{Z}_{k_n}$$

with $k \in \mathbb{N}^n$. Thus, if we pass to a $(k_1, \ldots, k_n)$-fold covering of $\mathcal{P}$, the action τ becomes a free[8] action and regular reduction can again be applied.

---

[8] Strictly speaking this is only true locally around the lift of the torus $\mathbb{T}^n(z)$.

On the covering space the isotropy group $\mathbb{T}^n_z$ acts as the group of deck transformations, fixing the lift of $\mathbb{T}^n(z)$. This induces a symplectic $\mathbb{T}^n_z$-action on the reduced phase space, which we locally identify with $\mathbb{R}^2$. Here the origin is the image of the lift of $\mathbb{T}^n(z)$ under the reduction mapping and, by Bochner's theorem we may assume that $\mathbb{T}^n_z$ acts linearly on $\mathbb{R}^2$. Recall that the only finite subgroups of $SL_2(\mathbb{R})$ are the cyclic groups $\mathbb{Z}_\ell$. This yields an epimorphism from the deck group onto $\mathbb{Z}_\ell$, the kernel of which we denote by N.

Identifying all points on the $(k_1, \dots, k_n)$-fold covering space of $\mathcal{P}$ that are mapped to each other by elements of N we pass to an ℓ-fold covering of $\mathcal{P}$. This has no influence on the reduced phase space $\mathbb{R}^2$, in particular the image of the lift of $\mathbb{T}^n(z)$ remains a regular point. In this way the action of the deck group $\mathbb{Z}_\ell = \mathbb{T}^n_z/N$ on $\mathbb{R}^2$ becomes faithful.

Only if we go further and also identify points within the $\mathbb{Z}_\ell$-orbit on the ℓ-fold covering space do we introduce a singularity on the reduced phase space. In particular, if we reduce the n-torus action τ directly on $\mathcal{P}$ we are led to a singularity of type $\mathbb{R}^2/\mathbb{Z}_\ell$ of the reduced phase space. This has been used in [49] to study n-tori with a normal-internal resonance; the necessary action τ was introduced by means of normalization.

1.1.2 Distinguished Parameters

Torus bifurcations occur in families of invariant tori, and the necessary parameters enter Hamiltonian systems in various fashions. This leads to a hierarchical structure where some parameters are *distinguished* with respect to others. To explain the basic mechanism let us start with a family of Hamiltonian systems that depends on an external parameter α. Then co-ordinate transformations $z \mapsto \tilde{z}$ on the phase space[9] $\mathcal{P}$ may clearly depend on the parameter α, while re-parametrisations $\alpha \mapsto \tilde{\alpha}$ are not allowed to depend on the phase space variable z. This ensures that after re-parametrisation and co-ordinate transformation the distinction between phase space variables $\tilde{z}$ and external parameters $\tilde{\alpha}$ remains valid.

Let the Hamiltonian system now be symmetric with respect to a symplectic action of a compact Lie group G. According to Noether's theorem every (continuous) symmetry induces a conserved quantity. If we divide out the group action, then the latter become Casimirs. Hence, we can treat their value μ as a parameter the reduced system depends upon. In the hierarchy the place of μ is "between" the external parameter α and the variable ζ on the reduced phase space. Indeed, while co-ordinate transformations $\zeta \mapsto \tilde{\zeta}$ now may depend on both α and μ, re-parametrisations $\alpha \mapsto \tilde{\alpha}$ are not allowed to depend on either ζ or μ – recall that (μ, ζ) constitute together with the reduced variable along the orbit of the Lie group G the "original" variable z on the phase space $\mathcal{P}$. While a re-parametrisation $\mu \mapsto \tilde{\mu}$ may (still) depend on the external parameter α, there should be no dependence on ζ. We say

---

[9] For simplicity we let the phase space be the same for all parameter values α.

that the (internal) parameter μ is *distinguished* with respect to the (external) parameter α, cf. [288]. If the reduction of the G-action is not regular, but a singular reduction, then the re-parametrisation $\mu \mapsto \tilde{\mu}(\alpha, \mu)$ has to be restricted to preserve the singular values, cf. [43]. A typical example is that μ is the value of angular momenta and the restriction $\tilde{\mu}(\alpha, 0) = 0 \; \forall_\alpha$ imposes that the zero level be preserved.

In applications the existing symmetries often do not suffice to render the system integrable. A possible approach is then to introduce additional symmetries by means of a normal form. After a co-ordinate transformation the Hamiltonian is split into an integrable part and a small perturbation. The first step then is to understand the dynamics defined by the integrable part of the Hamiltonian.

Typically the additional symmetry introduced by normalization is a torus symmetry. Dividing out the group action turns the actions conjugate to the toral angles into Casimirs, the value I of which again plays the rôle of parameter. Clearly I is distinguished with respect to α and a re-parametrisation $I \mapsto \tilde{I}$ should not depend on the variable of the twice reduced phase space. But we also want I to be distinguished with respect to μ, i.e. our parameter changes should be of the form

$$(\alpha, \mu, I) \;\mapsto\; \left(\tilde{\alpha}(\alpha), \tilde{\mu}(\alpha, \mu), \tilde{I}(\alpha, \mu, I) \right) . \tag{1.3}$$

In this way the new $\tilde{I}$ is still the value of the momentum mapping of the approximate symmetry, and when adding the small perturbation to the integrable part of the normal form the perturbation analysis may be performed for fixed $\tilde{\alpha}$ and $\tilde{\mu}$. Where the symmetry reduction is singular the re-parametrisation (1.3) should preserve the singular values.

Our aim is to understand what happens to the ramified torus bundle defined by a single integrable Hamiltonian system under generic perturbations. In that setting there are no external parameters, and the perturbation does not leave part of the symmetry of the unperturbed system intact. However, in applications one easily encounters simultaneously two or even all three hierarchical levels of parameters. This leads to changes in the unfolding properties, cf. [288, 43, 188, 53]. Nevertheless, the starting point for such modifications would be a theory with a single class of parameters.

1.2 Outline

Bifurcations of invariant tori are to a large extent governed by their normal dynamics. In the following *Chapter 2* we therefore study bifurcations of equilibria in their own right. To this end we let the system depend on external parameters.

We first concentrate on bifurcating equilibria in Hamiltonian systems with one degree of freedom. This is indeed the situation one is led to when studying bifurcations of invariant n-tori in $n + 1$ degrees of freedom. In one degree of freedom the symplectic form becomes an area form, the Hamiltonian is a planar function and the equilibria correspond to planar *singularities*. Morse singularities lead to *centres* and *saddles*. Local bifurcations are in turn governed by unstable singularities and their *universal unfoldings*.

Next to the simple planar singularities, which form two infinite series $(A_k)_{k \geq 1}$, $(D_k)_{k \geq 4}$ and a finite series E_6, E_7, E_8, there are various series of planar singularities with *moduli*. In *Chapter 2* we address the latter only sporadically and leave a more systematic approach to *Appendices A and B*. It turns out that the moduli of planar singularities do not lead to moduli of bifurcations of equilibria in Hamiltonian systems with one degree of freedom.

Motivated by the reduction of the toral symmetry τ in Section 1.1.1 we also study bifurcations of equilibria at singular points of 2-dimensional Poisson spaces. There are two possibilities. Similar to bifurcations of regular equilibria the Hamiltonian may change under parameter variation. Alternatively, the bifurcation may be triggered by local changes of the phase space, e.g. leading to a singular point when the parameter attains the bifurcation value. In multiparameter systems there may also be combinations of these two mechanisms.

Next to the cyclic symmetry groups $\mathbb{Z}_\ell$ which lead to singular phase spaces there are other (discrete) symmetries of one-degree-of-freedom systems, sometimes reversing. The main example for the latter is the reflection

$$(q, p) \quad \mapsto \quad (q, -p) \ .$$

Such symmetries strongly influence the bifurcations that degenerate equilibria can undergo. The ensuing possibilities are detailed in *Chapter 2* as well.

The local bifurcations of one-degree-of-freedom systems can occur in more degrees of freedom as well. Indeed, for an equilibrium in d degrees of freedom that has a linearization with $2d - 2$ eigenvalues off the imaginary axis this hyperbolic part can be dealt with by means of a centre manifold. The flow on the latter is that of a one-degree-of-freedom Hamiltonian system, and the equilibrium undergoes one-degree-of-freedom bifurcations where the remaining 2 eigenvalues vanish. Where a zero eigenvalue with (algebraic) multiplicity 2 coexists with further purely imaginary pairs of eigenvalues the situation is much more complicated, cf. [43, 122].

We focus on two degrees of freedom and also content ourselves with bifurcations of regular equilibria, leaving aside a systematic study of local bifurcations of singular points in two (or more) degrees of freedom. In fact, already a complete understanding of co-dimension 2 bifurcations of regular equilibria in two degrees of freedom is beyond our present possibilities.

A new phenomenon in two degrees of freedom is that one may have two pairs of purely imaginary eigenvalues in resonance. The most important of these is the 1:−1 resonance. In generic 1-parameter families this resonance triggers a Hamiltonian Hopf bifurcation. The double pair of imaginary eigenvalues leads to an S^1-symmetry, and reduction yields a one-degree-of-freedom problem where the bifurcating equilibrium is a singular point of the phase space.

Here and also for other resonant equilibria normalization is an important tool. This procedure allows to "push a toral symmetry through the Taylor series" whence the system can be approximated by the integrable part of a normal form. For the convenience of the reader this well-known method is recalled in *Appendix C.*

In families of two-degree-of-freedom systems with at least 2 parameters one may encounter equilibria with nilpotent linearization. In case the system has an S^1-symmetry one can again reduce to one degree of freedom and can proceed as for the $1:-1$ resonance. However, in the absence of symmetry the phenomena become much more complicated. For instance, all forms of resonant equilibria occur in an unfolding of nilpotent equilibria.

In *Chapter 3* we consider bifurcations of periodic orbits. Here the *Floquet multipliers* play a rôle similar to that of the eigenvalues of the linearization of an equilibrium. One Floquet multiplier is always equal to 1 as it corresponds to the direction tangential to the periodic orbit. All other multipliers are in 1-1 correspondence with the eigenvalues of (the linearization of) the Poincaré mapping. In the present case of Hamiltonian systems one of these eigenvalues is equal to 1. The (generalized) eigenvector of this Floquet multiplier spans the direction conjugate to that of the "first" multiplier 1. Correspondingly, periodic orbits of Hamiltonian systems form 1-parameter families. Occurring bifurcations are determined by the distribution of the remaining Floquet multipliers.

In contrast to our treatment of equilibria we concentrate on a single Hamiltonian system, without dependence on external parameters. Therefore, the bifurcations of periodic orbits we encounter are of co-dimension 1. In this way we recover the well-known three types of bifurcations triggered by an additional double Floquet multiplier 1, by a double Floquet multiplier -1 and by a double pair of Floquet multpliers on the unit[10] circle. These are the periodic centre-saddle bifurcation, the *period-doubling bifurcation* and the periodic Hamiltonian Hopf bifurcation, respectively.

For all these bifurcations the key information is already contained in the behaviour of the corresponding bifurcation of equilibria. For the period-doubling bifurcation this is the *Hamiltonian flip bifurcation* treated in Section 2.1.2 in which a singular equilibrium loses its stability. Since we use a similar strategy for bifurcations of invariant tori the reasons that allow to carry the bifurcations of equilibria over to bifurcations of periodic orbits are presented in detail, although the results on periodic orbits themselves are well documented in the existing literature, cf. [208, 38] and references therein. Specifically, in [38] also multiparameter bifurcations with one distinguished parameter are considered; this allows to understand bifurcations of periodic orbits in families of Hamiltonian systems.

---

[10] This double pair is different from 1 or -1.

Invariant tori and their bifurcations are then studied in *Chapter 4*. Since the n actions $y_1, \ldots, y_n$ conjugate to the toral angles of an invariant n-torus serve as (internal) parameters, we may encounter bifurcations of arbitrary co-dimension already in a single Hamiltonian system, provided the number $d > n$ of degrees of freedom is sufficiently large. We therefore abstain again from including external parameters into this setting.

An important assumption we make, which is automatically fulfilled for lower dimensional invariant tori of integrable systems, is that the torus $y = y_0$ be reducible to Floquet form

$$\dot{x} = \omega(y_0) + \mathcal{O}(y - y_0, z^2) \tag{1.4a}$$

$$\dot{y} = \mathcal{O}(y - y_0, z^3) \tag{1.4b}$$

$$\dot{z} = \Omega(y_0) z + \mathcal{O}(y - y_0, z^2) \tag{1.4c}$$

where the matrix $\Omega(y_0) \in \mathfrak{sp}(2m, \mathbb{R})$, $m = d - n$ is independent of the toral angles $x_1, \ldots, x_n$. The eigenvalues of this matrix are called Floquet exponents. Their distribution determines occurring bifurcations.

In the integrable case where there is no dependence at all on x we can reduce (1.4) to m degrees of freedom and end up with the Hamiltonian system defined by (1.4c). Here the origin $z = 0$ is an equilibrium, which undergoes a bifurcation as the parameter y passes through y_0. This puts us in the framework of Chapter 2 – and the main purpose of that chapter is indeed to address this problem independent of where it originates from. In this way the results obtained there carry over to bifurcations of invariant tori in integrable Hamiltonian systems.

We therefore concentrate on those bifurcations that could be satisfactorily treated in Chapter 2. This means we mainly restrict to $m = 1$ normal degree of freedom and consider $m = 2$ only insofar as there is an S^1-symmetry that again allows reduction to one normal degree of freedom. In this way we clarify the structure of the ramified d-torus bundle around invariant n-tori for integrable systems with $d = n + 1$ degrees of freedom and also for some cases with $d = n + 2$ degrees of freedom.

The remaining question then is what happens to this integrable picture under small Hamiltonian perturbations. Inevitably, where perturbations of quasi-periodic motions are concerned, small denominators enter the scene. Correspondingly, Diophantine conditions are needed to obtain the necessary estimates. The persistence of the bifurcation scenario is obtained by a combination of KAM theory and singularity theory.

To prove persistence of invariant tori one often uses a Kolmogorov-like condition

$$\det D\omega(y) \neq 0 \tag{1.5}$$

to let the actions $y_1, \ldots, y_n$ control the frequencies $\omega_1, \ldots, \omega_n$. In the present bifurcational setting we already need the actions to control the unfolding parameters $\lambda_1, \ldots, \lambda_k$. In particular, if the co-dimension k of the bifurcation is equal to the dimension n of the invariant torus, then the most degenerate torus is isolated and may disappear in a resonance gap. We therefore restrict to co-dimensions $k \leq n - 1$ where even the most degenerate tori still form continuous families in the unperturbed integrable system. Replacing (1.5) by a Rüssmann-like condition that involves also higher derivatives of the frequency mapping then yields a *Cantor family* of invariant tori in the perturbed system. In this way one can decouple the frequencies from the Hamiltonian and obtain persistence of invariant tori of the latter by treating the former as independent parameters. This strategy was already very successful in the study of normally elliptic lower dimensional tori, cf. [55] and references therein.

When proving a persistence result for a whole bifurcation scenario, the difficult part is to keep track of the most degenerate "object" in the perturbed system. To this end a KAM iteration scheme is used, performing two operations at each iteration step. First the lower[11] order terms are made x-independent. Here one has to deal with small denominators to solve a (linear) homological equation. Then these lower order terms are transformed into the universal unfolding of the central singularity. This is achieved by explicit co-ordinate changes known from singularity theory. The technical details of this procedure are deferred to *Appendices D and E*, where we also discuss in how far this proof is still open to generalizations.

In the final *Chapter 5* we put the results obtained into context to describe the dynamics in integrable and nearly integrable Hamiltonian systems. A completely integrable system with d degrees of freedom has d commuting integrals $G_1 = H, G_2, \ldots, G_d$ and according to Liouville's theorem [3, 13, 16] bounded motions starting at regular points of $G : \mathcal{P} \longrightarrow \mathbb{R}^d$ are conditionally periodic. Singular values of G give rise to lower dimensional invariant subsets and yield the whole hierarchy of singular fibres of the ramified torus bundle defined by G. Excitation of normal modes of non-hyperbolic equilibria generates periodic orbits (this is Lyapunov's theorem, see [3, 16, 208]) and the same mechanism explains how families of n-tori shrink down to k-tori, $k < n$. In Chapter 2 we encounter many more mechanisms how the various families of invariant tori fit together.

Under small non-integrable perturbations the ramified torus bundle is "Cantorised" as the smooth action manifolds parametrising invariant tori are replaced by Cantor sets of large relative measure. In the non-degenerate case this implies that most motions of the perturbed system are quasi-periodic, and the question arises how the various Cantor families of tori fit together. For the excitation of normal modes it has been shown in [164, 261] that the persistent k-tori consist of Lebesgue density points of persistent n-tori. Similar results are obtained in Chapter 4 for all the cases treated in Chapter 2. The destruction of maximal tori with a single resonance exemplifies that "Cantorised"

---

[11] This notion is defined by means of the singularity at hand.

bifurcations of lower dimensional tori ocur in virtually every nearly integrable Hamiltonian system.

In case there are more integrals than degrees of freedom the system is superintegrable. The G_i no longer commute, but the compact connected components of their level sets are still invariant tori carrying a conditionally periodic flow. In important cases it is possible to construct an "intermediate system" that is still integrable, but non-degenerately so. It is the ramified torus bundle defined by this "intermediate system" that gets "Cantorised" when passing to the original perturbed dynamics.

2

Bifurcations of Equilibria

We are given a family of Hamiltonian systems, defined by a family of Hamiltonian functions H_α on a family of phase spaces $\mathcal{P}_\alpha$. Bifurcations of equilibria are localised both in phase space and parameter space. After a translation in the latter we may assume that the bifurcation occurs at $\alpha = 0$. If the bifurcating equilibrium has a neighbourhood in $\mathcal{P}_0$ on which the rank of the Poisson structure is constant, we may choose local co-ordinates φ^α on $U_\alpha \subseteq \mathcal{P}_\alpha$ in such a way that $V = \varphi^\alpha(U_\alpha)$ is independent of α. In two and more degrees of freedom such regular equilibria are the only equilibria we consider. Also in one degree of freedom we restrict the parameter dependence to the Hamiltonian and assume $\mathcal{P}_\alpha = \mathcal{P}_0 =: \mathcal{P}$ throughout the bifurcation. Next to singular equilibria of 2-dimensional Poisson spaces we furthermore study bifurcations in $\mathcal{P} = \mathbb{R}^3$ where the level sets of the Casimir are symplectic manifolds except at the bifurcation. This latter situation could also be phrased in terms of parameter-dependent 2-dimensional phase spaces.

2.1 Equilibria in One Degree of Freedom

In one degree of freedom Hamiltonian systems are always integrable. The phase curves coincide with the energy "shells" whence the phase portrait is given by the level sets of the Hamiltonian function. This allows us to define an equivalence relation on the set $\mathcal{A}$ of Hamiltonian functions.

Definition 2.1. Let $\mathcal{P}$ be a 2-dimensional symplectic manifold. Two Hamiltonian systems on $\mathcal{P}$ are (topologically) equivalent if there is a homeomorphism η on $\mathcal{P}$ that maps[1] phase curves on phase curves.

---

[1] We do *not* require the direction of time to be preserved, a time reversal of a Hamiltonian system can always be obtained by multiplying the Hamiltonian function with -1.

In the more general case of Poisson spaces, where the orientable surface $\mathcal{P}$ may have singular points, we have to be more careful. Indeed, while singular points are always equilibria, a mere homeomorphism may map these also to e.g. regular equilibria. We therefore explicitly require a *topological equivalence* η to preserve singular points.

We also need a topology on the set $\mathcal{A}$ of Hamiltonian functions. If $\mathcal{P}$ is a compact symplectic manifold then the C^k-topology, for $k \in \mathbb{N}$ fixed, is defined by the semi-norm

$$\|H\|_k = \sum_{i=1}^{k} \sup_{x \in \mathcal{P}} |D^i H(x)| .$$

Note that two Hamiltonians have distance 0 with respect to this semi-norm if and only if they define the same Hamiltonian vector field. In [191] the corresponding quotient space is called the set of normalized Hamiltonians.

The projective limit on $C^\infty(\mathcal{P})$ defines the C^∞-topology, which is equivalently given by the semi-metric

$$\mathrm{d}(H, K) = \sum_{k=1}^{\infty} \frac{2^{-k} \|H - K\|_k}{1 + \|H - K\|_k} .$$

On the set of normalized Hamiltonians this defines a metric. If $\mathcal{P}$ is not compact we instead resort to the Whitney C^k-topologies, $k \in \mathbb{N} \cup \{\infty\}$ defined in e.g. [126, 191]. Finally, for real analytic Hamiltonians we use the compact-open topology defined by the supremum norm of DH on holomorphic extensions, cf. [55].

Definition 2.2. Let $\mathcal{P}$ be a 2-dimensional symplectic manifold. A Hamiltonian $H : \mathcal{P} \longrightarrow \mathbb{R}$ defines a *structurally stable Hamiltonian system* if for every Hamiltonian $K : \mathcal{P} \longrightarrow \mathbb{R}$ close to H the two Hamiltonian systems are topologically equivalent.

The set of structurally stable Hamiltonians on $\mathcal{P}$ is not only open, but also dense since it contains the set of *Morse functions*. The latter are functions $H : \mathcal{P} \longrightarrow \mathbb{R}$ such that all critical points have a non-degenerate quadratic form and no two values that H takes on these coincide. Thus, the equilibria of a Hamiltonian system defined by a Morse function are centres and saddles, and there are no heteroclinic connections between the latter. In fact, there are no dynamic consequences if a Hamiltonian function takes the same value on a centre and another equilibrium.

Homoclinic orbits are commonplace for Hamiltonian systems. In one degree of freedom all Hamiltonian systems are integrable and there is no splitting of separatrices. Where separatrices of two saddles coincide to form a heteroclinic orbit we speak of a *connection bifurcation*. This is a global bifurcation and a necessary condition is that the two saddles have the same energy.

2.1.1 Regular Equilibria

In one degree of freedom the regular equilibria of a Hamiltonian system are given by the planar singularities[2] (or critical points) of the Hamiltonian function. Local bifurcations occur where these singularities are not (structurally) stable. The whole bifurcation scenario is then included in the universal unfolding of the unstable singularity.

To study planar singularities we consider *germs* $\mathcal{H} : (\mathbb{R}^2, 0) \longrightarrow (\mathbb{R}, 0)$. This notion formalizes that we may always restrict to a small(er) neighbourhood of the singularity at hand, which we translate to the origin. Similarly $\mathcal{H}(0) = 0$ is no restriction since adding a constant to the Hamiltonian has no dynamical consequences. In this way we choose a preferred normal form in the equivalence class of all Hamiltonians defining the same Hamiltonian system.

Definition 2.3. Two singularities $\mathcal{H}, \mathcal{K} \in C^k(\mathbb{R}^2, 0)$, $k \in \mathbb{N} \cup \{\infty\}$ are C^k-left-right equivalent if there are C^k-diffeomorphisms η on $\mathbb{R}^2$ and h on $\mathbb{R}$ with $\mathcal{K} = h \circ \mathcal{H} \circ \eta$. If we can choose $h = \mathrm{id}$ they are C^k-right equivalent. A singularity is finitely determined if it is C^k-right equivalent to every singularity with the same ℓ-jet for some $\ell < k$.

The following result from [43] clarifies the relation between Hamiltonian systems that are defined by equivalent planar Hamiltonian functions.

Proposition 2.4. Let $\mathcal{H}, \mathcal{K} \in C^k(\mathbb{R}^2, 0)$ be two C^k-left-right equivalent Hamiltonian functions. Then the corresponding Hamiltonian vector fields satisfy

$$X_{\mathcal{K}}(z) = h'(\mathcal{H}(\eta(z))) \cdot \det D\eta(z) \cdot D\eta^{-1}(\eta(z)) (X_{\mathcal{H}}(\eta(z)))$$

for all z in a sufficiently small neighbourhood of the origin.

Proof. First let $\mathcal{H}$ and $\mathcal{K}$ be right equivalent. Writing $z = (q, p)$ the symplectic form becomes the area element $\mathrm{d}q \wedge \mathrm{d}p$ and the effect of the diffeomorphism η reads

$$(\mathrm{d}q \wedge \mathrm{d}p \circ \eta) \bullet (D\eta, D\eta) = \Lambda^2 D\eta^*(\mathrm{d}q \wedge \mathrm{d}p) = \det D\eta \cdot (\mathrm{d}q \wedge \mathrm{d}p) .$$

The $\bullet$ in the expression on the left hand side means

$$[(\mathrm{d}q \wedge \mathrm{d}p \circ \eta) \bullet (D\eta, D\eta)](z) = (\mathrm{d}q \wedge \mathrm{d}p(\eta(z))) \circ (D\eta(z), D\eta(z))$$

for all z in a sufficiently small neighbourhood of the origin. Fixing another vector field Y on $\mathbb{R}^2$ we compute

---

[2] Note the distinction between *singular points* of the phase space and *singularities* of the Hamiltonian.

$$
\begin{aligned}
\mathrm{d}q \wedge \mathrm{d}p(X_{\mathcal{K}}, Y) \; &= \; \mathrm{d}\mathcal{K}(Y) \; = \; \mathrm{d}(\mathcal{H} \circ \eta)(Y) \\
&= \; ((\mathrm{d}\mathcal{H} \circ \eta) \bullet D\eta)(Y) \; = \; (\mathrm{d}\mathcal{H} \circ \eta)(D\eta(Y)) \\
&= \; (\mathrm{d}q \wedge \mathrm{d}p \circ \eta)\left(X_{\mathcal{H} \circ \eta}, D\eta(Y)\right) \\
&= \; (\mathrm{d}q \wedge \mathrm{d}p \circ \eta)\left(D\eta(D\eta^{-1}(X_{\mathcal{H}}) \circ \eta), D\eta(Y)\right) \\
&= \; \det D\eta \cdot \mathrm{d}q \wedge \mathrm{d}p \left(D\eta^{-1}(X_{\mathcal{H}}) \circ \eta, Y\right) \; .
\end{aligned}
$$

In case $\mathcal{K} = h \circ \mathcal{H}$ we similarly obtain

$$
\begin{aligned}
\mathrm{d}q \wedge \mathrm{d}p(X_{\mathcal{K}}, Y) \; &= \; \mathrm{d}\mathcal{K}(Y) \; = \; \mathrm{d}(h \circ \mathcal{H})(Y) \\
&= \; (h'(\mathcal{H}) \cdot \mathrm{d}\mathcal{H})(Y) \; = \; h'(\mathcal{H}) \cdot (\mathrm{d}q \wedge \mathrm{d}p)(X_{\mathcal{H}}, Y) \; . \qquad \square
\end{aligned}
$$

The two factors $h' \circ \mathcal{H} \circ \eta$ and $\det D\eta$ express that X_H and $X_{\mathcal{K}}$ are (topologically) equivalent since a vector field transforms under the co-ordinate change η as

$$
\eta^* Y \; = \; D\eta^{-1}(Y) \circ \eta \; .
$$

In the particular case of right equivalence by an area-preserving co-ordinate change η we recover (a special case of) Jacobi's result

$$
\eta^* X_{\mathcal{H}} \; = \; X_{\mathcal{H} \circ \eta}
$$

for symplectomorphisms, see [3]. Note that it is in general not possible to turn the equivalence of Hamiltonian vector fields provided by a right equivalence of the corresponding Hamiltonian functions into a conjugacy by combining it with a local diffeomorphism $h : (\mathbb{R}, 0) \longrightarrow (\mathbb{R}, 0)$. Indeed, the resulting time factor $h'(\mathcal{H}(\eta(z)))$ is the same throughout a whole orbit. We therefore keep this extra freedom for later use[3] and work as long as possible with C^k-right equivalences.

Proposition 2.4 implies in fact more strongly that Hamiltonian vector fields with right equivalent Hamiltonian functions have flows that are equivalent by means of a diffeomorphism. This allows us to relax Definition 2.3 to C^0-equivalence of C^k-germs in Definition 2.12 below. However, for the moment it is quite convenient that η and its inverse are both smooth.

Simple Singularities

All (planar) harmonic oscillators are right equivalent to

$$
\mathcal{H}(q, p) \; = \; \frac{p^2 + q^2}{2} \tag{2.1}
$$

and locally this easily extends to anharmonic oscillators as well. By the latter we mean a centre with invertible linearization, i.e. we now allow for higher

---

[3] Where convenient we do use $h = -\mathrm{id}$ and $h = \mathrm{id} + const$, though.

order terms. Similarly, every saddle with invertible linearization is (locally) right equivalent to

$$\mathcal{H}(q,p) = \frac{p^2 - q^2}{2} . \tag{2.2}$$

This covers the stable planar singularities

$$A_1^{\pm} : \qquad \mathcal{H}(q,p) = \frac{a}{2}p^2 + \frac{b}{2}q^2$$

where the sign $\pm$ equals sgn(ab). Here and below we refrain[4] from scaling away the non-zero coefficients a and b. In applications it is much easier to simply check the relative sign of these coefficients than to perform the actual transformations to achieve one of the forms (2.1) and (2.2).

The singularities $A_1^{\pm}$ describe the motion of a 1-dimensional particle in a quadratic potential, and such an interpretation holds true for the singularities

$$A_{d-1}^{\pm} : \qquad \mathcal{H}(q,p) = \frac{a}{2}p^2 + \frac{b}{d!}q^d$$

as well. Here the sign $\pm$ = sgn(ab) is only needed for even d since $q \mapsto -q$ allows to replace b by $-b$ when d is odd. For $d \geq 3$ the singularity $A_{d-1}^{\pm}$ is no longer stable and the universal unfolding

$$\mathcal{H}_{\lambda}(q,p) = \frac{a}{2}p^2 + \frac{b}{d!}q^d + \sum_{j=1}^{d-2} \frac{\lambda_j}{j!}q^j \tag{2.3}$$

describes the 1-dimensional motion in the family of potentials

$$V_{\lambda}(q) = \frac{b}{d!}q^d + \sum_{j=1}^{d-2} \frac{\lambda_j}{j!}q^j .$$

For $d = 3$ this yields the centre-saddle[5] bifurcation, related to the fold catastrophe, the sole local bifurcation that one may encounter in generic 1-parameter families of Hamiltonian systems on orientable smooth surfaces.

Remark 2.5. The situation changes if one forces the origin $(q,p) = 0$ to be an equilibrium for the unfolding as well. Then the potential has no linear term, but the monomial q^{d-1} can no longer be transformed away. For $d = 3$ we get the 1-parameter family

$$\mathcal{H}_{\lambda}(q,p) = \frac{a}{2}p^2 + \frac{b}{6}q^3 + \frac{\lambda}{2}q^2 \tag{2.4}$$

---

[4] However, we do transform away the mixed term pq.

[5] In the literature this is also called Hamiltonian saddle node bifurcation. Not to be confounded with a "saddle-centre" which is a hypo-elliptic equilibrium in two degrees of freedom that combines the normal behaviour of a saddle and a centre.

displaying the transcritical[6] bifurcation, where the equilibria at the origin and at $(q, p) = (-2\lambda/b, 0)$ exchange stability. The assumption that certain invariant subsets are not affected by a perturbation is often helpful in theoretical considerations – a famous example is the construction in [7] of a transition chain of hyperbolic tori – but rarely justified *per se*. We therefore do not especially give "origin-fixing" universal unfoldings (the necessary changes are easily performed, though). In Sections 2.1.3 and 2.1.4 we encounter equilibria that are forced to remain equilibria for intrinsic reasons.

If U_μ is a family of potentials with $U_0^{(d)}(0) \neq 0$ and $U_0^{(j)}(0) = 0$ for $j = 0, \ldots, d - 1$ then there is a re-parametrisation $\mu \mapsto \lambda(\mu)$ such that U_μ is right equivalent to $V_{\lambda(\mu)}$, and the family of co-ordinate transformations η_μ in q depends smoothly on μ as well, see [9, 257, 40]. The splitting lemma, see again [9, 40], yields that same result for more general perturbations of (2.3) as well. It is this property of universal unfoldings that we are interested in.

Example 2.6. Let us consider the universal unfolding of the singularity A_3^- describing the 1-dimensional motion with kinetic energy $\frac{1}{2}p^2$ in the 2-parameter family of potentials

$$U_{\lambda,\mu}(q) = -\frac{1}{24}q^4 + \lambda q + \frac{\mu}{2}q^2$$

in more detail, see Fig. 2.1. The equations of motion generated by this dual cusp catastrophe read

$$\dot{q} = p$$
$$\dot{p} = \frac{1}{6}q^3 - \lambda - \mu q .$$

There is always a saddle and additionally a centre and a saddle are born in a centre-saddle bifurcation at the cusp line

$$\left\{ (\lambda, \mu) \in \mathbb{R}^2 \;\middle|\; 9\lambda^2 = 8\mu^3 \right\} .$$

Furthermore, the two saddles have the same energy at the half line

$$\left\{ (\lambda, \mu) \in \mathbb{R}^2 \;\middle|\; \lambda = 0, \, \mu > 0 \right\}$$

and get connected by heteroclinic orbits. This connection bifurcation is an example of a global co-dimension 1 bifurcation subordinate to a local co-dimension 2 bifurcation. As the phase portraits in Fig. 2.1 show the motion is generally unbounded, except if the 1-dimensional particle is trapped between the two maxima of the potential that coexist for $9\lambda^2 < 8\mu^3$.   □

---

[6] Also this bifurcation is sometimes called a Hamiltonian saddle-node bifurcation in the literature.

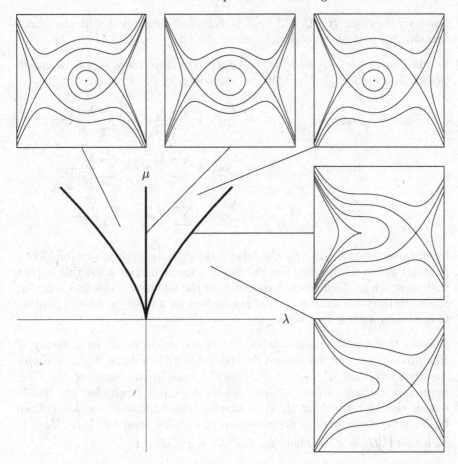

Fig. 2.1. Bifurcation diagram of the universal unfolding of the singularity A_3^-.

For the planar singularities $\mathcal{H}(q,p) = (a/2)p^2q + (b/k!)q^k$ of type $D_{k+1}^\pm$, $k \geq 3$, the sign $\pm = \mathrm{sgn}(ab)$ is only needed for odd k since otherwise b can be replaced by $-b$ after a combination of $q \mapsto -q$ and $\mathcal{H} \mapsto -\mathcal{H}$. The universal unfolding is given by

$$D_{k+1}^\pm : \qquad \mathcal{H}_\lambda(q,p) = \frac{a}{2}p^2q + \frac{b}{k!}q^k + \sum_{j=1}^{k-1}\frac{\lambda_j}{j!}q^j + \lambda_k p \,, \qquad (2.5)$$

see [9]. Again this means that the k-parameter family of Hamiltonians (2.5) is structurally stable, cf. [287]. The bifurcation diagrams of the unfoldings (2.5) of $D_4^\pm$ are given in [43].

Definition 2.7. A planar singularity $\mathcal{H} : (\mathbb{R}^2, 0) \longrightarrow (\mathbb{R}, 0)$ is *simple* if there are finitely many germs $\mathcal{K}_1, \ldots, \mathcal{K}_\ell$ such that every unfolding $\mathcal{H}_\lambda$ of $\mathcal{H} = \mathcal{H}_0$

has a representant $\widetilde{\mathcal{H}} : U \times \Lambda \longrightarrow \mathbb{R}$ such that for every $\lambda \in \Lambda$ the planar function $\widetilde{\mathcal{H}}_\lambda$ is right equivalent to one of the $\mathcal{K}_j$, $j = j(\lambda) \in \{1, \ldots, \ell\}$.

In addition to $(A_\mu^{\pm})_{\mu \geq 1}$ and $(D_\mu^{\pm})_{\mu \geq 4}$ there are three more simple singularities $(E_\mu)_{\mu = 6,7,8}$, see [9]. The universal unfoldings of these are given by

$$E_6 : \quad \mathcal{H}_y(q,p) = \frac{a}{6}p^3 + \frac{b}{24}q^4 + \sum_{j=1}^{2} \frac{\lambda_j}{j!}q^j + \sum_{j=0}^{2} \frac{\lambda_{j+3}}{j!}pq^j$$

$$E_7 : \quad \mathcal{H}_y(q,p) = \frac{a}{6}p^3 + \frac{b}{6}pq^3 + \sum_{j=1}^{4} \frac{\lambda_j}{j!}q^j + \sum_{j=0}^{1} \frac{\lambda_{j+5}}{j!}pq^j$$

$$E_8 : \quad \mathcal{H}_y(q,p) = \frac{a}{6}p^3 + \frac{b}{5!}q^5 + \sum_{j=1}^{3} \frac{\lambda_j}{j!}q^j + \sum_{j=0}^{3} \frac{\lambda_{j+4}}{j!}pq^j \ .$$

In all three series A_μ, D_μ, E_μ the label μ denotes the *multiplicity* (or Milnor number) of the singularity. For the simple singularities at hand this implies that there are $\mu - 1$ unfolding parameters in the universal unfoldings and that one needs the same number $\mu - 1$ of parameters for a family in general position to encounter such a singularity.

Theorem 2.8. On the neighbourhood U of the origin in $\mathbb{R}^2$ let a family of Hamiltonian systems be defined by the universal unfolding N_λ of a simple singularity N_0. Let $H_\mu = N_\mu + P_\mu$ be a C^∞-small perturbation of N_μ. Then there is a re-parametrising diffeomorphism $\lambda \mapsto \mu(\lambda)$ such that the Hamiltonian system defined by $H_{\mu(\lambda)}$ is topologically equivalent to that defined by N_λ, after both systems are restricted to suitable open sets $V_{\mu(\lambda)}, U_\lambda \subseteq U$ such that $\bigcup_\lambda U_\lambda$ is a neighbourhood of $\{0\} \times \Lambda \subseteq U \times \Lambda$. .

Proof. The implicit mapping theorem yields $(q_0, p_0) \in U$ and $\mu_0 \in \Lambda$ both close to zero such that the singularity of H_{μ_0} at (q_0, p_0) is right equivalent to N_0 at $(0,0)$. This defines $\mu(0) = \mu_0$ and neighbourhoods V_{μ_0} of (q_0, p_0) and U_0 of $(0,0)$ with $\eta_0(U_0) = V_{\mu_0}$ where the diffeomorphism η_0 effects $H_{\mu_0} = N_0 \circ \eta_0$. Since N_λ unfolds N_0 universally this extends to $\lambda \mapsto \mu(\lambda)$ and η_λ with $H_{\mu(\lambda)} = N_\lambda \circ \eta_\lambda$ and $\eta_\lambda(U_\lambda) = V_{\mu(\lambda)}$. Because of Proposition 2.4 this family of right equivalences between the Hamiltonian functions provides an equivalence between the Hamiltonian systems. $\qquad \square$

As the proof shows, Theorem 2.8 remains true if we replace "C^∞-small" perturbation by "C^d-small", where d is the degree of the (polynomial) singularity at hand.

Remark 2.9. The list of simple singularities remains the same if we require $\widetilde{\mathcal{H}}_\lambda$ only to be left-right equivalent to $\mathcal{K}_{j(\lambda)}$. It is also not necessary to explicitly require the singularity to be planar, cf. [9]; but we are only interested in planar (Hamiltonian) functions.

Moduli

It is instructive to understand why there are no further simple singularities, so we now delve a bit more into the theory. A deeper study of singularities is contained in Appendix A (or, of course, in any book on that topic). We continue to exclusively consider planar singularities.

Definition 2.10. Let $\mathcal{H} : (\mathbb{R}^2, 0) \longrightarrow (\mathbb{R}, 0)$ be a planar singularity. The *multiplicity (or Milnor number)* of $\mathcal{H}$ is the dimension μ of the *local algebra*

$$Q_{\mathcal{H}} \;=\; \mathcal{E}_2/\mathcal{I}_{\mathcal{H}}$$

where $\mathcal{E}_2$ denotes the ring of all germs at the origin of $\mathbb{R}^2$ and

$$\mathcal{I}_{\mathcal{H}} \;:=\; <\frac{\partial \mathcal{H}}{\partial q}, \frac{\partial \mathcal{H}}{\partial p}>$$

is the so-called *Jacobi ideal*, generated by the partial derivatives. The singularity $\mathcal{H}$ is *non-degenerate* if the multiplicity μ is finite.

A singularity is non-degenerate if and only if it is finitely determined, see [15]. To be precise, if μ is the multiplicity of $\mathcal{H}$, then every singularity $\mathcal{K}$ that has the same $(\mu+1)$-jet as $\mathcal{H}$ is right equivalent to $\mathcal{H}$. If $\mathcal{H}$ is non-degenerate, then representants $\mathbb{1} = h_0, h_1, \ldots, h_{\mu-1}$ of a basis of the local algebra $Q_{\mathcal{H}}$ yield a universal unfolding

$$\mathcal{H}_\lambda \;=\; \mathcal{H} + \sum_{j=1}^{\mu-1} \lambda_j h_j$$

of $\mathcal{H} = \mathcal{H}_0$. A non-degenerate polynomial singularity contains a monomial of the form p^β or $p^\beta q$ and a monomial of the form pq^γ or q^γ. Indeed, otherwise one could factor q^2 or p^2. This allows to distribute weights α_p and α_q as detailed in Table 2.1, thus defining a *gradation* on $\mathbb{R}[q, p]$.

Definition 2.11. A polynomial $\mathcal{H} \in \mathbb{R}[q, p]$ is *quasi-homogeneous* of order d with weight (α_q, α_p) if

Table 2.1. Non-degenerate quasi-homogeneous polynomials. To make the cases of the second and third row unique one may require $\beta \leq \gamma$, cf. the choice of unfolding terms in D_μ and E_7.

case	α_q	α_p	order	basis monomials for the local algebra
$p^\beta + q^\gamma$	β	γ	$\beta\gamma$	$p^k q^j$, $j = 0, \ldots, \gamma - 2$, $k = 0, \ldots, \beta - 2$
$p^\beta + pq^\gamma$	$\beta - 1$	γ	$\beta\gamma$	$p^k q^j$, $j = 0, \ldots, \gamma - 2$, $k = 0, \ldots, \beta - 2$ and q^j, $j = \gamma - 1, \ldots, 2\gamma - 2$
$p^\beta q + q^\gamma$	β	$\gamma - 1$	$\beta\gamma$	$p^k q^j$, $j = 0, \ldots, \gamma - 1$, $k = 0, \ldots, \beta - 2$ and $p^{\beta-1}$
$p^\beta q + pq^\gamma$	$\beta - 1$	$\gamma - 1$	$\beta\gamma + 1$	$p^k q^j$, $j = 0, \ldots, \gamma - 1$, $k = 0, \ldots, \beta - 1$

$$\bigwedge_{\tau \in \mathbb{R}} \bigwedge_{(q,p) \in \mathbb{R}^2} \mathcal{H}(e^{\alpha_q \tau} q, e^{\alpha_p \tau} p) = e^{d\tau} \mathcal{H}(q,p) \ .$$

A singularity is *semi-quasi-homogeneous* of order d with weight (α_q, α_p) if it is the sum $\mathcal{H} + \mathcal{K}$ of a non-degenerate quasi-homogeneous polynomial $\mathcal{H}$ of order d with weight (α_q, α_p) and a germ $\mathcal{K} \in \mathcal{E}_2$ of (weighted) order strictly greater than d.

All simple singularities are obviously (non-degenerate) quasi-homogeneous polynomials. The situation can best be visualized by means of the *Newton diagram*. Here the integer points $(j, k) \in \mathbb{N}_0^2$ stand for the monomial $p^k q^j$ and those monomials that have weighted order d all lie on a straight line with slope $-\alpha_q / \alpha_p$. For $A_{2\ell}, D_{2\ell+1}, E_6, E_7, E_8$ there are only two monomials on that straight line. The singularity $A_{2\ell-1}^{\pm}$ is quasi-homogeneous of order $d = 4\ell$ with weight $(\alpha_q, \alpha_p) = (2, 2\ell)$ and the monomial pq^ℓ has the same weighted order 4ℓ as well. This term can be transformed away by means of the (symplectic) co-ordinate change

$$(q, p) \mapsto (q, p - \gamma q^\ell)$$

with an appropriately chosen constant γ (thereby also changing the coefficient of the term $q^{2\ell}$ of the singularity). Similarly, the monomial $pq^{\ell+1}$ has order 4ℓ with respect to the weight $(\alpha_q, \alpha_p) = (2, 2\ell - 1)$ and can be transformed away when added to the singularity $D_{2\ell}^{\pm}$. A special case is the homogeneous polynomial

$$A p^3 + B p^2 q + C p q^2 + D q^3$$

of degree 3. For $(A, B, C, D) \in \mathbb{R}^4$ in general position the three roots are different from each other and a suitable (symplectic) co-ordinate transformation

$$(q, p) \mapsto (q - \beta p, (1 + \beta \gamma) p - \gamma q) \tag{2.6}$$

leads to the singularity D_4^+ if there is a complex conjugate pair of roots and to D_4^- if all three roots are real.

For the simple singularities there are monomial representants $h_{jk} = p^k q^j$ of a basis of the local algebra that all have (weighted) order below that of the singularity (i.e. the corresponding integer points (j, k) in the Newton diagram lie below the straight line characterising the quasi-homogeneous singularity). More generally, as shown in [10], a non-degenerate quasi-homogeneous polynomial has a basis of the local algebra with the monomial representants given in Table 2.1. Correspondingly, semi-quasi-homogeneous singularities cease to be simple when there are basis monomials $p^k q^j$ of (weighted) order $\alpha_p k + \alpha_q j$ equal or bigger than the order d of the quasi-homogeneous part.

There are two weights that yield a basis monomial of weighted order d, but none of higher order, see Fig. 2.2. These are $(\alpha_q, \alpha_p) = (3, 6)$ and $(\alpha_q, \alpha_p) = (4, 4)$ and correspond to the quasi-homogeneous singularities

$$J_{10} \ : \qquad \mathcal{H}(q,p) \ = \ \frac{a}{6}p^3 + \frac{m}{24}pq^4 + \frac{b}{6!}q^6$$

$$X_9^{\pm} \ : \qquad \mathcal{H}(q,p) \ = \ \frac{a}{24}p^4 + \frac{\mu}{4}p^2q^2 + \frac{b}{24}q^4 \ .$$

To ensure non-degeneracy, next to $a, b \neq 0$ the conditions $200\,m^3 + ab^2 \neq 0$ and $9\mu^2 \neq ab$ are needed. While it is still possible to transform a to 1 and b to 1 or ± 1 the coefficients m and μ are moduli that cannot be further simplified by right equivalences.

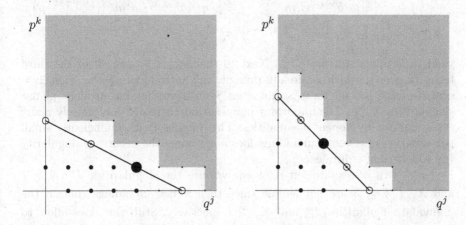

Fig. 2.2. Newton diagrams for the unimodal quasi-homogeneous planar singularities J_{10} and X_9.

The modulus μ derives directly from the grandfather of all moduli, the cross ratio. Considering $\mathcal{H}$ as a complex function defined on $\mathbb{C}^2$ this is immediate; in the real category things are a bit more subtle as pointed out in [289] (see also [93]). The modulus m stems from the fact that it is impossible to simultaneously transform three given parabolas to three predescribed normal form parabolas. In particular both m and μ are also moduli with respect to left-right equivalences.

The two quasi-homogeneous singularities J_{10} and X_9 start two series of *unimodal* planar singularities

$$J_{i+4} \ : \qquad \mathcal{H}(q,p) \ = \ \frac{a}{6}p^3 + \frac{m}{4}p^2q^2 + \frac{b}{i!}q^i \ , \quad i \geq 7$$

$$X_{j+5}^{\pm} \ : \qquad \mathcal{H}(q,p) \ = \ \frac{a}{24}p^4 + \frac{\mu}{4}p^2q^2 + \frac{b}{j!}q^j \ , \quad j \geq 5 \ .$$

The conditions $m \neq 0$ and $\mu \neq 0$ imply that these are not semi-quasi-homogeneous, depending on the chosen weights the quasi-homogeneous part is divisible by p^2 or q^2. Non-degeneracy is then ensured by $a \neq 0$ and $b \neq 0$.

For these singularities the moduli disappear if we use left-right equivalences instead of right equivalences. Indeed, the scalings

$$\eta(q,p) = \left(\pm \left[a^2 b^{6-(12/i)} m^{-6} \right]^{1/(i^2-2i)} \cdot q, \ \left[a^{6-(12/i)} b^2 m^{-6} \right]^{1/(3i-6)} \cdot p \right)$$

$$h(\mathcal{H}) = \pm \left[a^{-4} b^{-(12/i)} m^6 \right]^{1/(i-2)} \cdot \mathcal{H}$$

get rid of the coefficients in J_{i+4}; for even i this introduces a sign $\pm = \operatorname{sgn}(abm)$ of e.g. the third term. The scalings

$$\eta(q,p) = \left(\pm \left[a^4 b^{8-(16/j)} m^{-8} \right]^{1/(j^2-2j)} \cdot q, \ \left[a^{8-(16/j)} b^4 m^{-8} \right]^{1/(4i-8)} \cdot p \right)$$

$$h(\mathcal{H}) = \pm \left[a^{-4} b^{-(16/j)} m^8 \right]^{1/(i-2)} \cdot \mathcal{H}$$

yield the same result for $X_{j+5}^{\pm}$. Next to the sign $\pm = \operatorname{sgn}(ab)$ of the third term for even j which is already present, this introduces another sign $\pm = \operatorname{sgn}(\mu)$ of the second term, so for even j we have four inequivalent planar singularities $X_{j+5}^{\pm,\pm}$. The important point is that there are only finitely many cases, defined by "open" inequalities. This implies that a sufficiently small perturbation of a versal unfolding does not change the occurring singularity (up to left-right equivalence).

Even with respect to left-right equivalence the singularities $J_{i+4}^{\pm}, i \geq 7$ and $X_{j+5}^{\pm,\pm}, j \geq 5$ are not simple since their versal unfoldings contain the unimodal singularities J_{10}^m and $X_9^{\pm,\mu}$, respectively. Still, this does allow to prove a result similar to Theorem 2.8 on simple singularities by taking care that the re-parametrising diffeomorphism $\lambda \mapsto \lambda(\mu)$ preserves the modulus occurring in the unfolding. This strategy would also work for the unimodal singularities

$$E_{12} : \qquad \mathcal{H}_y(q,p) = \frac{a}{6}p^3 + \frac{b}{7!}q^7 + \frac{\mu}{5!}pq^5$$

$$E_{13} : \qquad \mathcal{H}_y(q,p) = \frac{a}{6}p^3 + \frac{b}{5!}pq^5 + \frac{\mu}{8!}q^8$$

$$E_{14} : \qquad \mathcal{H}_y(q,p) = \frac{a}{6}p^3 + \frac{b}{8!}q^8 + \frac{\mu}{6!}pq^6$$

or the remaining unimodal planar singularities, detailed in Appendix A. However, as shown in [263], the left diffeomorphism $h : \mathbb{R} \longrightarrow \mathbb{R}$ can only take care of one modulus (of right equivalence). To extend Theorem 2.8 to all non-degenerate planar singularities we therefore use that it is not necessary for the topological equivalence of Hamiltonian flows searched for to be differentiable.

Definition 2.12. Two singularities $\mathcal{H}, \mathcal{K} \in C^k(\mathbb{R}^2, 0)$, $k \geq 4$ are C^0-left-right equivalent if there are homeomorphisms η on $\mathbb{R}^2$ and h on $\mathbb{R}$ with $\mathcal{K} = h \circ \mathcal{H} \circ \eta$. If $h = \operatorname{id}$ they are C^0-right equivalent.

Note that the stable singularity $A_1^+ : p^2 + q^2$ is C^0-right equivalent to $p^4 + 2p^2q^2 + q^4$ which is not only unstable, but even fails to be non-degenerate.

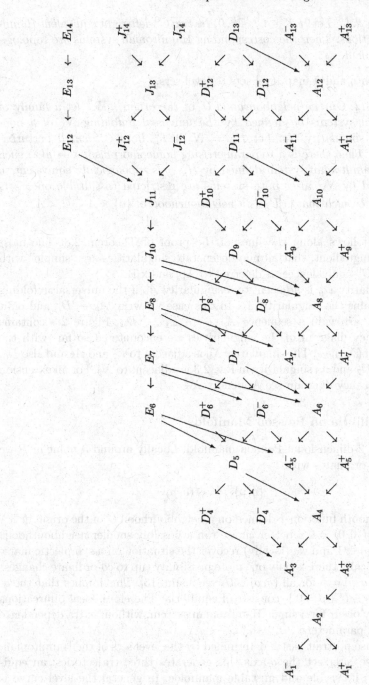

Fig. 2.3. Adjacency diagram of planar singularities with non-vanishing 3-jet.

Proposition 2.13. Let $\mathcal{H}, \mathcal{K} \in C^k(\mathbb{R}^2, 0)$ *be two C^0-left-right equivalent Hamiltonian functions. Then the corresponding Hamiltonian systems are topologically equivalent.*

Proof. Both η and h map level sets to level sets. $\square$

Theorem 2.14. On the neighbourhood U of the origin in $\mathbb{R}^2$ let a family of Hamiltonian systems be defined by the universal unfolding N_λ of a non-degenerate singularity N_0. Let $H_\mu = N_\mu + P_\mu$ be a C^∞-small perturbation of N_μ. Then there is a re-parametrising homeomorphism $\lambda \mapsto \mu(\lambda)$ such that the Hamiltonian system defined by $H_{\mu(\lambda)}$ is topologically equivalent to that defined by N_λ, after both systems are restricted to suitable open sets $V_{\mu(\lambda)}, U_\lambda \subseteq U$ such that $\bigcup_\lambda U_\lambda$ is a neighbourhood of $\{0\} \times \Lambda \subseteq U \times \Lambda$.

The proof follows along the lines of the proof of Theorem 2.8. The most important ingredient, that all non-degenerate singularities are "simple" with respect to C^0-equivalences, is deferred to Appendix B.

A singularity D_μ is *adjacent* to a singularity A_ν if the universal unfolding of D_μ contains the singularity A_ν. In this case we write $A_\nu \leftarrow D_\mu$ and omit the explicit arrow in a sequence $A_\nu \leftarrow A_{\nu+1} \leftarrow D_\mu$. Figure 2.3 contains the adjacency diagram of the singularities we encountered so far, with the exception of $(X_\mu)_{\mu \geq 9}$. The singularity X_9 is adjacent to E_7 and A_7 and also D_6, but not to D_7 and no singularity in Fig. 2.3 is adjacent to X_9. For an extension of this adjacency diagram see Appendix A.

2.1.2 Equilibria on Poisson Manifolds

Let $\mathcal{P}$ be a 2-dimensional Poisson manifold. Locally around a point in $\mathcal{P}$ we can find co-ordinates with

$$\{q, p\} = f(q, p)$$

for some smooth function f defined on a neighbourhood U of the origin in $\mathbb{R}^2$, see [13]. If $f(0,0) \neq 0$ a better choice (on a possibly smaller neighbourhood) yields $f(q, p) \equiv 1$ and we (locally) recover the situation on a symplectic manifold. Generically there is only one more possibility (up to co-ordinate changes) given by $f(q, p) = p$ for all $(q, p) \in U$, see again [13]. This implies that the q-axis $\{(q, p) \in U \mid p = 0\}$ consists of equilibria. Therefore, local bifurcations may already occur for a single Hamiltonian system, without extra dependence on external parameters.

The phase portrait is still determined by the level sets of the Hamiltonian. Where these intersect the q-axis this generates three trajectories: an equilibrium and its stable and unstable manifolds. In general the level curve is transverse to the q-axis. On a sufficiently small neighbourhood of such an equilibrium (which we translate to the origin) the Hamiltonian system is (topologically) equivalent to the Hamiltonian vector field

$$\dot{q} = 0$$
$$\dot{p} = -p$$

defined by the Hamiltonian function $\mathcal{H}(q,p) = q$. Along the 1-parameter family of equilibria the level curves may become tangent (without vanishing). Here the Hamiltonian system is locally (topologically) equivalent to that defined by the Hamiltonian

$$\mathcal{H} = p + \frac{q^2}{2} .$$

If the Hamiltonian function depends on external parameters, we may arrange for singularities of $\mathcal{H}_\lambda$ to occur on the q-axis for special values of λ. In this way we encounter again the whole hierarchy of the previous subsection. Notably the co-dimension is increased by 1: it is a co-dimension one phenomenon for a singularity $A_1^{\pm}$ to occur on the q-axis, for a singularity $D_4^{\pm}$ we need 4 parameters and so on.

For a family of Hamiltonian systems depending on external parameters it may as well be the Poisson manifold that undergoes changes leading to bifurcations. In [13] an A,D,E-scheme is used to classify the local Poisson structures up to co-dimension 7 with A_0 labelling $\{q,p\} = p$. In this context moduli already occur at co-dimension one: there is no co-ordinate change to transform away the modulus $m \neq 0$ of

$$A_1^{\pm,m} : \qquad \{q,p\} = m\,(p^2 \pm q^2) .$$

However, for the Hamiltonian system

$$\dot{q} = m\,(p^2 \pm q^2)\,\frac{\partial \mathcal{H}}{\partial p}$$
$$\dot{p} = -m\,(p^2 \pm q^2)\,\frac{\partial \mathcal{H}}{\partial q}$$

the scalar m can be taken care of by the left equivalence $\mathcal{H} \mapsto m^{-1}\mathcal{H}$ (i.e. by a rescaling of time). Where there are sufficiently many parameters, the two phenomena of parameter dependent Hamiltonian function and parameter dependent Poisson structure can interact to generate further local bifurcations.

2.1.3 Singular Equilibria

Motivated by the example of Section 1.1.1 we allow for the phase space to have singular points, ceasing to be a smooth manifold. We formalize this as follows.

Definition 2.15. A Poisson space is a topological manifold $\mathcal{P}$ together with a subring $\mathcal{A}$ of $C(\mathcal{P})$ that is a Poisson algebra: there is an alternating bilinear mapping

$$\{\,,\,\} : \mathcal{A} \times \mathcal{A} \longrightarrow \mathcal{A}$$

satisfying the *Leibniz rule*

$$\bigwedge_{f,g,h\in\mathcal{A}} \{f\cdot g,h\} = f\cdot\{g,h\} + \{f,h\}\cdot g$$

and the *Jacobi identity*

$$\bigwedge_{f,g,h\in\mathcal{A}} \{\{f,g\},h\} = \{f,\{g,h\}\} + \{\{f,h\},g\} .$$

For a Poisson manifold one has $\mathcal{A} = C^{\infty}(\mathcal{P})$.

Proposition 2.16. *On the Poisson space $\mathcal{P}$ consider the action of a compact group Γ that satisfies*

$$\bigwedge_{\gamma\in\Gamma}\bigwedge_{f,h\in\mathcal{A}} \{f\circ\gamma, h\circ\gamma\} = \{f,h\}\circ\gamma \qquad (2.7)$$

and thus defines a Poisson symmetry. Then the quotient space $\mathcal{B} = \mathcal{P}/\Gamma$ is again a Poisson space with Poisson algebra

$$\left\{ f\in\mathcal{A} \;\middle|\; \bigwedge_{\gamma\in\Gamma} f\circ\gamma = f \right\}$$

of Γ-invariant Hamiltonian functions.

In this way $\mathbb{R}^2/\mathbb{Z}_\ell$ is a Poisson space for every $\ell \in \mathbb{N}$.

Proof. By definition of the quotient topology the continuous functions on $\mathcal{B}$ correspond to the Γ-invariant continuous functions on $\mathcal{P}$. For Γ-invariant functions f, h in $\mathcal{A}$ we have

$$\{f,h\} = \{f\circ\gamma, h\circ\gamma\} = \{f,h\}\circ\gamma$$

which yields the desired Poisson bracket.   □

Singular points of 2-dimensional Poisson spaces are automatically equilibria. Correspondingly, a nontrivial $\mathbb{Z}_\ell$-symmetry forces the origin of $\mathbb{R}^2$ to be an equilibrium. In this way we are led to $\mathbb{Z}_\ell$-symmetric planar singularity theory, cf. [271, 128, 61, 253]. As shown in [238] one can obtain the $\mathbb{Z}_\ell$-symmetric universal unfolding of a $\mathbb{Z}_\ell$-symmetric singularity $\mathcal{H}_0$ from the universal unfolding

$$\mathcal{H}_\lambda(q,p) = \mathcal{H}_0(q,p) + \sum_{j=1}^{k}\lambda_j\mathcal{E}_j(q,p)$$

of the latter by symmetrisation

$$\frac{1}{\ell}\sum_{\gamma\in\mathbb{Z}_\ell}\mathcal{H}_\lambda\circ\gamma = \mathcal{H}_0 + \sum_{j=1}^{k}\frac{\lambda_j}{\ell}\sum_{\gamma\in\mathbb{Z}_\ell}\mathcal{E}_j\circ\gamma . \qquad (2.8)$$

Here the *Hilbert basis*, a set of generators of the ring $\mathcal{A}$ of $\mathbb{Z}_\ell$-invariant smooth planar functions comes into play.

$\mathbb{Z}_2$-Symmetry

For $\ell = 2$ we choose the Hilbert basis $u = \frac{1}{2}q^2$, $v = \frac{1}{2}p^2$, $w = pq$ which are related by the *syzygy* $2uv = \frac{1}{2}w^2$. The planar singularities A_{2l} are not $\mathbb{Z}_2$-symmetric. By $\mathbb{Z}_2$-symmetrisation we obtain for odd $k = 2l - 1$ the universal $\mathbb{Z}_2$-symmetric unfolding

$$\mathcal{H}_\lambda(q,p) = \frac{a}{2}p^2 + \frac{b}{(2l)!}q^{2l} + \sum_{j=1}^{l-1} \frac{\lambda_j}{(2j)!}q^{2j}$$

of the $\mathbb{Z}_2$-symmetric singularity $A_k^\pm$. For $k = 3$ this yields the Hamiltonian flip and the *dual* Hamiltonian flip bifurcation. Let us consider these in more detail.

Instead of working in the co-ordinates q, p on the covering space $\mathbb{R}^2$ we want to understand the geometric situation on the quotient space $\mathcal{B}$. Our choice of the Hilbert basis allows to identify

$$\mathcal{B} = \mathbb{R}^2/\mathbb{Z}_2 = \left\{ (u,v,w) \in \mathbb{R}^3 \;\middle|\; 2uv = \frac{w^2}{2}, u \geq 0, v \geq 0 \right\} \tag{2.9}$$

where the Poisson bracket relations read

$$\{u, v\} = w \;, \quad \{u, w\} = 2u \;, \quad \{v, w\} = -2v \;.$$

On this cone the phase portrait is given by the intersection with the energy level sets $\mathcal{H}(u, v, w) = h$. While the tip $(u, v, w) = (0, 0, 0)$ of the cone is a singular point and hence always an equilibrium, the regular equilibria occur where the two surfaces $\mathcal{B}$ and $\mathcal{H}^{-1}(h)$ are tangent to each other. For all $\lambda \in \mathbb{R}$ the level set of

$$\mathcal{H}_\lambda(u, v, w) = av + \frac{b}{6}u^2 + \lambda u$$

is a parabolic cylinder along the w-axis. Since $\mathcal{B}$ is a surface of revolution all regular equilibria satisfy $w = 0$. Therefore we restrict our attention to the relative position of $\{av + (b/6)u^2 + \lambda u = h\}$ to the u- and v-axes, see Figs. 2.4 and 2.5.

For $ab > 0$ the singular equilibrium loses its stability as λ passes from positive to negative values, giving rise to a regular centre at $(u, v) = (-3\lambda/b, 0)$. On the covering space $\mathbb{R}^2$ the (un)stable manifold of the saddle at the origin forms the familiar figure eight encircling the two centres at $(q, p) = \pm(\sqrt{-6\lambda/b}, 0)$. At the bifurcation value $\lambda = 0$ the parabolic cylinder $\mathcal{H}^{-1}(0)$ touches the tip of the cone $\mathcal{B}$ from the outside.

In the second case $ab < 0$ the additional regular equilibrium comes into existence as λ passes from negative to positive values. The singular equilibrium turns from unstable to stable and gives rise to a regular saddle at $(u, v) = (3\lambda/b, 0)$. The phase portraits on the covering space $\mathbb{R}^2$ are those of Fig. 2.1 along $\mu = 0$. At the bifurcation value $\lambda = 0$ of this dual Hamiltonian flip bifurcation the tip of the cone $\mathcal{B}$ is unstable as the parabolic cylinder $\mathcal{H}^{-1}(0)$ intersects $\mathcal{B}$ along two lines, the stable and unstable manifolds.

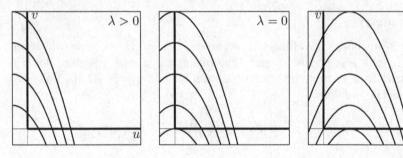

Fig. 2.4. Relative position of $\mathcal{B}$ and $\mathcal{H}_\lambda^{-1}(h)$ within the plane $w = 0$ in the case $ab > 0$.

Theorem 2.17. *Let the 2-dimensional phase space $\mathcal{P}$ have a conical singular point. Consider a 1-parameter family of Hamiltonian systems that has local level sets $(\mathcal{H}_\lambda^{-1}(h))_{h,\lambda}$ in general position. The level set $\mathcal{H}_\lambda^{-1}(0)$ of the singular point intersects the local cone for $\lambda < 0$ and for each $\lambda > 0$ there is a neighbourhood of the origin in which the local cone has no further intersection with $\mathcal{H}_\lambda^{-1}(0)$. Then a Hamiltonian flip bifurcation occurs if $\mathcal{H}_0^{-1}(0)$ touches the local cone from the outside and a dual Hamiltonian flip bifurcation occurs if there are additional intersection points of $\mathcal{H}_0^{-1}(0)$ with the local cone.*

Proof. The assumptions for $\lambda \neq 0$ imply that $\mathcal{H}_0^{-1}(0)$ touches the local cone when entering the singular point. Since the level sets $\mathcal{H}_\lambda^{-1}(0)$ are in general position the normal vectors e_λ of the tangent planes $T_0\mathcal{H}_\lambda^{-1}(0)$ satisfy the transversality condition

$$\frac{\mathrm{d}}{\mathrm{d}\lambda} e_\lambda \Big|_{\lambda=0} \neq 0$$

and the contact between $\mathcal{H}_0$ and the local cone is non-degenerate, i.e. of quadratic order.   □

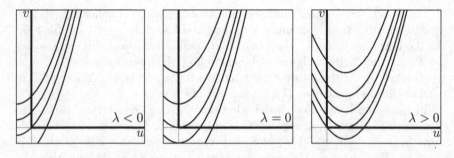

Fig. 2.5. Relative position of $\mathcal{B}$ and $\mathcal{H}_\lambda^{-1}(h)$ within the plane $w = 0$ in the case $ab < 0$.

The geometric situation is similar albeit increasingly complicated for the degenerate Hamiltonian flip bifurcations as the order of contact between the cone (2.9) and the level set $\mathcal{H}_0^{-1}(0)$ of

$$\mathcal{H}_\lambda(u,v,w) \;=\; av + \frac{2^l b}{(2l)!} u^l + \sum_{j=1}^{l-1} \frac{2^j \lambda_j}{(2j)!} u^j$$

becomes higher and higher. The various transversality conditions

$$\frac{\mathrm{d}}{\mathrm{d}\lambda} e_\lambda \bigg|_{\lambda=0} \;\neq\; 0$$

concern the orders of contacts from $l-1$ to quadratic and linear.

Example 2.18. Let us consider the reduced normal form of an axially symmetric perturbation of the so-called Penning trap. In [174] this perturbed ion trap has been treated as an axially symmetric perturbation of 1:1:1-resonant oscillators. Singular reduction of both the axial symmetry and the oscillator symmetry (introduced by normalization) leads to a one-degree-of-freedom problem, see [83]. The phase space depends on a parameter $\mu \in [-1,1]$ and reads

$$V_\mu \;=\; \left\{ (x,y,z) \in \mathbb{R}^3 \;\;\middle|\;\; R_\mu(x,y,z) = 0,\, |\mu| \leq x \leq 1 \right\}$$

with Poisson bracket

$$\{f,g\} \;=\; \langle \nabla f \times \nabla g, \nabla R_\mu \rangle$$

where $R_\mu(x,y,z) = y^2 + z^2 - (1-x)^2(x^2 - \mu^2)$. The parameter $\mu = N/L$ is the quotient of the third component N of the angular momentum and the unperturbed energy L of the resonant oscillator with three equal frequencies $\omega = 1$, see [83] for more details. For $\mu = 0$ the phase space has the form of a lemon, and the conical singularity at $(x,y,z) = (1,0,0)$ exists for all values $|\mu| < 1$, giving the phase space the form of a turnip (or balloon).

The (perfect) Penning trap describes the motion of a single ion in a constant magnetic and a quadrupole electric field. The relative strengths of these two fields can be adjusted to lead to a 1:1:1-resonant oscillator. We follow [174] and use a small detuning parameter δ to measure the difference between the third frequency (which derives directly from the electric field) and the first and second frequency (which are always equal to each other and are determined by magnetic and electric field together). Introducing the sextupolar terms of the electric field yields a further parameter β (where axial symmetry is still assumed). Normalization with respect to the resonant oscillator, reduction of the axial symmetry and the oscillator symmetry, dropping of constant terms and rescaling time leads to the Hamiltonian

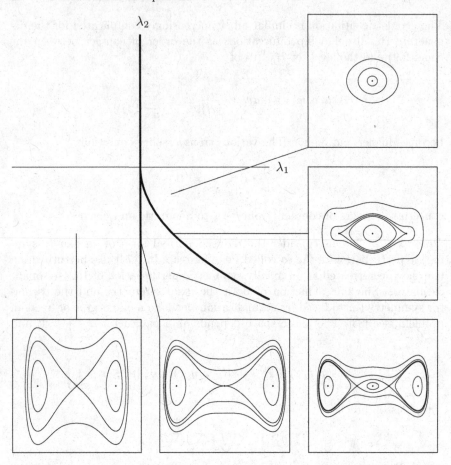

Fig. 2.6. Bifurcation diagram of the universal $\mathbb{Z}_2$-symmetric unfolding of A_5^+ displaying a degenerate Hamiltonian flip bifurcation.

$$\mathcal{H}_\lambda(x, y, z) = 16y + 41x^2 + (\lambda - 56)x$$

in one degree of freedom, cf. [174] for more details. The parameter $\lambda = \delta(4 - \delta)/(\beta^2 L)$ encodes the dependence on the external fields.

We are interested in bifurcations involving the singular point $(x, y, z) = (1, 0, 0)$ of V_μ. To this end we search for parameter values (λ, μ) where the energy level set $\mathcal{H}_\lambda^{-1}(\lambda - 15)$ through the singular point is tangent to the local cone. The energy level sets are parabolic cylinders along the z-axis and the phase spaces are surfaces of revolution around the x-axis. Therefore, such a tangency can only occur in the (x, y)-plane $\{z = 0\}$ which leads to the equations

$$-\frac{\lambda + 26}{16} = \pm\sqrt{1 - \mu^2}$$

whence we recover the ellipse

$$\frac{1}{256}(\lambda + 26)^2 + \mu^2 = 1 \qquad (2.10)$$

already obtained in [174]. Note that the tangency is for $\lambda > -26$ along the "upper" arc $y = (1 - x)\sqrt{x^2 - \mu^2}$ of the meridian section $V_\mu \cap \{z = 0\}$ and along the "lower" arc $y = (x - 1)\sqrt{x^2 - \mu^2}$ for $\lambda < -26$. At $\lambda = -26$ the ellipse (2.10) passes through $\mu = \pm 1$ where the phase space V_μ reduces to the point $(1, 0, 0)$. To establish quadratic contact between $\mathcal{H}_\lambda^{-1}(\lambda - 15)$ and V_μ for (λ, μ) satisfying (2.10) we compute second order derivatives and obtain the coefficient

$$-\frac{41}{8} + \frac{3}{\sqrt{1 - \mu^2}} \qquad \text{for } \lambda > -26$$

$$-\frac{41}{8} - \frac{3}{\sqrt{1 - \mu^2}} \qquad \text{for } \lambda < -26.$$

In the latter case the energy level set touches the local cone always from the outside – indeed, all parabolas $\mathcal{H}_\lambda^{-1}(h) \cap \{z = 0\}$ are "downward" oriented. We conclude that Hamiltonian flip bifurcations occur, confirming a conjecture in [174]. When $\lambda > -26$ the coefficient may vanish; this happens at $\lambda_* = -810/41$ (with $\mu_*^\pm = \pm\sqrt{1425/1681}$ determined from (2.10)). For $\lambda \in \,]-26, \lambda_*[$ the energy level set still touches V_μ from the outside and a Hamiltonian flip bifurcation takes place. In case $\lambda > \lambda_*$ we have a dual Hamiltonian flip bifurcation, again already conjectured in [174], since $\mathcal{H}_\lambda^{-1}(h) \cap V_\mu$ consists not only of the singular point.

For $\lambda = \lambda_*$ there is higher order contact between phase space and energy level set. We therefore compute the third derivative

$$\frac{3}{\sqrt{1 - \mu_*^2}}\left(1 + \frac{9}{1 - \mu_*^2}\right) = \frac{41}{8}\left(1 + \frac{41^2}{64}\right) > 0$$

at $x = 1$: the energy level set $\mathcal{H}_{\lambda_*}^{-1}(\lambda_* - 15)$ touches V_{μ_*} from the outside and we have a (once) degenerate Hamiltonian flip bifurcation. This proves that the bifurcation diagram locally around $(\lambda_*, \mu_*^\pm)$ is that defined by the $\mathbb{Z}_2$-symmetric universal unfolding of A_5^+ given in Fig. 2.6. Indeed, dividing out the $\mathbb{Z}_2$-symmetry yields the local phase portraits on Fig. 5 in [174] around the singular point. In particular there are two curves of centre-saddle bifurcations emanating from the two points $(\lambda_*, \mu_*^\pm)$. □

The universal $\mathbb{Z}_2$-symmetric unfolding of A_5^- has a subordinate global bifurcation of co-dimension 1 where the unstable singular point becomes connected to the regular saddle by heteroclinic orbits, cf. Fig. 2.7. Singularities with non-zero cubic terms are not $\mathbb{Z}_2$-symmetric, whence no further singularities of Fig. 2.3 occur in the present context. As shown in [21] the $\mathbb{Z}_2$-symmetric singularity $X_9^\pm$ is unimodal, has the universal $\mathbb{Z}_2$-symmetric unfolding

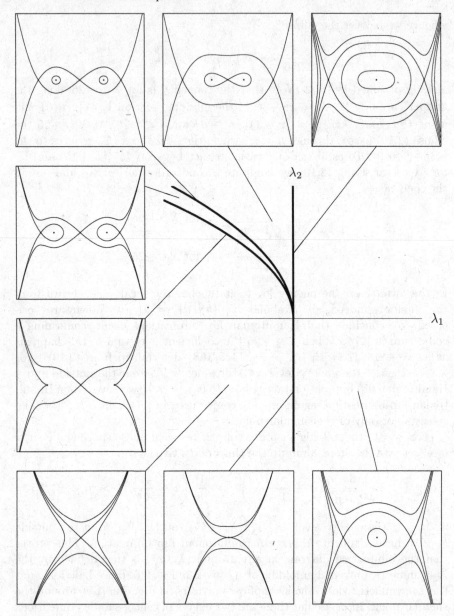

Fig. 2.7. Bifurcation diagram of the universal $\mathbb{Z}_2$-symmetric unfolding of A_5^- displaying a dual degenerate Hamiltonian flip bifurcation.

$$\mathcal{H}_\lambda(q,p) = \frac{a}{24}p^4 + \frac{\mu}{4}p^2q^2 + \frac{b}{24}q^4 + \frac{\lambda_1}{2}p^2 + \frac{\lambda_2}{2}q^2 + \lambda_3 pq \quad (2.11)$$

and thus yields a bifurcation of co-dimension 3, starting a series $X_{2i+5}^{\pm,\pm}$ of $\mathbb{Z}_2$-symmetric bifurcations of co-dimension $i+1$.

$\mathbb{Z}_3$-Symmetry

For $\ell = 3$ we need a Hilbert basis of the ring of $\mathbb{Z}_3$-invariant smooth planar functions and choose

$$u = \frac{p^2 q}{2} - \frac{q^3}{6} , \quad v = \frac{p^3}{6} - \frac{pq^2}{2} , \quad w = \frac{p^2 + q^2}{2} .$$

This allows us the identification

$$\mathcal{B} = \mathbb{R}^2/_{\mathbb{Z}_3} = \left\{ (u,v,w) \in \mathbb{R}^3 \;\middle|\; \frac{3}{2}(u^2 + v^2) = \frac{w^3}{3}, w \geq 0 \right\}$$

with Poisson bracket relations

$$\{u,v\} = w^2 , \quad \{u,w\} = 3v , \quad \{v,w\} = -3u .$$

Note that the two singularities A_1^+ and D_4^- are among the above generators. On the other hand the saddle A_1^- is not $\mathbb{Z}_3$-symmetric. It is therefore generic for a $\mathbb{Z}_3$-symmetric equilibrium to be a centre and hence dynamically stable. One needs at least one parameter to encounter the dynamically unstable "ape saddle" D_4^-, which has the universal $\mathbb{Z}_3$-symmetric unfolding[7]

$$3_3 : \qquad \mathcal{H}_\lambda(u,v,w) = au + \lambda w , \qquad (2.12)$$

see also [271, 128, 253]. Here two of the three unfolding terms vanish due to the $\mathbb{Z}_3$-symmetrisation. As shown in [289] there are no further simple $\mathbb{Z}_3$-symmetric planar singularities with respect to C^∞-right equivalences. We have already seen in Section 2.1.1 that passing to C^∞-left-right equivalences allows to remove one modulus (if the singularity is not quasi-homogeneous). In this way we obtain from [289] an additional simple $\mathbb{Z}_3$-symmetric planar singularity $(a/2)w^2 + (m/2)(u^2 - v^2)$ with non-degeneracy conditions $a, m \neq 0$. The co-dimension within the ring $\mathcal{A}$ of $\mathbb{Z}_3$-invariant functions is three and a universal $\mathbb{Z}_3$-symmetric unfolding is given by

$$2_{2,3} : \qquad \mathcal{H}_\lambda(u,v,w) = \frac{a}{2}w^2 + \frac{m}{2}(u^2 - v^2) + \lambda_1 u + \lambda_2 v + \lambda_3 w .$$

It is remarkable that there is no $\mathbb{Z}_3$-symmetric local bifurcation of co-dimension two.

---

[7] We use the labelling introduced in [289].

$\mathbb{Z}_4$-**Symmetry**

For $\ell = 4$ we choose the Hilbert basis

$$u = \frac{p^4}{24} - \frac{p^2 q^2}{4} + \frac{q^4}{24} \ , \quad v = -\frac{p^3 q - pq^3}{6} \ , \quad w = \frac{p^2 + q^2}{2}$$

of the ring of $\mathbb{Z}_4$-invariant smooth planar functions and thus identify

$$\mathcal{B} = \mathbb{R}^2/_{\mathbb{Z}_4} = \left\{ (u,v,w) \in \mathbb{R}^3 \ \Big| \ 2(u^2 + v^2) = \frac{w^4}{18} \, , \, w \geq 0 \right\}$$

with Poisson bracket relations

$$\{u,v\} = \frac{2}{9}w^3 \ , \quad \{u,w\} = 4v \ , \quad \{v,w\} = -4u \ .$$

The stable singularity A_1^+ is the sole simple $\mathbb{Z}_4$-symmetric planar singularity with respect to C^∞-left-right equivalences, cf. [289]. In the $\mathbb{Z}_4$-symmetric context the unimodal singularity $au + (m/2)w^2$ has the non-degeneracy conditions $a \neq 0$ and $3m \neq \pm a$. This latter condition expresses that the parabolic cylinder $au + (m/2)w^2 = 0$ (along the v-axis) does not coincide along a whole curve with the phase space $\mathcal{B}$. Therefore the universal $\mathbb{Z}_4$-symmetric unfolding

$$4_4 \ : \qquad \mathcal{H}_\lambda(u,v,w) = au + \frac{m}{2}w^2 + \lambda w$$

yields, after we pass again to C^0-equivalences, the two different types of co-dimension one bifurcations detailed in [271, 128, 253]. Both types appear as 1-parameter subfamilies in the universal $\mathbb{Z}_4$-symmetric unfolding

$$5_{3,4} \ : \qquad \mathcal{H}_\lambda(u,v,w) = a\left(3u \pm \frac{w^2}{2}\right) + \frac{m}{6}w^3 + \lambda_1 w + \lambda_2 \frac{w^2}{2} \quad (2.13)$$

given in [289]. Here the central singularity $\mathcal{H}_0$ has to satisfy the non-degeneracy conditions $a, m \neq 0$ and the modulus (with respect to C^∞-right equivalences) can be brought back to $m = \pm 1 = \text{sgn}(am)$ by allowing for C^∞-left-right equivalences. Since $3u + \frac{1}{2}w^2 = \frac{1}{4}(p^2 - q^2)^2$ and $3u - \frac{1}{2}w^2 = -p^2 q^2$ this leaves us with two cases, which are mapped into each other by $\lambda_2 \mapsto \lambda_2 \pm \frac{1}{2}$.

Example 2.19. Let us consider the 2-parameter family (2.13) in more detail, see Fig. 2.8. We work with $3u - \frac{1}{2}w^2$, rescale to $a = -\frac{1}{4}$ and choose $m = +1$; furthermore we write $\lambda = \lambda_1$ and $\mu = \lambda_2$ for the unfolding parameters. Instead of working on the Poisson space $\mathcal{B}$ we study

$$\mathcal{H}_{\lambda,\mu}(q,p) = \frac{p^2 q^2}{4} + \frac{(p^2 + q^2)^3}{48} + \frac{\lambda}{2}(p^2 + q^2) + \frac{\mu}{8}(p^2 + q^2)^2 \quad (2.14)$$

on $\mathbb{R}^2$ with the $\mathbb{Z}_4$-symmetry still in place. This yields the equations of motion

$$\dot{q} = p\left(\tfrac{1}{2}q^2 + f(p^2 + q^2)\right)$$
$$\dot{p} = -q\left(\tfrac{1}{2}p^2 + f(p^2 + q^2)\right)$$

with

$$f(\varrho) = \frac{\varrho^2}{8} + \lambda + \frac{\mu}{2}\varrho \; .$$

One easily computes the equilibria and their linearizations:

$$DX_{\mathcal{H}_{\lambda,\mu}}(0,0) = \begin{pmatrix} 0 & \lambda \\ -\lambda & 0 \end{pmatrix} \qquad \text{for all } (\lambda,\mu) \in \mathbb{R}^2.$$

For $\lambda \le \tfrac{1}{2}\mu^2$ the roots $\varrho_\pm = -2\mu \pm \sqrt{4\mu^2 - 8\lambda}$ of $f(\varrho) = 0$ are real and may yield equilibria on the q- and p-axes by means of $q_\pm = p_\pm = \sqrt{\varrho_\pm}$:

$$DX_{\mathcal{H}_{\lambda,\mu}}(\pm q_\pm, 0) = \begin{pmatrix} 0 & \tfrac{1}{2}q_\pm^2 \\ -\mu q_\pm^2 - \tfrac{1}{2}q_\pm^4 & 0 \end{pmatrix}$$

$$DX_{\mathcal{H}_{\lambda,\mu}}(0, \pm p_\pm) = \begin{pmatrix} 0 & \mu p_\pm^2 + \tfrac{1}{2}p_\pm^4 \\ -\tfrac{1}{2}p_\pm^2 & 0 \end{pmatrix} \; .$$

The necessary condition $\varrho_+ > 0$ implies $\mu < 0$ or $\lambda < 0$ while $\varrho_- > 0$ if both $\mu < 0$ and $\lambda > 0$. For $\lambda \le \tfrac{1}{2}(\mu + \tfrac{1}{2})^2$ the roots $\rho_\pm = -(\mu + \tfrac{1}{2}) \pm \tfrac{1}{2}\sqrt{(2\mu+1)^2 - 8\lambda}$ of $\tfrac{1}{2}\rho + f(2\rho) = 0$ are real and may yield equilibria $(q,p) = (\pm r_\pm, \pm r_\pm)$ on the diagonals by means of $r_\pm = \sqrt{\rho_\pm}$:

$$DX_{\mathcal{H}_{\lambda,\mu}}(\pm(r_\pm, \epsilon r_\pm)) = \begin{pmatrix} \epsilon r_\pm^2(1 + \mu + r_\pm^2) & r_\pm^2(\mu + r_\pm^2) \\ -r_\pm^2(\mu + r_\pm^2) & -\epsilon r_\pm^2(1 + \mu + r_\pm^2) \end{pmatrix}$$

with $\epsilon = \pm 1$. Here $\rho_+ > 0$ implies $\mu < -\tfrac{1}{2}$ or $\lambda < 0$ and $\rho_- > 0$ holds if both $\mu < -\tfrac{1}{2}$ and $\lambda > 0$.

From this the bifurcation lines are easily obtained, cf. Fig. 2.8. On the μ-axis $\{\lambda = 0\}$ the origin ceases to be a centre and for $\mu \ne -\tfrac{1}{2}, 0$ one of the two types of co-dimension one $\mathbb{Z}_4$-symmetric bifurcations occurs as the μ-axis is crossed. For $\mu > 0$ this leads to four centres $(\pm q_+, 0), (0, \pm p_+)$ and four saddles $(\pm r_+, \pm r_+)$ when λ becomes negative. In case $-\tfrac{1}{2} < \mu < 0$ four saddles $(\pm q_-, 0), (0, \pm p_-)$ exist for positive λ and turn into four saddles $(\pm r_+, \pm r_+)$ as λ changes sign. At $(\lambda, \mu) = (0, -\tfrac{1}{2})$ we have the second parameter point of co-dimension two. For $\mu < -\tfrac{1}{2}$ four saddles $(\pm q_-, 0), (0, \pm p_-)$ and four centres $(\pm r_-, \pm r_-)$ bifurcate off from the origin as λ passes from negative to positive values.

The remaining bifurcations are secondary and do not involve the origin. Since they concern regular equilibria on $\mathcal{B}$ they form $\mathbb{Z}_4$-related quartets on $\mathbb{R}^2$. In this way four simultaneous centre-saddle bifurcations occur at

$$\left\{ (\lambda, \mu) \in \mathbb{R}^2 \;\middle|\; \lambda = \frac{\mu^2}{2}, \mu < 0 \right\}$$

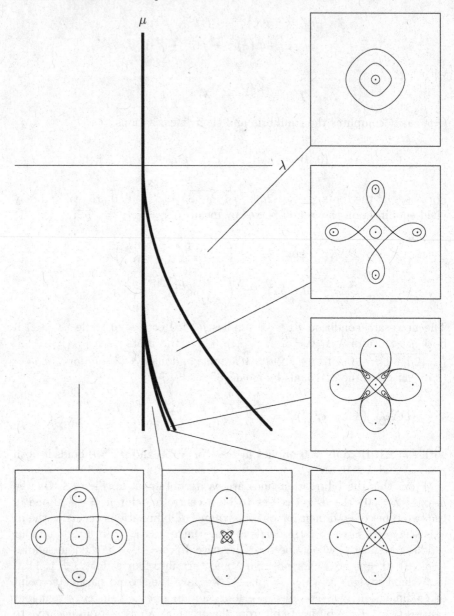

Fig. 2.8. Bifurcation diagram of (2.14) with $\mathbb{Z}_4$-symmetric phase portraits.

where the equilibria on the axes meet at parabolic equilibria and vanish. Similarly, the centres $(\pm r_-, \pm r_-)$ and the saddles $(\pm r_+, \pm r_+)$ meet at the four parabolic equilibria $(\pm\sqrt{-\mu - (1/2)}, \pm\sqrt{-\mu - (1/2)})$ and vanish in centre-saddle bifurcations as (λ, μ) passes through

$$\left\{ (\lambda, \mu) \in \mathbb{R}^2 \;\middle|\; \lambda = \frac{1}{2}(\mu + \frac{1}{2})^2, \; \mu < -\frac{1}{2} \right\} .$$

Finally a connection bifurcation takes place for $(\lambda, \mu) \in \mathbb{R}^2$ satisfying

$$12\mu^2 + 6\mu + 1 - 12\lambda = \left(\sqrt{4\mu^2 - 8\lambda}\right)^3 + \left(\sqrt{(2\mu + 1)^2 - 8\lambda}\right)^3$$

$$\mu < -\frac{1}{2}, \qquad 0 < \lambda < \frac{1}{2}(\mu + \frac{1}{2})^2$$

when the eight occurring saddles all have the same energy.   □

As shown in [289] there are two $\mathbb{Z}_4$-symmetric planar singularities of C^∞-left-right co-dimension three with universal unfoldings

$$2_{2,4} \;:\; \mathcal{H}_\lambda(u, v, w) = \frac{a}{2}w^2 + \frac{m}{2}(u^2 - v^2) + \lambda_1 u + \lambda_2 v + \lambda_3 w$$

$$5_{4,4} \;:\; \mathcal{H}_\lambda(u, v, w) = a\left(3u \pm \frac{w^2}{2}\right) + \frac{m}{24}w^4 + \lambda_1 w + \frac{\lambda_2}{2}w^2 + \frac{\lambda_3}{6}w^3$$

and non-degeneracy conditions $a, m \neq 0$.

$\mathbb{Z}_\ell$-Symmetry with $\ell \geq 5$

Following the pattern of $\ell = 3, 4$ we choose the Hilbert basis

$$u_\ell = \frac{1}{\ell!}\mathrm{Re}\,(q + ip)^\ell = \sum_{j=0}^{[\ell/2]} (-1)^j \frac{p^{2j}q^{\ell-2j}}{(2j)!(\ell - 2j)!}$$

$$v_\ell = \frac{1}{\ell!}\mathrm{Im}\,(q + ip)^\ell = \sum_{j=0}^{[(\ell-1)/2]} (-1)^j \frac{p^{2j+1}q^{\ell-2j-1}}{(2j + 1)!(\ell - 2j - 1)!}$$

$$w = \frac{p^2 + q^2}{2}$$

and identify

$$\mathcal{B} = \mathbb{R}^2/\mathbb{Z}_\ell = \left\{ (u_\ell, v_\ell, w) \in \mathbb{R}^3 \;\middle|\; R_\ell(u_\ell, v_\ell, w) = 0, \; w \geq 0 \right\}$$

with $R_\ell(u_\ell, v_\ell, w) = \frac{2^{\ell-1}}{(\ell - 1)!\,\ell!}w^\ell - \frac{\ell}{2}(u_\ell^2 + v_\ell^2)$ and Poisson bracket

$$\{f, g\} = \langle \nabla f \times \nabla g, \nabla R_\ell \rangle .$$

Next to the stable singularity A_1^+ we always have the unimodal $\mathbb{Z}_\ell$-symmetric planar singularity $au_\ell + (m/2)w^2$ with non-degeneracy conditions $a, m \neq 0$ and universal $\mathbb{Z}_\ell$-symmetric unfolding

$$1_{2,\ell} : \quad \mathcal{H}_\lambda(u_\ell, v_\ell, w) = au_\ell + \frac{m}{2}w^2 + \lambda w . \qquad (2.15)$$

Using C^∞-left-right equivalences the modulus m can be removed for odd ℓ and reduced to $\pm 1 = \operatorname{sgn}(am)$ if ℓ is even. For

$$2_{2,\ell} : \quad \mathcal{H}_\lambda(u_\ell, v_\ell, w) = \frac{a}{2}w^2 + \frac{m}{2}(u_\ell^2 - v_\ell^2) + \lambda_1 u_\ell + \lambda_2 v_\ell + \lambda_3 w$$

we are always left with $\pm 1 = \operatorname{sgn}(am)$.

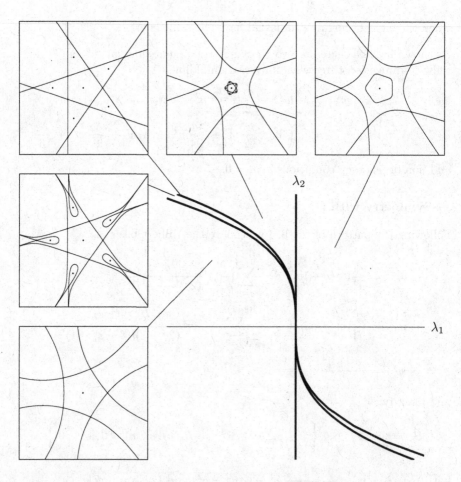

Fig. 2.9. Bifurcation diagram of (2.16) with $\mathbb{Z}_5$-symmetric phase portraits.

We follow [289] and consider further singularities for odd ℓ and even ℓ separately. If $\ell = 2k - 1$ we have

$$3_\ell \quad : \quad \mathcal{H}_\lambda(u_\ell, v_\ell, w) = au_\ell + \sum_{j=k}^{\ell-2} \frac{\mu_j}{j!} w^j + \sum_{j=1}^{k-1} \frac{\lambda_j}{j!} w^j$$

with for $\ell = 5$ the particular case

$$3_5 \quad : \quad \mathcal{H}_\lambda(u_5, v_5, w) = au_5 + \frac{\mu_3}{6} w^3 + \lambda_1 w + \frac{\lambda_2}{2} w^2 \qquad (2.16)$$

where the modulus μ_3 can be reduced to $\mu_3 \in \{0, 1\}$ by C^∞-left-right equivalences and removed completely using C^0-equivalences. The bifurcation diagram of (2.16) is straightforwardly obtained and given in Fig. 2.9. In case $\ell = 2k$ we need, next to $a \neq 0$, the non-degeneracy condition $\ell! \cdot \mu_k \neq 2^k k! \cdot a$ for

$$4_\ell \quad : \quad \mathcal{H}_\lambda(u_\ell, v_\ell, w) = au_\ell + \sum_{j=k}^{\ell-2} \frac{\mu_j}{j!} w^j + \sum_{j=1}^{k-1} \frac{\lambda_j}{j!} w^j \qquad (2.17)$$

and correspondingly also have

$$5_{m,\ell} \quad : \quad \mathcal{H}_\lambda(u_\ell, v_\ell, w) = a\left(u_\ell \pm \frac{2^k}{\ell!} w^k\right) + \sum_{j=m}^{m+k-2} \frac{\mu_j}{j!} w^j + \sum_{j=1}^{m-1} \frac{\lambda_j}{j!} w^j$$

for every integer $m \geq k + 1$.

2.1.4 Equilibria in the Phase Space $\mathbb{R}^3$

When studying the bifurcations of singular equilibria in the previous section we kept the phase space fixed and in particular had a singular equilibrium for every parameter value. Let now the phase space be parameter dependent as well. In 1-parameter families this typically results in the singular point only occurring at isolated parameter values.

All Poisson spaces of the previous section could be embedded into $\mathbb{R}^3$ and for this reason we consider Poisson structures

$$\{f, g\} = \langle \nabla f \times \nabla g, \nabla R \rangle \qquad (2.18)$$

with $R \in C^\infty(\mathbb{R}^3)$ in more detail. Where $\nabla R \neq 0$ the symplectic leaves can locally be identified with $\mathbb{R}^2$ parametrised by the value of R.

Let now $\nabla R(0) = 0$. Generically this singularity is stable and we can find, multiplying R by -1 if necessary, local co-ordinates u, v, w such that

$$R(u, v, w) = u^2 + v^2 \pm w^2 .$$

The two Poisson structures (2.18) on $\mathbb{R}^3$ defined[8] by these two quadratic functions are the *Lie–Poisson* structures on $\mathfrak{so}(3)^*$ and $\mathfrak{so}(2,1)^* \cong \mathfrak{sl}_2(\mathbb{R})^*$, respectively.

The symplectic leaves $\{R = \rho\}$ of $\mathfrak{so}(3)^*$ are the concentric spheres of radius $\rho > 0$ and the origin which is therefore always a (dynamically) stable equilibrium. The Hamiltonian (sub)systems on the spheres may undergo bifurcations as ρ varies, but this concerns isolated values. To obtain bifurcations that extend to the origin one needs the Hamiltonian on $\mathfrak{so}(3)^*$ to depend on external parameters. Then ρ will be distinguished with respect to α and re-parametrisations

$$(\alpha, \rho) \;\mapsto\; (\tilde{\alpha}(\alpha), \tilde{\rho}(\alpha, \rho))$$

also have to satisfy $\tilde{\rho}(\alpha, 0) = 0$ for all α.

Phase Space $\mathfrak{sl}_2(\mathbb{R})^*$

The Lie–Poisson structure on $\mathfrak{sl}_2(\mathbb{R})^*$ has the bracket relations

$$\{u, v\} \;=\; w \;, \quad \{u, w\} \;=\; 2u \;, \quad \{v, w\} \;=\; -2v$$

which is of the form (2.18) with

$$R(u, v, w) \;=\; \frac{w^2}{2} - 2uv \;. \tag{2.19}$$

The symplectic leaves are the one-sheeted hyperboloids $\{R = \rho\}$ with $\rho > 0$, the two sheets of the hyperboloids $\{R = \rho\}$ with $\rho < 0$, the double cone $\{R = 0, (u, v, w) \neq 0\}$ and the origin. Note that $\{R = 0, u \geq 0, v \geq 0\}$ is the Poisson space (2.9) of Section 2.1.3.

For a generic Hamiltonian system on $\mathfrak{sl}_2(\mathbb{R})^*$ the gradient $\nabla H(0)$ of the Hamiltonian function is nonzero. If $R(\nabla H(0)) > 0$ then the gradient $\nabla H(0)$ points inside the double cone and the origin is (dynamically) stable. The flow on the one-sheeted hyperboloids consists (locally) of periodic orbits (that are not contractible to a point) and the flow on a sheet of $\{R = \rho < 0\}$ consists of a family of periodic orbits that shrinks down to an elliptic equilibrium. In case $R(\nabla H(0)) < 0$ the gradient $\nabla H(0)$ points outside the double cone and the origin is unstable with stable and unstable manifolds

$$H^{-1}(0) \,\cap\, R^{-1}(0) \backslash \{0\} \;.$$

If $U \subseteq \mathbb{R}^3$ is a sufficiently small neighbourhood of the origin, then all other trajectories on the double cone leave U as $t \to \pm\infty$ and the same is true for all orbits with $R < 0$. On a one-sheeted hyperboloid $\{R = \rho > 0\}$ there are two saddles and no periodic orbit in U.

---

[8] The above does *not* provide a linearization of generic Poisson structures that vanish at the origin of $\mathbb{R}^3$. In fact, an example in [292] shows that in case of $\mathfrak{sl}_2(\mathbb{R})^*$ the symplectic leaves need not be level sets of a Casimir R.

It is a co-dimension one phenomenon to have a gradient $\nabla H(0)$ with $R(\nabla H(0)) = 0$. Here the energy level set $H^{-1}(0)$ is tangent to the cone $R^{-1}(0)$ and we may choose co-ordinates (u, v, w) such that, next to (2.19), we have

$$\nabla H(0) \parallel \begin{pmatrix} 0 \\ 1 \\ 0 \end{pmatrix} .$$

Multiplying H by -1 if necessary, we may assume that the factor $a = \partial H / \partial v(0)$ is positive. On the linear level the universal unfolding reads $av + \lambda u$ – for $\lambda > 0$ the origin is stable and for $\lambda < 0$ the origin is unstable. Without nonlinear terms the cone $R^{-1}(0)$ has infinite order of contact to the level set of the origin at $\lambda = 0$. Therefore we require the second derivative

$$b = \frac{\mathrm{d}^2}{\mathrm{d}u^2} H_0(0) > 0$$

to be nonzero. If $b < 0$ we perform the π-rotation

$$(u, v) \longmapsto (-u, -v) ,$$

thereby interchanging the $u, v \geq 0$ symplectic leaves $\{R = \rho \leq 0\}$ with those satisfying $u, v \leq 0$.

Within the (u, v)-plane $\{w = 0\}$ the relative position of the zero level sets of

$$H_\lambda(u, v, w) = av + \frac{b}{2}u^2 + \lambda u \tag{2.20}$$

as λ passes through 0 and the cone $\{R = 0, u \geq 0, v \geq 0\}$ is as depicted in Fig. 2.4, while Fig. 2.5 shows the relative positions of $H_\lambda^{-1}(0)$ to the cone $\{R = 0, u \leq 0, v \leq 0\}$ (after the above π-rotation, equivalently one may keep the cone and take b negative). The standard form (2.20) is w-independent, whence all equilibria (where ∇H and ∇R are parallel) lie in the (u, v)-plane. This leaves us with the relative position of the family $H_\lambda^{-1}(h) \cap \{w = 0\}$ of parabolas to the family $R^{-1}(\rho) \cap \{w = 0\}$ of hyperbolas, cf. Fig. 2.10.

Theorem 2.20. *The set of parameters $(\rho, \lambda) \in \mathbb{R}^2$ for which the one-degree-of-freedom system defined by (2.20) on $R^{-1}(\rho)$ is not structurally stable consists of the cubic line*

$$\left\{ (\rho, \lambda) \in \mathbb{R}^2 \;\middle|\; 8\lambda^3 + 27ab^2\rho = 0 \right\}$$

where parabolic equilibria occur.

Proof. We first fix $\lambda < 0$ whence $\nabla H_\lambda(0)$ lies outside the double cone. There are two saddles on evey one-sheeted hyperboloid $\{R = \rho\}$ with $\rho > 0$ sufficiently small and furthermore one centre. As ρ grows and passes through

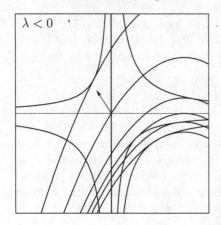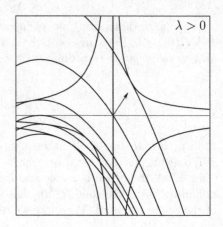

Fig. 2.10. From these intersections of parabolas and hyperbolas the phase potraits of the Hamiltonian sytem defined by (2.20) on $\mathfrak{sl}_2(\mathbb{R})^*$ can be easily obtained.

$$\rho = \frac{-8\lambda^3}{27ab^2} > 0$$

a centre and one of the saddles meet and vanish in a centre-saddle bifurcation. As ρ shrinks to zero the two saddles both tend to the (dynamically) unstable origin, while the family of centres tends to the centre

$$(u, v, w) = \left(-\frac{\lambda}{b}, 0, 0\right)$$

on the double cone $R^{-1}(0)$. From this point originates the family of centres on the positive sheet $\{R = \rho < 0, u \geq 0, v \geq 0\}$ while there are no equilibria on the negative sheets $\{R = \rho < 0, u \leq 0, v \leq 0\}$ of the two-sheeted hyperboloids.

To obtain the situation for $\lambda > 0$ we merely have to reflect the parabolas at the v-axis, see Fig. 2.10. However, since $R^{-1}(\rho)$ is a surface of revolution around the $(u = v)$-axis this very similar picture of relative positions of hyperbolas and parabolas leads to quite different phase portraits. The origin is now dynamically stable and serves as limit to the families of centres on both the positive sheet and the negative sheet of the two-sheeted hyperboloids $\{R = \rho < 0\}$. Furthermore there is a saddle

$$(u, v, w) = \left(-\frac{\lambda}{b}, 0, 0\right)$$

on the double cone which is part of a family of saddles starting on the one-sheeted hyperboloids and ending on the negative sheets of the two-sheeted hyperboloids. On the latter they meet the family of centres in a centre-saddle bifurcation at

$$\rho = \frac{-8\lambda^3}{27ab^2} < 0$$

and vanish.

For $\lambda = 0$ the only equilibria are a family of saddles on the one-sheeted hyperboloids and a family of centres on the positive sheets of the two-sheeted hyperboloids, which both converge to the origin. The latter is unstable, its stable and unstable manifolds lie within the negative cone $\{R = 0, u \leq 0, v \leq 0\}$ of the double cone.   □

It is instructive to study structural stability of the Hamiltonian systems defined by (2.20) on the 3-dimensional Poisson manifold $\mathfrak{sl}_2(\mathbb{R})^*$. This amounts to considering the 2-parameter family of Theorem 2.20 as a 1-parameter family, parametrised by λ, of 1-parameter families of Hamiltonian systems, parametrised by ρ; the value ρ of the Casimir R is distinguished with respect to λ. For all $\lambda \neq 0$ the centre-saddle bifurcation at $\rho = -8\lambda^3/(27ab^2)$ is universally unfolded under variation of ρ and the 1-parameter family of one-degree-of-freedom systems is structurally stable. Thus, $\lambda = 0$ is the only bifurcation value.

Example 2.21. We return to the Penning trap considered in Example 2.18. Recall that the Hamiltonian

$$\mathcal{H}_\lambda(x, y, z) = 16y + 41x^2 + (\lambda - 56)x$$

is defined on the phase space

$$V_\mu = \left\{ (x, y, z) \in \mathbb{R}^3 \;\middle|\; R_\mu(x, y, z) = 0, |\mu| \leq x \leq 1 \right\}$$

with Poisson bracket (2.18) and

$$R_\mu(x, y, z) = y^2 + z^2 - (1 - x)^2(x^2 - \mu^2) .$$

We concentrate on a small neighbourhood of the origin $(x, y, z) = (0, 0, 0)$. Putting $\mu = 0$ the origin is the singular point of the (local) double cone $R_0^{-1}(0)$ while $R_\mu^{-1}(0)$ is for $\mu \neq 0$ (locally) diffeomorphic to a two-sheeted hyperboloid. The restriction $x \geq 0$ singles out the "positive" sheets and the "positive" cone.

Once again the energy level sets $\mathcal{H}_\lambda^{-1}(h)$ are parabolic cylinders and can touch the surfaces of revolution V_μ only in the (x, y)-plane. The bifurcation of Theorem 2.20 occurs where the parabola

$$16y = h - 41x^2 - (\lambda - 56)x$$

touches one of the two lines

$$y = \pm x(1 - x)$$

at the origin, i.e. when

$$\lambda - 56 = \pm 16$$

which yields the two values $\lambda = 40$ and $\lambda = 72$.

At $\lambda = 72$ the parabolic cylinder remains outside the local cone and we recover the situation of Theorem 2.20 at the positive sheets (cone) $u \geq 0, v \geq 0$ with $\rho = -\mu^2 \leq 0$. This yields (local) structural stablity for $(\lambda, \mu) \neq (72, 0)$.

At $\lambda = 40$ the parabolic cylinder intersects the local cone and we recover the situation of Theorem 2.20 at the negative sheets/cone $u \leq 0, v \leq 0$ with $\rho = -\mu^2 \leq 0$. This yields a local cusp $2048(40 - \lambda)^3 = 2187\mu^2$ of centre-saddle bifurcations. As shown in [174] these two lines extend to the two values

$$(\lambda_*, \mu_*^\pm) = \left(-\frac{810}{41}, \pm\sqrt{\frac{1425}{1681}} \right)$$

where $(x, y, z) = (1, 0, 0)$ undergoes the degenerate Hamiltonian flip bifurcation studied in Example 2.18. The results of [83] imply that these are indeed two lines of centre-saddle bifurcations.

The bifurcation diagram given in Fig. 2 of [174] also contains the segment

$$\left\{ (\lambda, \mu) \in \mathbb{R}^2 \;\middle|\; 40 < \lambda < 72 , \mu = 0 \right\}$$

where the singular point $(0, 0, 0)$ of V_0 is dynamically unstable. While this situation is structurally stable with respect to perturbations of the Hamiltonian on V_0, there is no corresponding equilibrium on the (locally) smooth phase space V_μ with $\mu \neq 0$. □

Reconstructing the (normalized) Penning trap of Examples 2.18 and 2.21 to three degrees of freedom yields a supercritical Hamiltonian Hopf bifurcation at $\lambda = 72$ and a subcritical Hamiltonian Hopf bifurcation at $\lambda = 40$, cf. Section 2.2.2 below. In fact, the standard form of the Hamiltonian Hopf bifurcation leads to (2.20) on

$$\left\{ R = \rho = -\sigma^2 , u \geq 0, v \geq 0 \right\}$$

with Poisson bracket (2.18) and R given by (2.19). The restriction to the positive sheets/cone makes it necessary to distinguish between the supercritical case $b > 0$ and the subcritical case $b < 0$ of the Hamiltonian Hopf bifurcation; a scaling yields $b = \pm 1$.

Let us now consider the degenerate case of vanishing coefficient of u^2. On $\mathfrak{sl}_2(\mathbb{R})^*$ we can normalize the family of Hamiltonians H_λ with respect to the linear part $H_0(u, v, w) = av$ in such a way that $H_\lambda(u, v, w)$ does not depend on w and only linearly on v, see Appendix C. This yields the standard form

$$H_\lambda^\pm = av + \frac{b}{\ell!}u^\ell + \sum_{j=1}^{\ell-1} \frac{\lambda_j}{j!}u^j$$

where $\pm = \mathrm{sgn}(ab)$ distinguishes two cases if ℓ is odd. The bifurcation diagram for $\ell = 3$ with $a = b = 1$ is given in Fig. 2.11. Instead of 3-dimensional phase portraits only the relative positions of the cubics

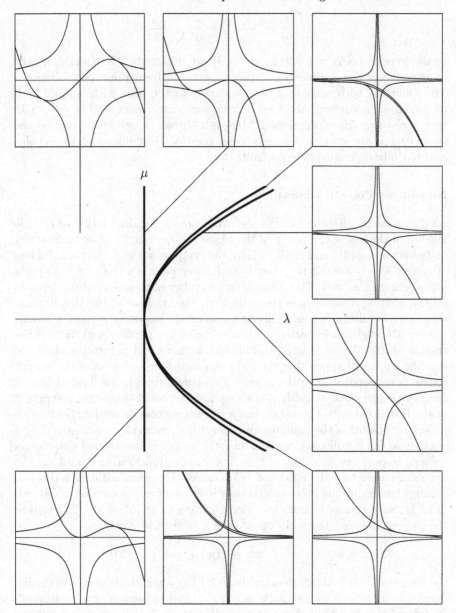

Fig. 2.11. Bifurcation diagram of the family of Hamiltonian systems on $\mathfrak{sl}_2(\mathbb{R})^*$ defined by $H_\lambda(u, v, w) = v + \frac{1}{6}u^3 + \lambda u + \frac{1}{2}\mu u^2$.

$$v = \frac{h}{a} - \frac{b}{6a}u^3 - \frac{\lambda}{a}u - \frac{\mu}{2a}u^2$$

to the hyperbolas $R = \rho$ within $\{w = 0\}$ are depicted. It is straightforward to reconstruct the 3-dimensional phase portraits from these two foliations. The foliations with $\mathrm{sgn}(ab) < 0$ are related to the ones with $\mathrm{sgn}(ab) > 0$ by simultaneous reflections $\lambda \mapsto -\lambda$ in parameter space and $u \mapsto -u$ in the (u, v)-plane. However, since $R^{-1}(\rho)$ is a surface of revolution around the $(u = v)$-axis the corresponding relative positions of cubics and hyperbolas lead to quite different phase portraits.

Non-linear Poisson structures

As discussed in [193] the Lie–Poisson structure on the dual $\mathfrak{se}(2)^*$ of the Lie algebra of the Euclidean group of the plane can be thought of as an intermediate between $\mathfrak{so}(3)^*$ and $\mathfrak{sl}_2(\mathbb{R})^*$. Here the symplectic leaves are the cylinders of radius $\rho > 0$ around the w-axis, and every point on the w-axis forms a symplectic leaf as well. The (dynamical) stability of these equilibria depends on the limiting dynamics on the cylinders. The gradient of the Hamiltonian at such an equilibrium generically has a nonzero component along the w-axis whence all neighbouring orbits on the cylinders are periodic and the equilibrium is stable. At an isolated equilibrium within the 1-parameter family of equilibria it may happen that the (still non-vanishing) gradient of the Hamiltonian is perpendicular to the w-axis. This equilibrium is the limit of both a family of centres and of saddles on the cylinders, provided that the restriction of the Hamiltonian to the w-axis has a definite second derivative. Under this genericity condition the equilibrium is (again) dynamically stable, see [235].

In case the Hamiltonian system on $\mathfrak{se}(2)^*$ depends on external parameters we also expect the gradient of the Hamiltonian function to vanish on the w-axis, ensuing more complicated bifurcations. However, without additional assumptions fixing the Poisson structure on $\mathbb{R}^3$ the bracket relation $\{u, v\} = 0$ of $\mathfrak{se}(2)^*$ may change as well, in general leading to $\mathfrak{so}(3)^*$ or $\mathfrak{sl}_2(\mathbb{R})^*$. Within the class of Poisson structures on $\mathbb{R}^3$ of the form (2.18) with

$$R(u, v, w) = u^2 + v^2 + r(w) , \quad r(0) = r'(0) = 0$$

the case $r(w) \equiv 0$ is highly degenerate. If $r^{(\ell)}(0) \neq 0$ is the first non-vanishing derivative at $w = 0$ we can achieve $r(w) = \pm w^\ell$ by means of a co-ordinate transformation in w; the sign $\pm = \mathrm{sgn}(r^{(\ell)}(0))$ can be chosen negative if ℓ is odd. We consider this in more detail below.

In the remaining cases $r(w) = w^{2k}$ with positive sign the symplectic leaves $\{R = \rho\}$ are compact which forces the origin to be a (dynamically) stable equilibrium. Similar to the case $k = 1$ of the Lie–Poisson structure $\mathfrak{so}(3)^*$ the Hamiltonian (sub)systems on the symplectic leaves may undergo bifurcations at leaves isolated from the origin. The singularities A_ℓ also enter the Poisson structure (2.18) on $\mathbb{R}^3$ if

$$R(u, v, w) = u^2 - v^2 + r(w)$$

and more generally the germ $\mathcal{R} : (\mathbb{R}^3, 0) \longrightarrow (\mathbb{R}, 0)$ may be any spatial singularity. The resulting equations of motion

$$\frac{\mathrm{d}}{\mathrm{d}t} \begin{pmatrix} u \\ v \\ w \end{pmatrix} = \nabla \mathcal{H} \times \nabla \mathcal{R}$$

are further complicated where $\mathcal{H} : (\mathbb{R}^3, 0) \longrightarrow (\mathbb{R}, 0)$ is a spatial singularity as well. Moreover, the Poisson structure need not be of the form (2.18), cf. [184].

Cuspoidal Singular Points

The $\mathfrak{sl}_2(\mathbb{R})^*$–Poisson structure on $\mathbb{R}^3$ can be thought of as an extension of the Poisson structure on the quotient $\mathcal{B} \subseteq \mathbb{R}^3$ of $\mathbb{R}^2$ by $\mathbb{Z}_2$ that we encountered in Section 2.1.3. Similarly, we can extend the Poisson structure on the quotient $R^{-1}(0) \subseteq \mathbb{R}^3$ of $\mathbb{R}^2$ by $\mathbb{Z}_3$ to a Poisson structure of the form (2.18) on all of $\mathbb{R}^3$ by putting

$$R(u, v, w) = \frac{w^3}{3} - \frac{3}{2}(u^2 + v^2) . \tag{2.21}$$

Here all level sets $R^{-1}(\rho)$, $\rho \neq 0$ are connected and form the regular symplectic leaves. Generically the gradient $\nabla H(0)$ will have a nonzero component along the w-axis. For $|\rho| \neq 0$ sufficiently small this yields a centre on $R^{-1}(\rho)$ and as $\rho \to 0$ these centres tend to the origin which is (dynamically) stable as well.

Within 1-parameter families the gradient of the Hamiltonian at the origin may become perpendicular to the w-axis whence the singular equilibrium becomes dynamically unstable. Under parameter variation also the Poisson structure defined by (2.21) may change and e.g. get defined according to

$$R_\lambda(u, v, w) = \frac{w^3}{3} - \frac{3}{2}(u^2 + v^2) + \lambda w .$$

The level sets $R_\lambda^{-1}(\rho)$ are surfaces of revolution about the w-axis. For $\lambda > 0$ these are all diffeomorphic to $\mathbb{R}^2$ and with the exception of $R_0^{-1}(0)$ this holds true for $\lambda = 0$ as well. When $\lambda < 0$ there are two singular points $(u, v, w) = (0, 0, \pm\sqrt{-\lambda})$. Next to symplectic leaves diffeomorphic to $\mathbb{R}^2$ the symplectic leaves

$$\left\{ R_\lambda = \rho \in \left] -\frac{2}{3}\sqrt{-\lambda^3}, +\frac{2}{3}\sqrt{-\lambda^3} \right[, w \leq \sqrt{-\lambda} \right\}$$

are compact and diffeomorphic to S^2. The different families are separated by the level set $R_\lambda^{-1}(\frac{2}{3}\sqrt{-\lambda^3})$ which consists of the three symplectic leaves $\{R_\lambda = \frac{2}{3}\sqrt{-\lambda^3}, w < \sqrt{-\lambda}\}$, $\{(0, 0, \sqrt{-\lambda})\}$ and $\{R_\lambda = \frac{2}{3}\sqrt{-\lambda^3}, w > \sqrt{-\lambda}\}$.

A straightforward generalization is to embed the quotient of $\mathbb{R}^2$ by $\mathbb{Z}_\ell$ into the family of Poisson structures of the form (2.18) on $\mathbb{R}^3$ defined by

$$R_\lambda(u,v,w) = \frac{2^{\ell-1}}{(\ell-1)!\ell!}w^\ell - \frac{\ell}{2}(u^2 + v^2) + \sum_{j=0}^{\ell-2}\frac{\lambda_j}{j!}w^j \ .$$

In total this results in an $(\ell-1)$-parameter unfolding of the Poisson space $R_0^{-1}(0)$ with cuspoidal singularity. In Section 2.2.3 below we will encounter mechanisms that single out a 1-parameter sub-unfolding.

2.1.5 Reversibility

To generalize the notion of Poisson symmetries of a Poisson space $\mathcal{P}$ we consider the automorphisms of the additive real group. Assuming continuity, the latter are automatically linear and restricting to compact groups Γ we are led to the identity id and to $-\mathrm{id} : \xi \mapsto -\xi$. Writing the group homomorphism $\Gamma \longrightarrow \mathrm{Aut}(\mathbb{R}, +)$ as $\gamma \mapsto (-1)^\gamma$ we obtain the generalized condition

$$\bigwedge_{\gamma \in \Gamma} \bigwedge_{f,h \in \mathcal{A}} \{f \circ \gamma, h \circ \gamma\} = (-1)^\gamma \cdot \{f, h\} \circ \gamma \ .$$

The group Γ is the disjoint union

$$\Gamma = \Gamma_+ \cup \Gamma_-$$

where

$$\Gamma_\pm = \left\{ \gamma \in \Gamma \ \middle| \ (-1)^\gamma = \pm 1 \right\}$$

and the normal subgroup Γ_+ consists of Poisson symmetries. Passing to the quotient by Γ_+ we may assume that $\Gamma = \{\mathrm{id}, \gamma\}$ with $(-1)^\gamma = -1$. A typical example on $\mathcal{P} = \mathbb{R}^2$ is given by the reflection $\gamma(q, p) = (q, -p)$. We refrain from reducing γ as well as would be appropriate for a mere study of Γ-symmetric singularities, cf. [264]. The reason is that no Poisson structure is induced on the resulting quotient space which correspondingly does not carry a reduced dynamics. Reversing symmetries and Poisson symmetries share the property that a normalization procedure (introducing additional symmetries) does preserve those (reversing and Poisson) symmetries that the system already has, cf [41, 74].

Let now $H \in \mathcal{A}$ be a Hamiltonian function on $\mathcal{P}$ that is invariant under γ. Then the Hamiltonian system defined by H is *reversible* with respect to γ as we have

$$\dot{z} \circ \gamma = \{z \circ \gamma, H\} = -\{z, H\} \circ \gamma \ .$$

The Hamiltonian systems defined by standard forms of singularities are often reversible. The reason is that these standard forms only contain the terms of lowest (weighted) order whence one (or even both) variable(s) may only enter squared. The quasi-homogeneous singularity $A_{d-1}^\pm$ contains the monomial p^2 whence the universal unfolding is invariant under the reflection $p \mapsto -p$ as well. In case $d = 2l$ is even, the second monomial in $A_{d-1}^{\pm}$ is $q^d = (q^2)^l$ and the singularity is also invariant under $q \mapsto -q$. However, this latter symmetry is broken by the universal unfolding

$$\mathcal{H}_\lambda(q,p) = \mathcal{H}_0(q,p) + \sum_{j=1}^{2(l-1)} \lambda_j \mathcal{E}_j(q,p) .$$

Symmetrisation $\frac{1}{2}(\mathcal{H}_\lambda + \mathcal{H}_\lambda \circ \gamma)$ yields the unfolding

$$\mathcal{H}_\lambda(q,p) = \frac{a}{2}p^2 + \frac{b}{(2l)!}q^{2l} + \sum_{j=1}^{l-1} \frac{\lambda_j}{(2j)!}q^{2j}$$

of $A_{2l-1}^{\pm}$ that we already encountered as the universal $\mathbb{Z}_2$-symmetric unfolding in Section 2.1.3. Indeed, the two reversing symmetries

$$(q,p) \mapsto (q,-p) \tag{2.22}$$

and

$$(q,p) \mapsto (-q,p) \tag{2.23}$$

compose to the π-rotation

$$(q,p) \mapsto (-q,-p) \tag{2.24}$$

which is a Poisson symmetry. In the reversible context where (2.23) has to be preserved, but (2.24) may be broken by higher order terms, the bifurcation unfolded by the 1-parameter family

$$\mathcal{H}_\lambda(q,p) = \frac{a}{2}p^2 + \frac{b}{24}q^4 + \frac{\lambda}{2}q^2$$

is called a Hamiltonian pitchfork bifurcation (if $ab > 0$ and a dual[9] Hamiltonian pitchfork bifurcation if the two coefficients have opposite signs).

The singularity $D_{k+1}^{\pm}$ is invariant under (2.22). Symmetrisation of the universal unfolding with respect to this reflection leads to

$$\mathcal{H}_\lambda(q,p) = \frac{a}{2}p^2q + \frac{b}{k!}q^k + \sum_{j=1}^{k-1} \frac{\lambda_j}{j!}q^j$$

and thus lowers the co-dimension by one. For $k = 3$ the resulting bifurcation diagrams have been given in [138], see also [57, 188, 53]. The singularity E_6 defines a Hamiltonian system that is reversible with respect to (2.23) and symmetrisation with respect to this reflection lowers the co-dimension from 5 to 3. The unimodal singularity $X_9^{\pm}$ is reversible with respect to both (2.22) and (2.23) and the $(\mathbb{Z}_2 \times \mathbb{Z}_2)$-symmetrisation of its universal unfolding is obtained from (2.11) by dropping the last term $\lambda_3 pq$. For the five different bifurcation diagrams of this 2-parameter family see [57, 144].

---

[9] In the literature this latter case $ab < 0$ is also called the subcritical type and the former case the supercritical type.

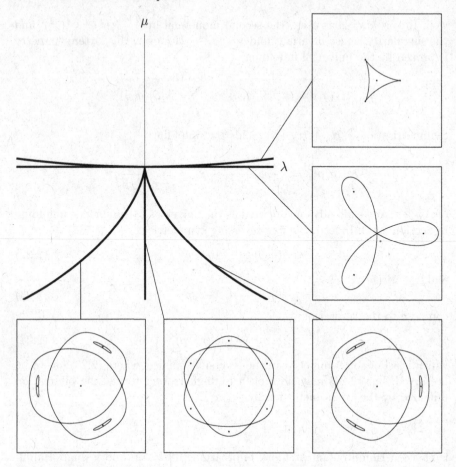

Fig. 2.12. Bifurcation diagram of the 2-parameter subfamily of Hamiltonian systems defined by the universal $\mathbb{Z}_3$-symmetric unfolding $2_{2,3}$ that is reversible with respect to (2.22).

Example 2.22. Let us consider a reversible subfamily of the 3-parameter family of $\mathbb{Z}_3$-symmetric Hamiltonian systems defined by $2_{2,3}$ that we encountered in Section 2.1.3. The central singularity

$$\mathcal{H}_0(q,p) \;=\; \frac{a}{8}(p^4 + 2p^2q^2 + q^4) \;-\; \frac{m}{72}(p^6 - 15p^4q^2 + 15p^2q^4 - q^6)$$

is invariant under both (2.22) and (2.23). Since symmetrisation with respect to (2.24) yields the $\mathbb{Z}_6$-symmetric 1-parameter family $1_{2,6}$ we require the subfamily to be invariant only under (2.22) and allow for the symmetry (2.23) to be broken. This yields the 2-parameter unfolding

$$\mathcal{H}_{\lambda,\mu}(q,p) \;=\; \mathcal{H}_0 \;+\; \frac{\lambda}{2}\Bigl(p^2q - \frac{1}{3}q^3\Bigr) \;+\; \frac{\mu}{2}(p^2 + q^2)$$

which has the bifurcation diagram given in Fig. 2.12. Along the q-axis $\{p = 0\}$ the symmetries enforce $\dot{q} = 0$ and we have

$$\dot{p}|_{p=0} \;=\; -qf(q)$$

$$D^2\mathcal{H}_{\lambda,\mu}(q,0) \;=\; \begin{pmatrix} g(q) & 0 \\ 0 & h(q) \end{pmatrix}$$

with

$$f(q) \;=\; \frac{m}{12}q^4 + \frac{a}{2}q^2 - \frac{\lambda}{2}q + \mu$$

$$g(q) \;=\; \frac{5m}{12}q^4 + \frac{3a}{2}q^2 - \lambda q + \mu$$

$$h(q) \;=\; -\frac{5m}{12}q^4 + \frac{a}{2}q^2 + \lambda q + \mu \; .$$

The origin $(q,p) = (0,0)$ is dynamically stable except for parameter values in

$$\left\{ (\lambda, \mu) \in \mathbb{R}^2 \;\middle|\; \lambda \neq 0, \, \mu = 0 \right\} \; .$$

When f and g have a common real root q_* a centre-saddle bifurcation takes place at $(q,p) = (q_*, 0)$ and its $\mathbb{Z}_3$-symmetric counterparts. The corresponding parameter values (λ, μ) are zeros of the resultant

$$R_{f,g}(\lambda, \mu) \;=\; -\frac{m\mu}{2^{10}3^4}\left[81m\lambda^4 + 72a\lambda^2(a^2 - 12m\mu) - 64\mu(4m\mu - 3a^2)^2\right]$$

of f and g, cf. [79]. Similarly, a Hamiltonian pitchfork bifurcation takes place at $(q,p) = (q^*, 0)$ and its $\mathbb{Z}_3$-symmetric counterparts for parameter values (λ, μ) where

$$R_{f,h}(\lambda, \mu) \;=\; -\frac{m\mu}{2^{10}}\left[3m\lambda^4 - 72a\lambda^2(a^2 + m\mu) - 64m^2\mu^3\right]$$

vanishes and f and h have a common real root q^*. One readily checks that in both cases there is a real root that tends to zero as $(\lambda, \mu) \to 0$. At

$$\left\{ (\lambda, \mu) \in \mathbb{R}^2 \;\middle|\; \lambda = 0, \, \mu < 0 \right\}$$

a connection bifurcation occurs.   □

A Hamiltonian system is also reversible with respect to $\gamma : \mathcal{P} \longrightarrow \mathcal{P}$ if γ preserves the Poisson structure and $H \circ \gamma = -H$. Combinations of two such reversing symmetries form the group Γ_+ of Poisson symmetries that leave H invariant and passing to the quotient by Γ_+ we obtain again $\gamma = \gamma^{-1}$. On $\mathbb{R}^2$ the π-rotation (2.24) is a reversing Poisson symmetry of the Hamiltonian systems defined by the singularities $D_{2l}^{\pm}$ and E_8. Since the quadratic terms vanish, there are no structurally stable equilibria at the origin. The (2.24)-reversible universal unfolding of $D_{2l}^{\pm}$ reads

$$\mathcal{H}_\lambda(q,p) = \frac{a}{2}p^2 q + \frac{b}{(2l-1)!}q^{2l-1} + \sum_{j=1}^{l-1} \frac{\lambda_j}{(2j-1)!}q^{2j-1} + \lambda_l p .$$

For the bifurcation diagrams of the 2-parameter unfoldings of $D_4^{\pm}$, the equilibria at the origin with lowest co-dimension, see [63]. Symmetrisation of the 7-parameter universal unfolding of E_8 yields a 4-parameter reversible unfolding.

The Hamiltonian system defined by the singularity $D_{2l}^{\pm}$ is also reversible with respect to the reflection (2.22) whence the combination (2.23) of the two reversing symmetries is a "true" symmetry of the Hamiltonian system that, however, preserves neither the Poisson bracket nor the Hamiltonian function. As always, the discrete symmetries of the standard form of the singularity may be broken by higher order terms and/or unfolding, but where the symmetry (2.23) is imposed the universal unfolding of $D_{2l}^{\pm}$ yields

$$\mathcal{H}_\lambda(q,p) = \frac{a}{2}p^2 q + \frac{b}{(2l-1)!}q^{2l-1} + \sum_{j=1}^{l-1} \frac{\lambda_j}{(2j-1)!}q^{2j-1} .$$

Similarly, the 6-parameter universal unfolding of the singularity E_7 contains the 2-parameter subfamily

$$\mathcal{H}_{\lambda,\mu}(q,p) = \frac{a}{6}p^3 + \frac{b}{6}pq^3 + \lambda p + \mu pq \tag{2.25}$$

of Hamiltonians satisfying $\mathcal{H}_{\lambda,\mu}(q,-p) = -\mathcal{H}_{\lambda,\mu}(q,p)$. Note that this reflectional symmetry forces the reflection axis to be invariant under the flow. In particular, the origin cannot be a centre. Hence, the only structurally stable equilibria are saddles – in the form $\mathcal{H}(q,p) = pq$ in case the reflection is given by (2.22) or (2.23).

2.2 Higher Degrees of Freedom

For Hamiltonian systems with more than one degree of freedom the concept of topological equivalence is too restrictive to be of importance. There are no structurally stable systems with recurrent dynamics, in particular only hyperbolic equilibria are (locally) structurally stable. Our aim is to coarsen Definition 2.2 in such a way that bifurcations of equilibria become again a co-dimension one phenomenon.

Let $z \in \mathcal{P}$ be an equilibrium of a Hamiltonian vector field X_H on a symplectic manifold $\mathcal{P}$ with linearization $A = DX_H(z)$. The eigenvalues of A form complex quartets $\pm\Re\pm i\Im$, next to the pairs of real and purely imaginary eigenvalues we already encountered in one degree of freedom. Zero eigenvalues come in pairs as well.

Bifurcations clearly occur where the number of hyperbolic eigenvalues changes. In Section 2.1.1 we have seen how a pair of zero eigenvalues triggers a bifurcation. A second mechanism is that two pairs of purely imaginary eigenvalues coincide and lead to a complex quartet. Since the eigenvalues do not vanish this does not lead to the destruction of equilibria. In 1-parameter families the generic bifurcation triggered by the second mechanism is the Hamiltonian Hopf bifurcation. Where the number of degrees of freedom is sufficiently high these two mechanisms of change in the eigenvalue configuration may also occur in a combined way.

While all Hamiltonian systems with one degree of freedom are integrable, this property becomes highly exceptional in two and more degrees of freedom. We mainly restrict to two degrees of freedom where most of the new phenomena can already be observed. Furthermore, we do not consider equilibria at singular points of the phase space. Since all symplectic manifolds locally look alike this allows us to restrict to $\mathbb{R}^4$ with symplectic structure

$$\mathrm{d}q_1 \wedge \mathrm{d}p_1 \; + \; \mathrm{d}q_2 \wedge \mathrm{d}p_2$$

whence the equations of motion read

$$\dot{q} \; = \; \nabla_p \mathcal{H} \; , \quad \dot{p} \; = \; -\nabla_q \mathcal{H} \; .$$

Again we can restrict our attention to Hamiltonians that are germs

$$\mathcal{H} \; : \; (\mathbb{R}^4, 0) \; \longrightarrow \; (\mathbb{R}, 0)$$

and freely use the left equivalence $h = -1$ where convenient. However, a smooth right equivalence $\eta : \mathbb{R}^4 \longrightarrow \mathbb{R}^4$ is helpful only when it preserves the symplectic structure – at least up to some factor, cf. the proof of Proposition 2.4. The Hamiltonian Hopf bifurcation of regular equilibria in two degrees of freedom is related to bifurcating equilibria at singular points in one degree of freedom. Similarly, bifurcations of regular equilibria in three or more degrees of freedom may lead to bifurcations of singular equilibria in two (or more) degrees of freedom. It would therefore be instructive to know whether all singular equilibria of two-degree-of-freedom systems can be obtained in this way.

2.2.1 Bifurcations Inherited from One Degree of Freedom

Let the linearization $DX_{H_0}(z)$ at $\lambda = 0$ of the equilibrium $z \in \mathcal{P}$ have a double eigenvalue zero and all other eigenvalues hyperbolic. Then a centre manifold allows to reduce the local situation to one degree of freedom, cf. [211]. In this way the bifurcations of the previous section occur in higher degrees of freedom as well.

Whenever there are hyperbolic eigenvalues of the linearization $DX_{H_0}(z)$ of $z \in \mathcal{P}$ a centre manifold reduction allows to reduce to a Hamiltonian system with less degrees of freedom where all eigenvalues lie on the imaginary axis. It only remains to treat the case of a doulbe zero eigenvalue combined with elliptic eigenvalues. In the unfolding the double zero eigenvalue may lead to a pair of (small) hyperbolic eigenvalues; these are not dealt with by means of the centre manifold.

Let us first consider such a normally elliptic bifurcation in *two* degrees of freedom and let $\pm i\alpha$ be the nonzero eigenvalues. In suitable Darboux coordinates (x, y, q, p) the linearization reads

$$\frac{\alpha}{2}(x^2 + y^2) + \frac{a}{2}p^2$$

with $a = 1$ or $a = -1$ if the two zero eigenvalues come from a nilpotent block and $a = 0$ in the semi-simple case, see [120, 165, 157]. Using the normalization procedure detailed in Appendix C we can construct a normal form $\mathcal{H}_\lambda$ of the original family H_λ where x and y only enter as a function of

$$I := \frac{x^2 + y^2}{2}$$

whence $\mathcal{H}_\lambda$ is invariant under the S^1-symmetry

$$
\begin{array}{ccc}
S^1 \times \mathbb{R}^4 & \longrightarrow & \mathbb{R}^4 \\[4pt]
(\varphi, \begin{pmatrix} x \\ y \\ q \\ p \end{pmatrix}) & \mapsto & \begin{pmatrix} \cos\varphi & -\sin\varphi & 0 & 0 \\ \sin\varphi & \cos\varphi & 0 & 0 \\ 0 & 0 & 1 & 0 \\ 0 & 0 & 0 & 1 \end{pmatrix} \begin{pmatrix} x \\ y \\ q \\ p \end{pmatrix}
\end{array} \tag{2.26}
$$

introduced by normalization. Reducing this symmetry yields a one-degree-of-freedom problem and amounts to considering I as a (distinguished) parameter in

$$\mathcal{H}_\lambda^I(q, p) = \mathcal{H}_\lambda(I, q, p) .$$

The group action (2.26) is not free at the plane

$$\left\{ (x, y, q, p) \in \mathbb{R}^4 \;\middle|\; x = y = 0 \right\} \tag{2.27}$$

which is (pointwise) fixed by the whole group. This makes the reduction singular at $I = 0$. Re-parametrisations to transformed parameters J and μ have to be of the form

$$(\lambda, I) \mapsto (\mu(\lambda), J(\lambda, I)) \tag{2.28a}$$

to ensure that J can again be interpreted as a phase space function in x and y while μ inherits from λ the rôle of an external parameter. In this way the parameter I is distinguished with respect to λ. Furthermore, we require $J(\lambda, 0) = 0 \; \forall_\lambda$ in order to preserve the singular zero level set. This amounts to

$$J(\lambda, I) \;=\; I \cdot j(\lambda, I) \tag{2.28b}$$

and to truly preserve $I = 0$ we add

$$\bigwedge_{I \geq 0} \; j(\lambda, I) \;>\; 0$$

as an extra[10] requirement – where $j(\lambda, 0) = 0$ the re-parametrisation (2.28) is not invertible.

The family $\mathcal{H}_\lambda^I$ of planar Hamiltonians with its bifurcating equilibrium at $q = p = 0 = \lambda$ can now be treated along the lines of Section 2.1. The necessary adjustments of singularity theory to the present context of boundary-preserving re-parametrisations (2.28) have been derived in [43]. In the co-dimension one case $\lambda \in \mathbb{R}$ it turns out that the unfolding parameter μ in the universal unfolding has to be replaced by either $\mu + J$ or by $\mu - J$. For higher co-dimensions one can still achieve $\mu_1 \pm J$ for the "first" unfolding parameter, while the others have to be replaced by coefficient functions

$$\begin{aligned}
\delta_i \;:\; \mathbb{R}^k \times \mathbb{R} &\longrightarrow \quad \mathbb{R} \\
(\mu, J) &\longmapsto \quad \delta_i(\mu, J)
\end{aligned}$$

for $i = 2, \ldots, k$. See also [188, 53] where this result is generalized to equivariant singularity theory.

Example 2.23. Let us consider the normal form

$$\mathcal{H}_\lambda \;=\; I + \frac{p^2}{2} + \frac{q^3}{6} + (\lambda - I)q$$

in more detail. The equations of motion are

$$\dot{x} = y - yq , \qquad \dot{q} = p ,$$
$$\dot{y} = -x + xq , \qquad \dot{p} = -\frac{1}{2}q^2 - \lambda + I .$$

The (q, p)-plane $I = 0$ is invariant and displays a centre-saddle bifurcation. For $I \neq 0$ this centre-saddle bifurcation is superposed by a periodic motion in the (x, y)-plane – as long as the value of q remains less than $+1$.

This yields a family of quasi-periodic motions for $I > \lambda$, while other generic motions are unbounded. Periodic orbits are given by

$$q = \pm\sqrt{2(I - \lambda)} , \quad p = 0$$

---

[10] For germs this is equivalent to $j(0,0) > 0$. A mild assumption as $J \leq 0$ would be the only alternative.

if $I \geq \lambda$ and by initial conditions

$$x = 0 , \quad y = 0 , \quad -\sqrt{-2\lambda} < q < \sqrt{-2\lambda} , \quad p = 0$$

if $\lambda < 0$. In this latter case there are furthermore two equilibria at

$$(x, y, q, p) \;=\; (0, 0, \pm\sqrt{-2\lambda}, 0)$$

which meet at the origin and vanish as λ passes through 0.

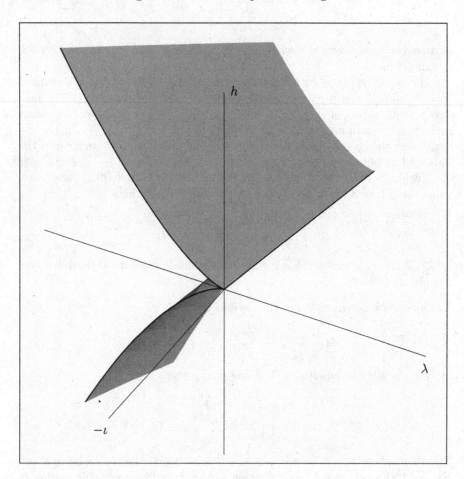

Fig. 2.13. Sketch of the set of singular values of the energy-momentum mapping of $\mathcal{H}_\lambda$. The front (λ, h)-plane $\iota = 0$ is part of that set (though not shaded).

Information on the global dynamics is collected in the (parameter dependent) set of singular values of the *energy-momentum mapping*

$$(\mathcal{H}_\lambda, I) \;:\; \mathbb{R}^4 \;\longrightarrow\; \mathbb{R} \times [0, \infty[$$

which consists of the values (λ, ι, h) taken on those points where $X_{\mathcal{H}_\lambda}$ and X_I are linearly dependent. Next to the (λ, h)-plane $\iota = 0$ these consist of the values $\lambda = h = \iota > 0$ on parabolic periodic orbits, the values

$$\left\{ (\lambda, \iota, h) \;\middle|\; h = \iota - \frac{1}{3}\left(\sqrt{2(\iota - \lambda)}\right)^3 , \, \iota > \max(\lambda, 0) \right\}$$

on elliptic periodic orbits and the values

$$\left\{ (\lambda, \iota, h) \;\middle|\; h = \iota + \frac{1}{3}\left(\sqrt{2(\iota - \lambda)}\right)^3 , \, \iota > \max(\lambda, 0) \right\}$$

on hyperbolic periodic orbits, see Fig. 2.13. Note that the latter two surfaces meet in a cusp-like shape at the straight line $\lambda = h = \iota \geq 0$. $\square$

It is straightforward to reconstruct the dynamics defined by the normal form $\mathcal{H}_\mu$ in two degrees of freedom. Let us be explicit in the co-dimension one case with $\mu - J$ unfolding the planar singularity. Where $J = \mu > 0$ the bifurcating planar equilibrium is superposed by the periodic flow generated by

$$\dot{x} = y \cdot \frac{\partial \mathcal{H}_\mu}{\partial J} , \quad \dot{y} = -x \cdot \frac{\partial \mathcal{H}_\mu}{\partial J}$$

with $\partial \mathcal{H}_0 / \partial J(0,0,0) = \alpha$ and yields a bifurcating periodic orbit with Floquet multipliers equal to 1. As detailed in Chapter 3 these periodic bifurcations are versally unfolded where $\{\mu = J\}$ is transversely crossed. In the limit $J \searrow 0$ this bifurcation scenario unfolded by μ converges to the invariant plane (2.27) within which the planar bifurcation described by $\mathcal{H}_\mu^J$ in one degree of freedom yields that same bifurcation versally unfolded by μ.

Bifurcations of higher co-dimensions are still versally unfolded by μ within the invariant plane (2.27). Indeed, the genericity condition

$$\det\left((\frac{\partial \delta_i}{\partial \mu_j}(0,0))_{i,j=2,\dots,k} \right) \neq 0$$

allows to write

$$\delta_i = \mu_i + J \cdot \Delta_i(\mu, J)$$

for $i = 2, \dots, k$. Here the $\Delta_i(\mu, J)$ can be any smooth functions in μ and J. In Section 2.1.1 we encountered moduli, which would correspond to the constant coefficients if the Δ_i were known a priori to be polynomials. In the present situation we thus have infinitely many moduli.

To draw conclusions concerning the dynamics defined by the original Hamiltonian H_μ from the approximating normal form $\mathcal{H}_\mu$ we use that the latter is as close as we want to the former – if we restrict to a sufficiently small neighbourhood of the bifurcating equilibrium, both in phase space and in parameter space. Recall that J is not a conserved quantity of X_{H_μ}, the co-ordinates x and y enter H_μ directly and not only as a function of J. However, the plane (2.27) is also X_{H_μ}-invariant and in this one-degree-of-freedom subsystem the results of Section 2.1 apply: the bifurcation is versally unfolded by μ.

The hyperbolic and elliptic periodic orbits of $X_{\mathcal{H}_\mu}$ survive the perturbation to X_{H_μ} as well. As we shall see in Chapter 3 the same holds true for the co-dimension one bifurcations. Where bifurcations of higher co-dimensions are concerned, already integrable perturbations of $\mathcal{H}_\mu$ could alter the functions $\Delta_i(\mu, J)$ which makes it very unlikely that such bifurcation scenarios of periodic orbits described by $\mathcal{H}_\mu$ survive the perturbation to H_μ.

Furthermore there are certain features of the dynamics that are not persistent under a perturbation from an integrable to a non-integrable Hamiltonian. Where e.g. the stable and unstable manifolds of hyperbolic periodic orbits may coincide for integrable systems, it is generic already in two degrees of freedom Hamiltonian systems for these to split, cf. [247, 248]. For compact energy level sets almost all motion is restricted to invariant 2-tori. After a small perturbation this is still true for the majority of motion since quasi-periodic 2-tori with Diophantine frequencies persist according to KAM theory, while resonant tori are expected to break up.

Under parameter variation the equilibrium at the origin bifurcates and gives rise to hypo-elliptic equilibria and to elliptic equilibria with two pairs $\pm i\alpha(\mu)$ and $\pm i\beta(\mu)$ of purely imaginary eigenvalues, where $\alpha(0) = \alpha$ and $\beta(\mu) \xrightarrow{\mu \to 0} 0$. This yields infinitely many resonances $(k, l) \in \mathbb{Z}^2$ with

$$k\,\alpha(\mu) + l\,\beta(\mu) = 0$$

as μ tends to zero. The order $|k| + |l|$ of these is quite high, and correspondingly there seem to be no dynamical implications. We come back to this in Section 2.2.3 below.

Normally elliptic bifurcations in $\ell \geq 3$ degrees of freedom that are triggered by a double zero eigenvalue have nonzero eigenvalues $\pm i\alpha_1(\lambda), \ldots, \pm i\alpha_{\ell-1}(\lambda)$. To be able to normalize H_λ up to order n there should be no resonances

$$k_1\alpha_1 + \ldots + k_{\ell-1}\alpha_{\ell-1} = 0$$

of order $|k_1| + \ldots + |k_{\ell-1}| \leq n$; see Appendix C for more details. Again the co-ordinates x_j and y_j enter the normal form $\mathcal{H}_\lambda$ only as a function of

$$I_j = \frac{x_j^2 + y_j^2}{2}, \quad j = 1, \ldots, \ell - 1$$

and reducing the symmetry generated by normalization yields a one-degree-of-freedom problem. The multiplicity of the resulting planar singularity of $\mathcal{H}_{\lambda=0}^{I=0}$ at $(q, p) = 0$ determines whether occuring resonances between the normal frequencies α_j are of sufficiently high order as we need $m + 1 \leq n$ for the determining $(m + 1)$-jet to be accounted for in the reduced normal form $\mathcal{H}_\mu^I$.

Again we have $I_j \geq 0$ and the reduction is singular at the boundary

$$\{I_1 = 0\} \cup \{I_2 = 0\} \cup \ldots \cup \{I_{\ell-1} = 0\} \ .$$

Re-parametrisations (2.28a) that respect this boundary lead to a very restrictive unfolding theory, cf. [43]. Therefore we follow [43, 188, 53] and only require

$$\{I_1 = 0\} \cap \{I_2 = 0\} \cap \ldots \cap \{I_{\ell-1} = 0\}$$

to be preserved whence the octand $\bigcap\{I_j \geq 0\}$ may be slightly distorted[11] by the re-parametrisations (2.28a). This allows to simplify the functions $\delta_i(\mu, J)$ to $\mu_i \pm J_i$ for $i = 2, \ldots, \min(k, \ell - 1)$ as well in the universal unfolding of the planar singularity of $\mathcal{H}_0^0$ at the origin. In particular if $k \leq \ell - 1$ then all the arbitrary functions $\Delta_i(\mu, J)$, $i = 2, \ldots, k$ can be turned into the constants 1 or -1 and we do not have (infinitely many) moduli. Thus, when encountering normally elliptic bifurcations of high co-dimensions it is in fact helpful if there are many (and not few) normal frequencies – provided these are not involved in a low order resonance.

2.2.2 The Hamiltonian Hopf Bifurcation

For Hamiltonian systems with one degree of freedom the sole generic co-dimension one bifurcation of regular equilibria is the centre-saddle bifurcation. In two or more degrees of freedom an equilibrium may also undergo a Hamiltonian Hopf[12] bifurcation. In the absence of extra symmetry there are no further local bifurcations in generic 1-parameter families of Hamiltonian systems on symplectic manifolds.

Transversality of Linear Terms

In a *Krein collision*[13] two pairs of purely imaginary eigenvalues meet in a double pair on the imaginary axis and split off to form a complex quartet $\pm\Re\pm$ i$\Im$ of hyperbolic eigenvalues. We concentrate here on the ensuing phenomena in two degrees of freedom. This eigenvalue movement may also occur in three or more degrees of freedom where it is superposed with additional hyperbolic and/or elliptic eigenvalues. Some of the resulting complications are discussed below.

Since no zero eigenvalues are involved the existence of the corresponding equilibrium is not in question and a simple (parameter dependent) translation ensures that the equilibrium is always at the origin. Let $\pm i\Omega$ be the double pair of imaginary eigenvalues. Then the linearization at the origin contains two harmonic oscillators with frequency $\Omega > 0$ (if $\Omega < 0$ we reverse time). There are two possible choices for the relative sign.

When the two oscillators are in 1:1 resonance the quadratic[14] part of the Hamiltonian is of the form

---

[11] Correspondingly, we must be careful when interpreting the resulting unfoldings.

[12] In the literature this is also called a Brown or Trojan bifurcation.

[13] In the literature this is also called complex instability or splitting case.

[14] This is the Hamiltonian of the vector field linearized at the origin.

$$H(x,y) \;=\; \Omega\frac{x_1^2 + y_1^2}{2} \;+\; \Omega\frac{x_2^2 + y_2^2}{2}$$

whence the origin is a local minimum of the Hamiltonian function. This is a structurally stable situation. In particular, small perturbations can only lead to two pairs of single purely imaginary eigenvalues and no Krein collision takes place. In case of

$$H(x,y) \;=\; \Omega\frac{x_1^2 + y_1^2}{2} \;-\; \Omega\frac{x_2^2 + y_2^2}{2}$$

one speaks of two oscillators in $1{:}{-}1$ resonance. The canonical co-ordinate change

$$\begin{pmatrix} x_1 \\ x_2 \\ y_1 \\ y_2 \end{pmatrix} \;\longmapsto\; \begin{pmatrix} q_1 \\ q_2 \\ p_1 \\ p_2 \end{pmatrix} \;=\; \frac{1}{\sqrt{2}} \begin{pmatrix} 1 & 0 & 0 & -1 \\ 0 & 1 & -1 & 0 \\ 0 & 1 & 1 & 0 \\ 1 & 0 & 0 & 1 \end{pmatrix} \begin{pmatrix} x_1 \\ x_2 \\ y_1 \\ y_2 \end{pmatrix}$$

turns the Hamiltonian into $H = \Omega \cdot S$ with

$$S(q,p) \;=\; q_1 p_2 \;-\; q_2 p_1 \;.$$

By construction $X_H = DX_H(0)$ is again semi-simple, but now we can add a nilpotent part

$$N(q,p) \;=\; \frac{q_1^2 + q_2^2}{2}$$

to the quadratic Hamiltonian. This is in fact what triggers the Krein collision in the 1-parameter family of quadratic Hamiltonians

$$H_\lambda(q,p) \;=\; \Omega S(q,p) \;+\; a N(q,p) \;+\; \lambda M(q,p) \tag{2.29}$$

where

$$M(q,p) \;=\; \frac{p_1^2 + p_2^2}{2}$$

is the nilpotent part adjoint to N. The eigenvalues of the linear vector field defined by H_λ are indeed given by $\pm\sqrt{-a\lambda} \pm i\Omega$.

The vector field X_S defined by the semi-simple part has the periodic flow

$$\begin{pmatrix} q_1 \\ q_2 \\ p_1 \\ p_2 \end{pmatrix} \;\longmapsto\; \begin{pmatrix} \cos\rho & -\sin\rho & 0 & 0 \\ \sin\rho & \cos\rho & 0 & 0 \\ 0 & 0 & \cos\rho & -\sin\rho \\ 0 & 0 & \sin\rho & \cos\rho \end{pmatrix} \begin{pmatrix} q_1 \\ q_2 \\ p_1 \\ p_2 \end{pmatrix} \tag{2.30}$$

and thus defines a (diagonal) S^1-action ϱ on $\mathbb{R}^4 = T^*\mathbb{R}^2 = \mathbb{R}^2 \times \mathbb{R}^2$. The ring of ϱ-invariant functions on $\mathbb{R}^4$ is generated by S, N, M and

$$P \;=\; p_1 q_1 \;+\; p_2 q_2 \;.$$

These invariants satisfy the syzygy

$$2NM = \frac{1}{2}P^2 + \frac{1}{2}S^2$$

and are restricted by the relations $N \geq 0$ and $M \geq 0$.

The unfolding (2.29) of the non semi-simple 1:$-$1 resonance is a 1-parameter unfolding, and a universal unfolding within the "universe" of all linear Hamiltonian systems with two degrees of freedom needs 2 parameters. Such a universal unfolding is e.g. given by the *linear centraliser unfolding*

$$H_\mu(q,p) = (\Omega + \mu_1)S(q,p) + aN(q,p) + \mu_2 M(q,p) , \qquad (2.31)$$

cf. [8, 120, 165, 198, 157, 153, 48]. The left equivalence $h(H) = \Omega \cdot H/(\Omega + \mu_1)$ re-parametrises time to fix the coefficient of S at Ω. An appropriate scaling $(q,p) \mapsto (\alpha q, (1/\alpha)p)$ makes a again the coefficient of N and reveals the unfolding parameter μ_1 to be mute. In fact, we recover (2.29) with $\lambda = \Omega^2 \mu_2/(\Omega + \mu_1)^2$. Time re-parametrisation and scaling could also be combined to achieve $\Omega = a = 1$ in (2.29) – in case the two coefficients have different signs the symplectic involution

$$(q_1, q_2, p_1, p_2) \mapsto (q_2, q_1, p_2, p_1) \qquad (2.32)$$

can be used to remedy this. While we refrain from explicitly transforming away Ω and a in (2.29), we do keep in mind that there is thus only one type of Krein collision.

Non-Degeneracy of Higher Order Terms

With a slight abuse of notation, let now H_λ denote a 1-parameter family of two-degree-of-freedom Hamiltonians such that for every parameter value $\lambda \in \mathbb{R}$ the origin is an equilibrium with linearization given by (2.29). Our first step is to approximate H_λ by a normal form $\mathcal{H}_\lambda$ with respect to the linear part $\Omega S + aN$, see [198, 208] or Appendix C. The resulting simplification is best described as a two-step procedure.

The first step is to normalize with respect to the semi-simple part S. This introduces the symmetry (2.30) whence $\mathcal{H}_\lambda$ depends on (q,p) only as a function of the invariants (S, N, M, P). The second step consists in a further normalization with respect to the nilpotent part N. Since the two parts of the Jordan–Chevalley decomposition commute,

$$[X_S, X_N] = X_{\{N,S\}} = 0 ,$$

this does not spoil the achievements of the first step. The normalization with respect to N allows to remove N and P in the higher order terms of $\mathcal{H}_\lambda$ and yields

$$\mathcal{H}_\lambda(q,p) = \Omega S(q,p) + aN(q,p) + \lambda M(q,p) + \frac{b}{2}(M(q,p))^2$$

$$+ cS(q,p)M(q,p) + \frac{d}{2}(S(q,p))^2 + \frac{e}{6}(M(q,p))^3 + \cdots .$$

We shall see that under the non-degeneracy condition $b \neq 0$ the terms in the second line do not influence the qualitative behaviour of the Hamiltonian Hopf bifurcation.

The S^1-symmetry (2.30) allows to reduce the Hamiltonian system defined by $\mathcal{H}_\lambda$ to one degree of freedom. Our choice of Hilbert basis of the ring of smooth S^1-invariant functions on $\mathbb{R}^4$ yields the identification of the reduced phase space $\mathcal{P}$ with

$$\mathbb{R}^4 /_{S^1} = \left\{ (S, N, M, P) \in \mathbb{R}^4 \ \middle| \ 2NM = \frac{1}{2}P^2 + \frac{1}{2}S^2 \ , \ N \geq 0, \ M \geq 0 \right\}$$

and Poisson bracket relations

$$\begin{array}{lll} \{S, N\} = 0 \ , & \{S, M\} = 0 \ , & \{S, P\} = 0 \ , \\ \{N, M\} = P \ , & \{N, P\} = 2N \ , & \{M, P\} = -2M \ . \end{array}$$

As expected, the generator S of the S^1-symmetry becomes a Casimir on the reduced phase space $\mathcal{P}$. The three remaining equations reproduce the Lie–Poisson bracket of $\mathfrak{sl}_2(\mathbb{R})$, see Section 2.1.4. Fixing $S = \sigma$ the one-degree-of-freedom dynamics takes place in

$$U_\sigma = \left\{ (N, M, P) \in \mathbb{R}^3 \ \middle| \ R_\sigma(N, M, P) = 0 \ , \ N \geq 0, \ M \geq 0 \right\}$$

where

$$R_\sigma(N, M, P) = \frac{1}{2}P^2 - 2NM + \frac{\sigma^2}{2} \ .$$

For $\sigma \neq 0$ the surface U_σ is one sheet of the two-sheeted hyperboloid $R_\sigma^{-1}(0)$ and in particular $U_\sigma = U_{-\sigma}$, while U_0 is the (positive) cone, see again Section 2.1.4. For $\lambda = 0$ the energy level set $\mathcal{H}_0^{-1}(0) \subseteq \mathfrak{sl}_2(\mathbb{R})$ is tangent to the cone U_0 and the relative sign of a and b determines whether $\mathcal{H}_0^{-1}(0) \cap U_0$ consists solely of the origin (locally around the origin) or contains an (un)stable manifold of that singular equilibrium. Indeed, U_0 is a surface of revolution and $\mathcal{H}_0$ does not depend on P, whence the geometric situation can be read off from

$$N = \frac{-b}{2a}M^2 - \frac{e}{6a}M^3 - \cdots \ ,$$

showing that the higher order terms locally have no influence. On the other hand, while we may rescale Ω and a to 1, we can reduce b only to ± 1.

Definition 2.24. Let H_λ be a family of Hamiltonians with normal form $\mathcal{H}_\lambda$ as above. If a and b have the same sign then H_λ undergoes a supercritical[15] Hamiltonian Hopf bifurcation. If $ab < 0$ the Hamiltonian Hopf bifurcation at $\lambda = 0$ is a subcritical[16] one.

---

[15] In the literature this case is also called direct, hyperbolic, or liberation.

[16] In the literature this dual case is also called inverse, elliptic, or the bubble.

Restricting in Fig. 2.10 to the first quadrant (supercritical case) or to the third quadrant (subcritical case) the phase portraits in one degree of freedom are easily obtained. In particular the proof of Theorem 2.20 shows that in the subcritical case subordinate centre-saddle bifurcations of regular equilibria occur where $16\lambda^3 \approx 27ab^2\sigma^2$. In case $(b/2)M^2$ is the sole non-quadratic term of $\mathcal{H}_\lambda$ this becomes an exact equality.

It is straightforward to reconstruct the full dynamics of $X_{\mathcal{H}_\lambda}$ in two degrees of freedom. At regular points of

$$\mathcal{P} = \bigcup_{\sigma \in \mathbb{R}} \{\sigma\} \times U_\sigma$$

we have to superpose a periodic orbit with frequency

$$\frac{\partial \mathcal{H}_\lambda}{\partial S}(q,p) = \Omega + cM(q,p) + dS(q,p) + \dots$$

and from the singular point we reconstruct the central equilibria. In the supercritical case almost all orbits of the reduced system are periodic, leading to invariant 2-tori in two degrees of freedom. In the subcritical case there are, for $\lambda > 0$, some invariant 2-tori as well, but the vast majority of orbits leaves a neighbourhood[17] of the origin. The flow is organized by the relative equilibria – from regular equilibria on $\mathcal{P}$ we reconstruct periodic orbits.

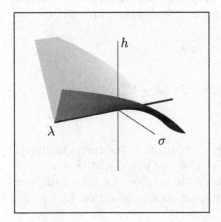 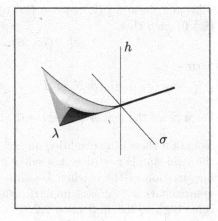

Fig. 2.14. Sketch of the set of singular values of the energy-momentum mapping of $\mathcal{H}_\lambda$ for (a) the supercritical case and (b) the subcritical case.

We also have to understand how the invariant sets fit together to form the phase space $\mathbb{R}^4$. The relevant information is collected in the (parameter dependent) set of singular values of the energy-momentum mapping

---

[17] It depends on the coefficients $c, d, e, \dots$ of the higher order terms whether there are indeed orbits to infinity.

$$(\mathcal{H}_\lambda, S) : \mathbb{R}^4 \longrightarrow \mathbb{R}^2$$

which consists of the values (λ, σ, h) taken on those points where $X_{\mathcal{H}_\lambda}$ and X_S are linearly dependent. Next to the λ-axis $h = \sigma = 0$ these consist of the values on periodic orbits. Where this latter surface detaches from the former line in the supercritical case, see Fig. 2.14a, monodromy is created, cf. [225, 96]. In the subcritical case the monodromy around the λ-axis for $\lambda < 0$ persists at $\lambda = 0$ and turns into "island monodromy" for $\lambda > 0$, cf. [280, 105, 104]. In Fig. 2.14b one can clearly recognize the cusp lines of parabolic periodic orbits where the surfaces of hyperbolic and of elliptic periodic orbits meet in a periodic centre-saddle bifurcation.

We have seen how only the "lowest" higher order term $(b/2)M^2$ determines the qualitative behaviour. This can be formalized in terms of co-ordinate changes determined by left-right equivalences of the energy-momentum mapping to the following result.

Theorem 2.25. Let $\mathcal{H}_\lambda = \mathcal{H}_\lambda(S, N, M, P)$ be an S^1-symmetric unfolding of the 1:−1 resonance satisfying

$$\frac{\partial \mathcal{H}_0}{\partial N} = a \neq 0 \,, \quad \frac{\partial \mathcal{H}_\lambda}{\partial M} = \lambda \quad \text{and} \quad \frac{\partial^2 \mathcal{H}_0}{\partial M^2} = b \neq 0$$

at the origin $(S, N, M, P) = (0, 0, 0, 0)$. Then there exists an S^1-equivariant diffeomorphism $\eta : (\mathbb{R}^4, 0) \longrightarrow (\mathbb{R}^4, 0)$ and a diffeomorphism $h : (\mathbb{R}^2, 0) \longrightarrow (\mathbb{R}^2, 0)$ such that

$$h \circ (\mathcal{H}_\lambda, S) \circ \eta \ = \ (\mathcal{G}_\lambda, S)$$

with

$$\mathcal{G}_\lambda(S, N, M, P) \ = \ aN \ + \ \lambda M \ + \ \frac{b}{2} M^2 \,.$$

For a proof the reader is referred to [198]. □

Note that there is no condition on $\Omega = \partial \mathcal{H}_0 / \partial S(0, 0, 0, 0)$ and correspondingly the semi-simple part does not enter in $\mathcal{G}_\lambda$. The sole purpose of $\Omega \neq 0$ in the linearization of the original Hamiltonian H_λ is to allow for a normal form approximation that is symmetric with respect to the S^1-action (2.30). We come back to this in Section 2.2.4.

It remains to show that the dynamics defined by the original Hamiltonian H_λ inherits the main qualitative features of its normal form approximation $\mathcal{H}_\lambda$. Here the condition $\Omega \neq 0$ on the frequency of nearby periodic orbits does become important. We can use for $\mathcal{H}_\lambda$ a normal form approximation as high as we wish, so the error $\|H_\lambda - \mathcal{H}_\lambda\|_V$ on a sufficiently small neighbourhood V of the origin can be made as small as needed.

Since H_λ starts with $\Omega S + aN + \lambda M$ as well, we already know that the origin is an elliptic equilibrium for $a\lambda > 0$ and hyperbolic for $a\lambda < 0$. The smallness of $\|H_\lambda - \mathcal{H}_\lambda\|_V$ allows to conclude that for $\lambda = 0$ and $ab < 0$ the origin is unstable as well, we formulate the corresponding statement for $\lambda = 0$ and $ab > 0$ below.

Furthermore, the elliptic and hyperbolic periodic orbits persist, and as addressed in Section 3.1 this holds true for the periodic centre-saddle bifurcation as well. In [198] Moser–Weinstein reduction is used to associate the periodic orbits of X_{H_λ} to those of $X_{\mathcal{H}_\lambda}$. In case $ab > 0$ and $a\lambda < 0$ the stable and unstable manifolds of the hyperbolic equilibrium of $X_{\mathcal{H}_\lambda}$ coincide and form a *pinched torus*. Under perturbations these are expected to split. This also applies to coinciding stable and unstable manifolds of hyperbolic periodic orbits, appearing when $ab < 0$ and $a\lambda < 0$.

Persistence of quasi-periodic motions is obtained by KAM theory. This is quite standard for $a\lambda > 0$ where the resulting Cantor family of invariant 2-tori extends to the elliptic equilibrium. Since the 2-dimensional tori divide the 3-dimensional energy shells this yields dynamical stability of the origin. The same holds true for $\lambda = 0$ and $ab > 0$, see [266, 197]. Where there is monodromy the results of [225, 245, 246, 103, 46] apply, and it is possible to extend these to globally yield the persistence of invariant 2-tori, cf. [47]. Interestingly, the iso-energetic non-degeneracy condition is necessarily violated for values (λ, σ, h) on a surface emanating from $\{(\sigma, h) = (0,0), a\lambda \le 0\}$, see [103, 99].

Example 2.26. For the planar circular restricted three body problem we consider two bodies of masses μ and $1 - \mu$ moving in circles about their common centre of mass and let a third particle move within that plane of circular motion. The "restricting" assumption is that the particle does not influence the Keplerian motion of the two "primary" bodies. It can be shown that this is a valid approximation of the full three body problem in the limit where the ratio of the particle mass to that of the primaries tends to zero, cf. [208, 268] and references therein.

The (full) three body problem has central configurations in the form of equilateral triangles, and in the restricted three body problem this yields a (relative) equilibrium at the so-called Lagrangian point $\mathcal{L}_4$. Passing to co-rotating co-ordinates and translating the origin from the centre of mass to $\mathcal{L}_4$ puts the Hamiltonian into the form

$$H_\mu(x, y) = \frac{1}{2}(y_1^2 + y_2^2) - (x_1 y_2 - x_2 y_1) - U_\mu(x)$$

with

$$U_\mu(x) = \frac{1}{2}x_1 + \frac{\sqrt{3}}{2}x_2 + \frac{1}{d_1} + \mu\left[\frac{1}{d_2} - \frac{1}{d_1} - x_1\right]$$

where

$$d_1^2 = \left(x_1 + \frac{1}{2}\right)^2 + \left(x_2 + \frac{\sqrt{3}}{2}\right)^2$$

$$d_2^2 \;=\; \left(x_1 - \frac{1}{2}\right)^2 + \left(x_2 + \frac{\sqrt{3}}{2}\right)^2$$

are the two distances of the particle to the primaries, cf. [198, 208, 110]. Since $\mathcal{L}_4$ is an equilibrium, the linear terms in U_μ cancel and the eigenvalues of the linearization DX_{H_μ} are easily computed to be

$$\pm\sqrt{-\frac{1}{2} \pm \frac{1}{2}\sqrt{1 - 27\mu(1-\mu)}} \; .$$

Hence, a Krein collision takes place as the mass μ of the smaller of the two primaries passes through the so-called Routh ratio

$$\mu_1 \;=\; \frac{1}{2} - \frac{\sqrt{69}}{18} \;\approx\; 3.85 \cdot 10^{-2} \; .$$

A linear co-ordinate change turns the quadratic part of H_μ into

$$H_{\lambda(\mu)}(q,p) \;=\; \frac{q_1 p_2 - q_2 p_1}{\sqrt{2}} + \frac{q_1^2 + q_2^2}{2} + \lambda(\mu)\frac{p_1^2 + p_2^2}{2}$$

with $\lambda(\mu_1) = 0$ and $\lambda'(\mu_1) > 0$. To determine the type of the Hamiltonian Hopf bifurcation the sign of the coefficient of $(p_1^2 + p_2^2)^2$ in the normal form of order 4 has to be determined. It is sufficient to compute this normal form at $\mu = \mu_1$ and this yields a positive coefficient, see [267, 197], whence the Hamiltonian Hopf bifurcation at the Routh ratio is supercritical.   □

For a Hamiltonian Hopf bifurcation in three or more degrees of freedom we can repeat the discussion of Section 2.2.1. Additional hyperbolic eigenvalues can again be dealt with by means of a centre manifold. Let $\pm i\alpha_1, \ldots, \pm i\alpha_{\ell-1}$ be $\ell - 1$ pairs of nonzero purely imaginary eigenvalues. We complete these with $\pm i\alpha_\ell = \pm i\Omega$ and require that there are no resonances

$$k_1\alpha_1 \;+\; \ldots \;+\; k_\ell\alpha_\ell \;=\; 0$$

of order $|k_1| + \ldots + |k_\ell| \leq 4$. This allows again to compute a normal form of order 4, see Appendix C for more details. The sign of the coefficient of $(p_1^2 + p_2^2)^2$ in this normal form determines whether the normally elliptic Hamiltonian Hopf bifurcation is supercritical or subcritical.

It is clear that the situation is much more difficult if $\pm i\Omega$ is a triple pair of purely imaginary eigenvalues. But also an additional pair of eigenvalues $\pm i\alpha$ in 2:1 or 3:1 resonance poses new problems. This is even more true if a double eigenvalue $\pm i\Omega$ is combined with a (double) zero eigenvalue.

Higher Co-Dimensions

In a generic 1-parameter family encountering a 1:−1 resonance the coefficient b of $\frac{1}{2}M^2 = \frac{1}{8}(p_1^2 + p_2^2)^2$ determines the qualitative behaviour and it is a co-dimension two phenomenon that b vanishes at the bifurcation. From the dynamics on $\mathfrak{sl}_2(\mathbb{R})$ studied in Section 2.1.4 we infer that the coefficient e of $\frac{1}{6}M^3$ should then be non-zero. The normal form of a 2-parameter unfolding of the $1{:}{-}1$ resonance does not cleanly reduce to a 2-parameter family of Hamiltonian systems on $\mathfrak{sl}_2(\mathbb{R})$, but rather defines a 3-parameter unfolding of one-degree-of-freedom problems. The third parameter is the value σ of the semi-simple part S and determines the symplectic leaf of $\mathfrak{sl}_2(\mathbb{R})$ on which the reduced dynamics takes place. It enters squared in $2NM = \frac{1}{2}P^2 + \frac{1}{2}\sigma^2$ whence the monomial SM in the normal form Hamiltonian becomes important in the degenerate case $b = 0$ and its coefficient c should not vanish. This has to be taken into account when interpreting the bifurcation diagram Fig. 2.11 in the present context. When trying to generalize Theorem 2.25 to the present degenerate case it turns out that c is a modal parameter with respect to C^∞-singularity theory, cf [35, 200]. As shown in [200] this modulus can be dealt with by allowing the left-right equivalence of energy-momentum mappings to be a homeomorphism. Next to $a \neq 0$ the necessary inequalities in the degenerate case $b = 0$ read

$$c,\, e,\, c^2 + e,\, e - 3c^2 \neq 0 \ . \qquad (2.33)$$

For an approach to these results relying on $\mathbb{Z}_2$-equivariant singularity theory see [35, 36].

Where one of the expressions in (2.33) does vanish the phenomena become of co-dimension three. Another co-dimension three bifurcation is the universal unfolding of the semi-simple $1{:}{-}1$ resonance, cf. [170].

2.2.3 Resonant Equilibria

The two basic mechanisms that alter the number of hyperbolic eigenvalues of an equilibrium are a passage through zero and the Krein collision. Hyperbolic eigenvalues can be dealt with by means of a centre manifold, in particular the flow is locally conjugate to its linearization if all eigenvalues are hyperbolic (i.e. the equilibrium is hyperbolic). The difficulties stem from recurrent behaviour, which is triggered by elliptic eigenvalues. We assume that all eigenvalues are nonzero and that there are no pairs of purely imaginary eigenvalues in $1{:}{\pm}1$ resonance. After reduction to a centre manifold the equilibrium is elliptic.

Letting $\pm i\alpha_1, \ldots, \pm i\alpha_\ell$ denote the nonzero and distinct eigenvalues the quadratic part of the Hamiltonian can be brought into the form

$$H_\alpha^2(x, y) = \sum_{j=1}^{\ell} \alpha_j \frac{x_j^2 + y_j^2}{2} \ . \qquad (2.34)$$

Note that it is not possible to enforce $\alpha_j > 0 \ \forall_j$ as the equilibrium need not be a local extremum of the Hamiltonian function. The linear centraliser unfolding of (2.34) consists of an unfolding $\alpha_j + \lambda_j$ of the ℓ frequencies. In $\ell = 2$ degrees of freedom a time scaling makes this a 1-parameter unfolding where the unfolding parameter detunes the frequency ratio. We first restrict to this setting and consider the extensions to $\ell \geq 3$ degrees of freedom at the end.

The dynamics of the linear Hamiltonian system defined by the quadratic Hamiltonian function

$$H_\alpha^2(x,y) \;=\; \alpha_1 \frac{x_1^2 + y_1^2}{2} \;+\; \alpha_2 \frac{x_2^2 + y_2^2}{2} \tag{2.35}$$

is readily analysed. The (x_1, y_1)-plane and the (x_2, y_2)-plane both consist of periodic orbits and the rest of the phase space is foliated by invariant 2-tori. Our aim is to understand what happens under addition $H_\alpha = H_\alpha^2 + H^3 + \dots$ of higher order terms.

It is evident that the origin remains an equilibrium. If α_1/α_2 is not an integer, Liapunov's Center Theorem guarantees the existence of a surface tangent to the (x_2, y_2)-plane at the origin that is filled by elliptic periodic orbits of X_{H_α} whose period tends near the origin to $2\pi/\alpha_2$, cf. [3, 208, 127]. We can always[18] arrange $|\alpha_2| < \alpha_1$ whence the similar family of periodic orbits with period tending to $2\pi/\alpha_1$ is unconditionally guaranteed.

Higher Order Resonances

Persistence of invariant 2-tori is addressed by KAM theory. The linear approximation H_α^2 of H_α has a constant frequency mapping and thus fails to satisfy the Kolmogorov condition. To obtain an integrable approxiation $\mathcal{H}_\alpha$ of H_α that does satisfy the Kolmogorov condition we compute a normal form with respect to H_α^2, cf. Appendix C. If there are no resonances $k_1\alpha_1 + k_2\alpha_2 = 0$ of order $|k_1| + |k_2| \leq n$, in particular if α_2/α_1 is irrational, then the normal form of order n depends on (x, y) only as a function of the invariants

$$I_1 \;=\; \frac{x_1^2 + y_1^2}{2} \quad \text{and} \quad I_2 \;=\; \frac{x_2^2 + y_2^2}{2} \;.$$

For such a *Birkhoff normal form*

$$\mathcal{H}_\alpha \;=\; \alpha_1 I_1 \;+\; \alpha_2 I_2 \;+\; \frac{\alpha_{11}}{2} I_1^2 \;+\; \alpha_{12} I_1 I_2 \;+\; \frac{\alpha_{22}}{2} I_2^2 \;+\; \dots$$

our analysis of the linearized system $X_{H_\alpha^2}$ essentially remains valid. The important difference is that the frequency mapping of the invariant tori

$$(I_1, I_2) \;\mapsto\; \left(\frac{\partial \mathcal{H}_\alpha}{\partial I_1}, \frac{\partial \mathcal{H}_\alpha}{\partial I_2} \right)$$

may now become invertible. If

---

[18] Here we need our assumption $\alpha_1 \neq \alpha_2$.

$$\det \begin{pmatrix} \frac{\partial^2 \mathcal{H}_\alpha}{\partial I_1^2}(0) & \frac{\partial^2 \mathcal{H}_\alpha}{\partial I_1 \partial I_2}(0) \\ \frac{\partial^2 \mathcal{H}_\alpha}{\partial I_1 \partial I_2}(0) & \frac{\partial^2 \mathcal{H}_\alpha}{\partial I_2^2}(0) \end{pmatrix} = \alpha_{11}\alpha_{22} - \alpha_{12}^2 \neq 0 \qquad (2.36)$$

one has immediately that a measure-theoretically large part of a neighbour-hood of the origin in the phase space $\mathbb{R}^4$ is foliated by a Cantor family of X_{H_α}-invariant 2-tori. The condition

$$\det \begin{pmatrix} \alpha_{11} & \alpha_{12} & \alpha_1 \\ \alpha_{12} & \alpha_{22} & \alpha_2 \\ \alpha_1 & \alpha_2 & 0 \end{pmatrix} = 2\alpha_{12}\alpha_1\alpha_2 - \alpha_{11}\alpha_2^2 - \alpha_{22}\alpha_1^2 \neq 0 \qquad (2.37)$$

of iso-energetic non-degeneracy ensures persistence of most 2-tori on the 3-dimensional enery shells and thus implies that the origin is dynamically stable also in the indefinite case $\alpha_1\alpha_2 < 0$ – when H_α^2 is positive or negative definite the Hamiltonian serves as a Liapunov function and the origin is dynamically stable independent of the higher order terms.

The conditions (2.36) and (2.37) require the computation of the Birkhoff normal form of order 4, whence our present considerations do not apply to resonances up to that order. Next to the 1:0 resonance of a double zero eigenvalue and the 1:±1 resonances this excludes the 1:±2 and 1:±3 resonances. Correspondingly, in [254] all other resonances are called higher order resonances. An alternative would be to reserve this notion to equilibria that do admit a Birkhoff normal form of sufficiently high order. For instance, if (2.37) is not satisfied one would compute higher and higher order Birkhoff normal forms $\mathcal{H}_\alpha$ until $\mathcal{H}_\alpha(\alpha_2, -\alpha_1) \neq 0$ to establish iso-energetic non-degeneracy, cf. [208]. Already where one has to normalize up to order 6 this would turn the 1:±4, 1:±5 and 2:±3 resonances into lower order resonances.

Example 2.27. We return to the planar circular restricted three body problem considered in Example 2.26. Recall that the eigenvalues of the linearization at the equilibrium $\mathcal{L}_4$ are given by

$$\pm \sqrt{-\frac{1}{2} \pm \frac{1}{2}\sqrt{1 - 27\mu(1-\mu)}}$$

where the parameter μ is the mass of the smaller of the two primaries. For μ between 0 and the Routh ratio μ_1 these are elliptic, but the linearization is indefinite. With the exception of the 1:−2 resonance at

$$\mu_2 = \frac{1}{2} - \frac{\sqrt{1833}}{90} \approx 2.43 \cdot 10^{-2}$$

and the 1:−3 resonance at

$$\mu_3 = \frac{1}{2} - \frac{\sqrt{213}}{30} \approx 1.35 \cdot 10^{-2}$$

one can compute the normal form $\mathcal{H}_\mu$ of order 4 and verify $\mathcal{H}_\mu(\alpha_2, -\alpha_1) \neq 0$, cf. [92, 209, 208]. It turns out that this expression does vanish at an isolated value $\mu_c \approx 1.09 \cdot 10^{-2}$ where the frequencies are

$$\alpha_1 \approx 0.96 \quad \text{and} \quad \alpha_2 \approx 0.28 \ .$$

Hence one may compute the normal form $\mathcal{H}_{\mu_c}$ of order 6 and indeed $\mathcal{H}_{\mu_c}(\alpha_2, -\alpha_1) \neq 0$, cf. [209, 208]. This proves stability of $\mathcal{L}_4$ for all but two masses $\mu = \mu_2, \mu_3$ in the interval $]0, \mu_1]$.   □

Low Order Resonances

In case the two frequencies α_1 and α_2 are in resonance the flow defined by (2.35) is periodic and induces an S^1-action ϱ on $\mathbb{R}^4 \cong \mathbb{C}^2$. Introducing complex co-ordinates $z_j = x_j + iy_j$, $j = 1, 2$ this action reads

$$\varrho : \begin{array}{ccc} S^1 \times \mathbb{C}^2 & \longrightarrow & \mathbb{C}^2 \\ (\rho, z_1, z_2) & \longmapsto & (e^{-ik_2\rho}z_1, e^{ik_1\rho}z_2) \end{array}$$

for the $k_1 : -k_2$ resonance, i.e. when $k_1\alpha_1 + k_2\alpha_2 = 0$. In this latter equation we can always arrange for k_1 to be positive and according to our convention $|\alpha_2| < \alpha_1$ we have $k_1 < |k_2|$. Note that we have a $k_1 : |k_2|$ resonance if H_α^2 is definite while in the indefinite case k_2 is positive as well.

The ring of ϱ-invariant functions on $\mathbb{R}^4$ is generated by

$$I_1 = \frac{z_1 \bar{z}_1}{2} = \frac{x_1^2 + y_1^2}{2}$$

$$I_2 = \frac{z_2 \bar{z}_2}{2} = \frac{x_2^2 + y_2^2}{2}$$

$$J = \frac{1}{k_1! |k_2|!} \mathrm{Re}\,(z_1^{k_1} \bar{z}_2^{|k_2|})$$

$$K = \frac{1}{k_1! |k_2|!} \mathrm{Im}\,(z_1^{k_1} \bar{z}_2^{|k_2|})$$

for $k_2 < 0$ while in the case $k_2 > 0$ the last two polynomials have to be replaced by

$$J = \frac{1}{k_1! k_2!} \mathrm{Re}\,(z_1^{k_1} z_2^{k_2}) \quad \text{and} \quad K = \frac{1}{k_1! k_2!} \mathrm{Im}\,(z_1^{k_1} z_2^{k_2}) \ .$$

In both cases they are constrained by $I_1 \geq 0$, $I_2 \geq 0$ and the syzygy

$$\frac{2^{k_1 + |k_2| - 1}}{(k_1!)^2 (|k_2|!)^2} I_1^{k_1} I_2^{|k_2|} = \frac{J^2}{2} + \frac{K^2}{2} \ .$$

It is advantageous to replace I_1, I_2 by the k-dependent combinations

$$I = k_1 I_1 + k_2 I_2$$
$$L = k_1 I_2 - k_2 I_1$$

which are better adapted to the resonance at hand. This turns the syzygy into

$$\frac{J^2 + K^2}{2} = \frac{2^{k_1 + |k_2| - 1}(k_1 I - k_2 L)^{k_1}(k_2 I + k_1 L)^{|k_2|}}{(k_1! |k_2|!)^2 (k_1^2 + k_2^2)^{k_1 + |k_2|}} \qquad (2.38)$$

and makes the quadratic part H_α^2 of the Hamiltonian a multiple[19] of the generator L of the S^1-action ϱ.

Again we pass from H_α to a normal form approximation $\mathcal{H}_\alpha$ with respect to the quadratic part H_α^2, cf. [233] or Appendix C. Being at a low order resonance means that the normal form $\mathcal{H}_\alpha$ we are interested in depends on (x, y) as a function not only of I and L but also of J and K. These four invariants serve as co-ordinates on the reduced phase space $\mathcal{P}$ where L is a Casimir. Fixing $L = \ell$ the one-degree-of-freedom dynamics takes place in

$$U_\ell = \left\{ (I, J, K) \in \mathbb{R}^3 \ \middle| \ R_\ell(I, J, K) = 0 \, ; \, I \in \mathcal{I} \right\}$$

with Poisson structure

$$\{f, g\} = \langle \nabla f \times \nabla g, \nabla R_\ell \rangle$$

where

$$R_\ell(I, J, K) = \frac{2^{k_1 + |k_2| - 1}(k_1 I - k_2 \ell)^{k_1}(k_2 I + k_1 \ell)^{|k_2|}}{(k_1! |k_2|!)^2 (k_1^2 + k_2^2)^{k_1 + |k_2|}} - \frac{J^2 + K^2}{2} .$$

The interval $\mathcal{I}$ of admissible values of I depends on the sign of k_2. In the definite case $k_2 < 0$ we have

$$\mathcal{I} = \left[\frac{k_2}{k_1} \ell, \frac{k_1}{-k_2} \ell \right]$$

and the surface of revolution U_ℓ is compact. For $k_2 > 0$ the value ℓ of L may be negative as well and

$$\mathcal{I} = \left[\max\{ \frac{k_2}{k_1} \ell, \frac{k_1}{-k_2} \ell \}, \infty \right[$$

is unbounded. The singular points at $I = (k_1/-k_2)\ell \neq 0$ are of the type $\mathbb{R}^2/_{|k_2|\mathbb{Z}}$. For $k_1 = 1$ the point $(I, J, K) = ((k_2/k_1)\ell, 0, 0) \neq 0$ is regular and becomes a singular point of type $\mathbb{R}^2/_{k_1\mathbb{Z}}$ for $k_1 \geq 2$. These singular points correspond to the normal modes guaranteed by Lyapunov's Center Theorem, cf. [3, 16, 208, 127]. In the definite case $k_2 < 0$ the origin $(I, J, K) = 0$ forms all of U_0 and for $k_2 > 0$ the origin is a singular point of type $\mathbb{R}^2/_{(k_1 + k_2)\mathbb{Z}}$.

---

[19] Note that in the indefinite case this multiple Ω is negative.

While a Birkhoff normal form involves only I and L, the normal form approximation $\mathcal{H}_\alpha$ of a low order resonance depends on J and K as well. For $|k_2| = 2$ the linear centraliser unfolding yields a 1-parameter family

$$\mathcal{H}_\alpha^\lambda(I, J, K, L) = \Omega L + \lambda I + a(\lambda)J + b(\lambda)K \qquad (2.39)$$

of normal forms with $\Omega = \alpha_2/k_1 \,(=-\alpha_1/k_2)$ and coefficient functions a, b that generically do not both vanish at $\lambda = 0$. By means of a λ-dependent rotation in the (J, K)-plane we can then achieve $b(\lambda) \equiv 0$, see [94, 171]. Fixing the value ℓ of the Casimir L, the phase curves are given by the intersections of the surface of revolution U_ℓ with the plane $\mathcal{H}_\alpha^\lambda = h$ whence (relative) equilibria satisfy $K = 0$. While I and L are quadratic in the original variables x, y, the invariant J is of order 3 and a λ-dependent scaling allows to achieve $a(\lambda) \equiv 1$. Correspondingly, one obtains a structurally stable 1-parameter family with respect to ϱ-invariant perturbations, see [94, 127, 188, 53].

For the low order resonances $1{:}{\pm}3$ normalizing the linear centraliser unfolding yields, again after a rotation in the (J, K)-plane,

$$\mathcal{H}_\alpha^\lambda = \Omega L + \lambda I + a(\lambda)J + b(\lambda)I^2 + c(\lambda)IL + d(\lambda)L^2 . \qquad (2.40)$$

Next to $a(0) \neq 0$ we need in case of the $1{:}{-}3$ resonance also the genericity condition $100\, b(0) \neq \pm\sqrt{3}\, a(0)$. In [94, 188, 53] the detuning parameter is accompanied by two moduli, while the approach in [127] leads to a single modulus.

Example 2.28. We return to the planar circular restricted three body problem considered in Examples 2.26 and 2.27. Recall that the parameter μ is the mass of the smaller of the two primaries and that the equilibrium $\mathcal{L}_4$ is in $1{:}{-}i$ resonance when $\mu = \mu_i$, $i = 2, 3$. We claim that for both these values of μ the equilibrium $\mathcal{L}_4$ is dynamically unstable.

For $\mu = \mu_2$ this is immediate. Indeed the normal form $\mathcal{H}_\alpha^\lambda = \Omega L + \lambda I + J$ yields at $L, \lambda = 0$ the stable and unstable manifolds

$$\left\{ (I, J, K) \in \mathbb{R}^3 \ \middle| \ J = 0,\ 125K^2 = 8I^3 \right\}$$

of the singular equilibrium at the origin.

For $\mu = \mu_3$ one needs the relative strength of the coefficients $a(0)$ of J and $b(0)$ of I^2 in the normal form (2.40). The computations in [64, 110] show $a(0) > -(100/\sqrt{3})b(0)$ whence the origin has at $L, \lambda = 0$ again stable and unstable manifolds, proving again dynamical instability. $\qquad\square$

The general normal form of the $k_1{:}{-}k_2$ resonance starting with the linear centraliser unfolding can be written as

$$\begin{aligned}
\mathcal{H}_\alpha^\lambda(I, J, K, L) = {}& \Omega L + \lambda I + \frac{a_1(\lambda)}{2}I^2 + a_2(\lambda)IL + \frac{a_3(\lambda)}{2}L^2 + \cdots \\
& + b(\lambda)J + c_1(\lambda)IJ + c_2(\lambda)IK + c_3(\lambda)JL + c_4(\lambda)KL + \cdots \\
& + \frac{d_1(\lambda)}{2}J^2 + d_2(\lambda)JK + \frac{d_3(\lambda)}{2}K^2 + \cdots
\end{aligned}$$

and our aim is to truncate at the lowest significant order. For $k_1 + |k_2| \geq 5$ the terms explicitly given in the first line constitute the Birkhoff normal form of order 4 and the iso-energetic non-degeneracy condition (2.37) reads $a_1(0) \neq 0$ in the present variables. In case this[20] condition is violated the terms starting with $b(\lambda)J$ become important. In the indefinite case $k_2 > 0$ this may[21] lead to dynamical instability.

The (relative) equilibria of $\mathcal{H}_\alpha^\lambda$ and their bifurcations are governed by the topology of the reduced phase space U_ℓ. The possible forms of the occurring singular points lead to the five cases

$$\begin{aligned}
k_1 &= 1 \quad \text{and} \quad |k_2| \geq 4 \\
k_1 &= 2 \quad \text{and} \quad |k_2| \geq 5 \\
k_1 &= 3 \quad \text{and} \quad |k_2| \geq 4 \\
k_1 &\geq 4 \quad \text{and} \quad |k_2| \geq 5 \\
k_1 &= 2 \quad \text{and} \quad k_2 = \pm 3
\end{aligned}$$

detailed in [254].

The 1:1 Resonance

The cases $\alpha_2 = 0$ and $\alpha_2 = -\alpha_1$ have been dealt with in Sections 2.2.1 and 2.2.2. In the remaining case $\alpha_1 = \alpha_2 =: \Omega$ the quadratic part

$$H_\Omega^2 = \Omega \frac{x_1^2 + y_1^2}{2} + \Omega \frac{x_2^2 + y_2^2}{2}$$

is semi-simple, although there are double eigenvalues $\pm i\Omega$. The reason is that H_Ω^2 is positive definite. This also excludes nilpotent terms in the linear centraliser unfolding, but does not reduce the number of parameters in the latter. The 1:1 resonance persistently occurs in 3-parameter families, cf. [169, 94].

Indeed, the ring of functions that are invariant under the flow of $X_{H_\Omega^2}$ is generated by the four quadratic polynomials

$$\begin{aligned}
I &= \frac{x_1^2 + y_1^2}{2} - \frac{x_2^2 + y_2^2}{2} \\
J &= x_1 x_2 + y_1 y_2 \\
K &= x_1 y_2 - x_2 y_1 \\
L &= \frac{x_1^2 + y_1^2}{2} + \frac{x_2^2 + y_2^2}{2}
\end{aligned}$$

and the linear centraliser unfolding

---

[20] For high $k_1 + |k_2|$ further non-degeneracy conditions on coefficients of monomials $I^i L^j$ have to be violated as well.

[21] For $k_1 + |k_2| = 5$ it is e.g. sufficient to have $b(0) \neq 0$ next to $a_1(0) = 0$.

$$(\Omega + \lambda_0)L + \lambda_1 I + \lambda_2 J + \lambda_3 K$$

has the eigenvalues $\pm i[\Omega + \lambda_0 \pm \sqrt{\lambda_1^2 + \lambda_2^2 + \lambda_3^2}]$. Rescaling time allows us to put $\Omega + \lambda_0 = 1$, leaving us with a 3-parameter unfolding of the linear vector field that is in 1:1 resonance only at $\lambda = 0$.

Even more parameters are needed for the full (nonlinear) problem. In [78] a universal unfolding with 7 parameters is computed. The Hénon–Heiles family is a 2-parameter unfolding of the 1:1 resonance that illustrates how extra integrability can lower the co-dimension of a bifurcation, see [149] for more details.

Three and More Degrees of Freedom

In $\ell \geq 3$ degrees of freedom the linear Hamiltonian system with Hamiltonian function (2.34) may have, next to resonances $k_i \alpha_i + k_j \alpha_j = 0$ that only involve two normal frequencies, also resonances $\sum k_i \alpha_i$ that couple three, four, ... up to all ℓ normal frequencies. The actual coupling would have to come from higher order term perturbations of the quadratic part (2.34) of the Hamiltonian, but it is a generic property for higher order terms to provide such couplings.

In case there are no resonances of order $\sum |k_i| \leq 4$ we can again normalize twice to obtain a normal form approximation

$$\overline{\overline{H}}_\alpha(x,y) = \sum_{i=1}^{\ell} \alpha_i I_i + \sum_{i,j=1}^{\ell} \frac{\alpha_{ij}}{2} I_i I_j$$

that depends on (x,y) only as a function of the invariants

$$I_i = \frac{x_i^2 + y_i^2}{2}, \quad i = 1, \ldots, \ell.$$

This Birkhoff normal form is integrable and has a single family of maximal invariant tori $I_i > 0 \; \forall_i$ near the equilibrium at the origin. Putting one or more $I_i = 0$ yields the families of elliptic lower-dimensional invariant tori until the ℓ "normal modes" are reached, the periodic orbits with $I_j > 0$ and all other $I_i = 0$.

From Lyapunov's Center Theorem we know that all ℓ normal modes are present in the original system as well if none of the quotients α_i/α_j is an integer, and if (2.34) is definite even this latter non-resonance condition is not needed, see [290, 217, 291, 3]. We have seen in the Hamiltonian Hopf bifurcation that normal modes may indeed vanish if $\alpha_i + \alpha_j = 0$ and [89, 22] show this for the 1:−2 resonance as well. While the normal modes are lacking in examples in [217], persistence of normal modes turns out to be a generic property in the remaining cases $\alpha_i + k\alpha_j = 0$ with $k \geq 3$, cf. [89].

A strong form of non-resonance in the form of Diophantine conditions is also needed to obtain persistence of (Cantor)-families of invariant tori. This is taken care of by genericity conditions on the coefficients of the Birkhoff normal form. For instance, most maximal tori near the origin survive provided that the Kolmogorov condition

$$\det D^2 \overline{\overline{H}}_\alpha (I = 0) \;=\; \det(\alpha_{ij}) \;\neq\; 0$$

is met and the iso-energetic non-degeneracy condition

$$\det \begin{pmatrix} D^2 \overline{\overline{H}}_\alpha (I = 0) & \nabla \overline{\overline{H}}_\alpha (I = 0) \\ D\overline{\overline{H}}_\alpha (I = 0) & 0 \end{pmatrix} \;=\; \det \begin{pmatrix} \alpha_{ij} & \alpha_i \\ \alpha_j & 0 \end{pmatrix} \;\neq\; 0$$

yields the same result on every energy shell.

The resonances of order ≤ 4 are the resonances

$$\alpha_i = \pm \alpha_j \;, \quad \alpha_i = \pm 2\alpha_j \;, \quad \alpha_i = \pm 3\alpha_j$$

we already encountered in two degrees of freedom and the resonances

$$\alpha_i + \alpha_j \pm \alpha_k = 0$$
$$\alpha_i \pm \alpha_j \pm 2\alpha_k = 0$$
$$\alpha_i + \alpha_j \pm \alpha_k \pm \alpha_l = 0 \;.$$

In case there is exactly one such resonance, normalizing the terms up to order 4 still yields an integrable approximation, cf. [16, 172]. In $\ell \geq 3$ degrees of freedom there may be two or more such resonances, in fact up[22] to $\ell - 1$ independent resonances. From now on we restrict to $\ell = 3$ whence 2 is already the maximal number of independent resonances.

With 2 independent resonances of order ≤ 4 all 3 frequencies have integer relations and, with two exceptions to which we come back below, the larger frequencies are integer multiples of the smallest frequency. Then, after a rescaling of time we may rewrite the quadratic part (2.34) of the Hamiltonian as

$$H^2_{k,l}(x,y) \;=\; \frac{x_1^2 + y_1^2}{2} + k\frac{x_2^2 + y_2^2}{2} + l\frac{x_3^2 + y_3^2}{2}$$

with $k, l \in \mathbb{Z}\backslash\{0\}$. Such an equilibrium is said to be in $1{:}k{:}l$ resonance. The linear Hamiltonian system defined by $H^2_{k,l}$ has a periodic flow and induces the S^1-action

$$\varrho : \quad \begin{array}{ccc} S^1 \times \mathbb{C}^3 & \longrightarrow & \mathbb{C}^3 \\ (\rho, z_1, z_2, z_3) & \mapsto & (e^{i\rho} z_1, e^{ik\rho} z_2, e^{il\rho} z_3) \end{array}$$

where we used compex co-ordinates $z_j = x_j + iy_j$, $j = 1, 2, 3$. When $H^2_{k,l}$ is positive definite, the monomials

---

[22] If there are ℓ independent resonances then all normal frequencies α_i vanish.

$$I_1 = \frac{z_1 \bar{z}_1}{2}, \qquad I_2 = \frac{z_2 \bar{z}_2}{2}, \qquad I_3 = \frac{z_3 \bar{z}_3}{2}$$

$$J_1 = \frac{z_1^k \bar{z}_2}{k!}, \qquad J_2 = \frac{z_1^l \bar{z}_3}{l!}, \qquad J_3 = \frac{z_2^l \bar{z}_3^k}{k! l!}$$

are among the generators[23] of the ring of ϱ-invariant functions on $\mathbb{C}^3$. In case l (or k) is negative we have to replace z_j^l by $\bar{z}_j^{|l|}$ (resp. z_1^k by $\bar{z}_1^{|k|}$ and $\bar{z}_3^k$ by $z_3^{|k|}$). From now on we concentrate on positive k and l and leave such adjustement to the reader. Note, however, that for two equal frequencies with opposite sign the semi-simple quadratic part (2.34) generically is completed by a nilpotent part as in the Hamiltonian Hopf bifurcation.

In addition, there are ϱ-invariant functions involving all three complex co-ordinates like

$$K_1 = \frac{z_1^{l-k} z_2 \bar{z}_3}{(l-k)!}, \qquad \text{assuming } 0 < k < l$$

that are not smooth functions of (I, J). Their precise number depends on the specific $1{:}k{:}l$ resonance at hand.

When $k = l$ all monomials involving all three complex co-ordinates are products of the I_i and[24] J_j. To obtain real generators of the ring $\left(C(\mathbb{R}^6) \right)^\varrho$ of real ϱ-invariant functions on $\mathbb{R}^6$ we have to take real and imaginary parts of the latter and end up with a set of 9 generators. Both for the $1{:}1{:}1$ and the $1{:}3{:}3$ resonance, polynomials of order 3 in the original variables (x, y) cannot be ϱ-invariant whence in the cubic normal form $\overline{H} = H^2 + \overline{H^3}$ obtained by normalizing once (and truncating after the terms of order three) the additional terms $\overline{H^3} = 0$ must vanish. The first nontrivial normal form is $\overline{\overline{H}} = H^2 + \overline{H^3} + \overline{H^4}$ and requires a second normalization; $\overline{\overline{H}}$ generically couples all three degrees of freedom. The general cubic normal form of the $1{:}2{:}2$ resonance reads

$$\overline{H}_{2,2} = I_1 + 2I_2 + 2I_3 + \text{Re}\,(A J_1 + B J_2) \qquad (2.41)$$

with complex coefficients $A, B \in \mathbb{C}$. A change of phases of z_2 and z_3 allows to make these coefficients real and a further[25] symplectic co-ordinate change rotating the complex (z_2, z_3)-plane yields $B = 0$, see [1, 171]. This separates (2.41) into an integrable two-degree-of-freedom system with Hamiltonian $I_1 + 2I_2 + A\text{Re}\,(J_1)$ and the harmonic oscillator defined by $2I_3$. In particular, the cubic normal form $\overline{H}_{2,2}$ of the $1{:}2{:}2$ resonance is integrable and allows for a detailed study based on the $1{:}2$ resonance, cf. [171, 131, 132, 133]. The second normalization yields $\overline{\overline{H}}_{2,2} = H_{2,2}^2 + \overline{H_{2,2}^3} + \overline{H_{2,2}^4}$ which generically couples the two subsystems and turns out to be non-integrable, see [131, 132, 133].

---

[23] Exceptions are specified below.

[24] Since $k = l$ we work with $J_3 = z_2 \bar{z}_3$ also for $k = l \neq 1$.

[25] This also works for the $1{:}{-}2{:}{-}2$ resonance, but not for the $1{:}2{:}{-}2$ resonance, which moreover generically has a nilpotent part added to (2.34) as well.

While (2.41) only seemingly couples all three degrees of freedom already in the cubic normal form, the 1:1:2, 1:2:3 and 1:2:4 resonance are *genuine first order resonances*. For the 1:1:2 resonance the invariants I_i and J_j are completed by $K_1 = z_1 z_2 \bar{z}_3$ and taking real and imaginary parts where appropriate yields a set of 11 generators of the ring $\left(C(\mathbb{R}^6)\right)^\varrho$ of ϱ-invariant functions. As shown in [95] the cubic normal form $\overline{H}_{1,2} = H_{1,2}^2 + \overline{H}_{1,2}^3$ is generically non-integrable. A large step towards the same result for the 1:2:3 resonance is made in [158, 156]. For this resonance the 13 generators of the ring $\left(C(\mathbb{R}^6)\right)^\varrho$ are given by the I_i and the real and imaginary parts of the J_j, K_1 and $K_2 = \frac{1}{2} z_1 \bar{z}_2^2 z_3$. The 1:2:4 resonance has again the set of 11 generators of the ring of ϱ-invariant functions coming from I_i, J_j and $K_1 = \frac{1}{2} z_1^2 z_2 \bar{z}_3$.

Additional discrete symmetries, including those that make the system reversible, may have great influence on these resonances. Indeed, since such symmetries are preserved under normalization, those generators of $\left(C(\mathbb{R}^6)\right)^\varrho$ that fail to be invariant under the discrete symmetry as well may no longer enter a cubic normal form. This renders both the cubic and the quartic normal form of some genuine first order resonances with discrete symmetry integrable, see [278, 256] for more details.

Table 2.2. Genuine first and second order semi-simple resonances in three degrees of freedom and the (minimal) number of generators of ϱ-invariant functions.

Resonance	# of generators (of degree ≤ 4)	coupling	generically not semi-simple
1:$\pm$1:$\pm$2	11	the cubic	1:$-$1:2
1:2:$-$2	9	normal form	1:2:$-$2
1:$\pm$2:$\pm$3	13	couples all	
1:$\pm$2:$\pm$4	11	three DOF	
1:1:$\pm$1	9	the cubic	1:1:$-$1
1:$\pm$1:$\pm$3	13	normal form	1:$-$1:3
1:$\pm$3:$\pm$3	9	is trivial	1:3:$-$3
1:$\pm$3:$\pm$5	15(9)		
1:$\pm$3:$\pm$7	15(7)		
1:$\pm$3:$\pm$9	13(7)		
$\pm$1:2:2	9	the cubic	
1:$\pm$2:$\pm$5	15(7)	normal form	
1:$\pm$2:$\pm$6	13(7)	is integrable	
1:$\pm$3:$\pm$4	15(7)		
1:$\pm$3:$\pm$6	11(7)		
2:$\pm$3:$\pm$4	11(7)		
2:$\pm$3:$\pm$6	9(7)		

Next to the 1:1:1 and 1:3:3 resonances there are 4 more resonances for which $\overline{H}^3 = 0$ and the first non-trivial normal form $\overline{H}_{k,l} = H_{k,l}^2 + \overline{H}_{k,l}^4$ is obtained only after the second normalization, see Table 2.2. It is to be expected that all these $\overline{\overline{H}}_{k,l}$ are generically non-integrable; a related (and for these resonances slightly weaker) conjecture states that in $\ell \geq 3$ degrees of freedom a sufficiently high normal form of any fully resonant equilibrium is non-integrable. In the first three cases one can force the normal form to be integrable by requiring $H_{k,l}$ to be axially symmetric, cf. [295, 83, 118] for the 1:1:1 resonance. The quadratic part of this latter resonance is even $SO(3)$-invariant, and while higher order terms that preserve this $SO(3)$-invariance do not lead to interesting dynamics, higher order terms that are invariant under a subgroup of $SO(3)$ are quite important in applications, see [107, 106, 104] for the case of tetrahedral symmetry. Similarly, the axially symmetric 1:1:2 resonance treated in [152, 98] to model the swing-spring (or elastic spherical pendulum) has integrable normal forms as well.

The additional generators of the ring $\left(C(\mathbb{R}^6)\right)^\varrho$ of the 1:1:3 resonance are the real and imaginary parts of $K_2 = \frac{1}{2}z_1 z_2^2 \bar{z}_3$. For the 1:3:5 resonance we work with $K_2 = \frac{1}{2}z_1 \bar{z}_2^2 z_3$ for that purpose, but $K_3 = \frac{1}{12}z_1 z_2^3 \bar{z}_2$, J_2 and J_3 are of orders 6 and 8, respectively, and thus do not enter $\overline{\overline{H}}_{3,5}^4$. The same is true for the generators K_1, J_2, J_3 and $K_2 = \frac{1}{12}z_1^2 \bar{z}_2^3 z_3$ of the ring $\left(C(\mathbb{R}^6)\right)^\varrho$ of the 1:3:7 resonance, but the generator $K_3 = \frac{1}{2}z_1 z_2^2 \bar{z}_3$ does enter $\overline{\overline{H}}_{3,7}^4$. For the 1:3:9 resonance the generator $J_3 = \frac{1}{6}z_2^3 \bar{z}_3$ of $\left(C(\mathbb{R}^6)\right)^\varrho$ is again of order 4, while J_2, K_1 and $K_2 = \frac{1}{12}z_1^3 z_2^2 \bar{z}_3$ are of orders $10, 8$ and 6, respectively,

The $1:\pm2:l$ resonances with $|l| \geq 5$ can be treated along the lines of the 1:2:2 resonance, see [2, 131, 132, 133]. For the 1:2:5 resonance the coupling term in $\overline{\overline{H}}_{2,5}^4$ is provided by $K_2 = \frac{1}{2}z_1 z_2^2 \bar{z}_3$, while J_2, J_3, K_1 and $K_3 = \frac{1}{6}z_1 z_2^3 z_3$ are of orders 5 to 7. In case of the 1:2:6 resonance the generator $J_3 = \frac{1}{6}z_2^3 \bar{z}_3$ is again of order 4, the orders of J_2, K_1 and $K_2 = \frac{1}{4}z_1^2 z_2^2 \bar{z}_3$ are $7, 6$ and 5, respectively.

The cubic normal form of the 1:3:4 resonance reads

$$\overline{H}_{3,4} = I_1 + 3I_2 + 4I_3 + \mathrm{Re}(AK_1) \tag{2.42}$$

with $A \in \mathbb{R}$ after a phase change in, say, z_3. Similar to the 1:2:2 resonance it is to be expected that the integrability of $\overline{H}_{3,4}$ is broken by the coupling term J_1 when passing to $\overline{\overline{H}}_{3,4} = \overline{H}_{3,4} + \overline{\overline{H}}_{3,4}^4$. The additional generators J_2, J_3, $K_2 = \frac{1}{4}z_1^2 \bar{z}_2^2 z_3$ and $K_3 = \frac{1}{12}z_1 \bar{z}_2^3 z_3^2$ have orders 5 to 7.

The remainig three resonances can again be treated along the lines of the 1:2:2 resonance, which becomes more obvious by writing them as the 3:6:1, 3:6:2 and 2:4:3 resonances, cf. [131, 132, 133]. Thus, for the 1:3:6 resonance the cubic generator is $J_3 = \frac{1}{2}z_2^2 z_3$ and J_1 provides the coupling in $\overline{\overline{H}}_{3,6}$, while K_1 and J_2 are of order 5 and 7, respectively. The cubic and quartic generators of the 2:3:6 resonance are $J_3 = \frac{1}{2}z_2^2 \bar{z}_3$ and $J_2 = \frac{1}{6}z_1^3 \bar{z}_3$ with $J_1 = \frac{1}{12}z_1^3 \bar{z}_2^2$ of order 5. In case of the 2:3:4 resonance the cubic and quartic generators are

$J_2 = \frac{1}{2}z_1^2 \bar{z}_3$ and $K_1 = \frac{1}{2}z_1 \bar{z}_2^2 z_3$, while $J_1 = \frac{1}{12}z_1^3 \bar{z}_2^2$ and $J_3 = \frac{1}{288}z_2^4 \bar{z}_3^3$ have again order 5 and 7, respectively.

2.2.4 Nilpotent 4 × 4 Matrices

Higher co-dimensions may not only result in bifurcations triggered by 1:−1 or 1:0 resonances to have degenerate terms of higher and higher order. Already in 2-parameter families one may encounter combinations of such resonances, e.g. a 1:−1:0 resonance in three degrees of freedom or two simultaneous 1:−1 resonances (with two different double frequencies) in four degrees of freedom. If we restrict to two degrees of freedom then the only additional possibility is to have a fourfold zero eigenvalue. At such an equilibrium the linearization is given by a nilpotent 4 × 4 matrix.

Unfolding of the Linear Part

A nilpotent matrix with exactly one eigenvector is of co-dimension 1 in one degree of freedom and of co-dimension 2 in two degrees of freedom. We have already seen that two eigenvectors are of co-dimension 3 in one degree of freedom and they are of co-dimension 4 in two degrees of freedom. Since Jordan blocks of uneven lengths must come in pairs in Hamiltonian systems, a 3 × 3 nilpotent matrix cannot occur in two degrees of freedom and that latter configuration must consist of two blocks of 2 × 2 nilpotent matrices.

A linear Hamiltonian system on $\mathbb{R}^4$ is given by a quadratic Hamiltonian function. In [120] the versal linear unfolding of a nilpotent system with one eigenvector is given by the linear centraliser unfolding in normal form

$$\mathcal{H}_{\lambda,\mu}(q,p) = \frac{p_1^2}{2} - q_1 q_2 - p_1 q_2 + \lambda p_1 p_2 + \mu \frac{p_2^2}{2} . \qquad (2.43)$$

The 4 eigenvalues of the equilibrium at the origin are $\pm\sqrt{\lambda \pm \sqrt{\mu}}$ and yield all conceivable eigenvalue configurations. Within the wedge $\lambda < 0 \leq \mu \leq \lambda^2$ we encounter the 1:−k resonances, $k \in \mathbb{N}$, at the semi-parabolas

$$\mu = \left(\frac{k^2 - 1}{k^2 + 1}\right)^2 \lambda^2 , \quad \lambda < 0$$

with the extreme cases of 1:0 and 1:−1 resonance at the boundary of the wedge.

Nilpotent linear two-degree-of-freedom systems with 2 eigenvectors are considered as an illustrating example in [157]. Imposing in addition reversibility the co-dimension is brought down from 4 to 3, and in two of the four resulting cases a further reduction to a versal 2-parameter unfolding is possible. However, in such an unfolding all possible k:l resonances are met, including the 1:1 resonance.

The Nonlinear Problem

At present a "universal unfolding" of $\frac{1}{2}p_1^2 - q_1 q_2 - p_1 q_2$ or, equivalently, of

$$\mathcal{H}_0(q,p) = \frac{p_2^2}{2} - p_1 q_2 \tag{2.44}$$

within the "universe" of all two-degree-of-freedom Hamiltonian systems is not known. A first step towards this goal is to look at a well-chosen standard form and consider the coefficients of the additional linear and quadratic terms as unfolding parameters. Following [86] such a normal form Hamiltonian is obtained by classical invariant theory, using representation theory of $\mathfrak{sl}_2(\mathbb{R})$.

Example 2.29. For the cubic normal form

$$\mathcal{H}_{\lambda,\mu}(q,p) = \frac{p_2^2}{2} - p_1 q_2 + \lambda q_1 + \mu\left(\frac{q_2^2}{2} + \frac{3}{4}p_2 q_1\right) + \frac{q_1^3}{6} \tag{2.45}$$

it is still feasable to study the equilibria and their bifurcations. The equations of motion read

$$\dot{q}_1 = -q_2$$
$$\dot{q}_2 = p_2 + \frac{3\mu}{4}q_1$$
$$\dot{p}_1 = -\lambda - \frac{3\mu}{4}p_2 - \frac{q_1^2}{2}$$
$$\dot{p}_2 = p_1 - \mu q_2$$

and have the two equilibria $(q,p) = (q_1, 0, 0, -\frac{3}{4}\mu q_1)$ with

$$q_1 = \frac{9\mu}{16} \pm \sqrt{\frac{3^4}{4^8}\mu^2 - \lambda} \, ,$$

if $\lambda \leq (3^4/4^8)\mu^2$. The linearization reads

$$\begin{pmatrix} 0 & -1 & 0 & 0 \\ \frac{3}{4}\mu & 0 & 0 & 1 \\ -q_1 & 0 & 0 & -\frac{3}{4}\mu \\ 0 & -\mu & 1 & 0 \end{pmatrix}$$

and has the eigenvalues

$$\pm\sqrt{-\frac{5}{4}\mu \pm \sqrt{\frac{25}{16}\mu^2 \pm \sqrt{\frac{3^4}{4^8}\mu^2 - \lambda}}}$$

depicted in Fig. 2.15. Again all $k{:}{-}l$ resonances, $k,l \in \mathbb{N}$, are met for $\mu > 0$ as λ varies from the 1:0 resonance at $\lambda = (3^4/4^8)\mu^2$ to the 1:$-$1 resonance at $\lambda = -(544/4^8)\mu^2$.   □

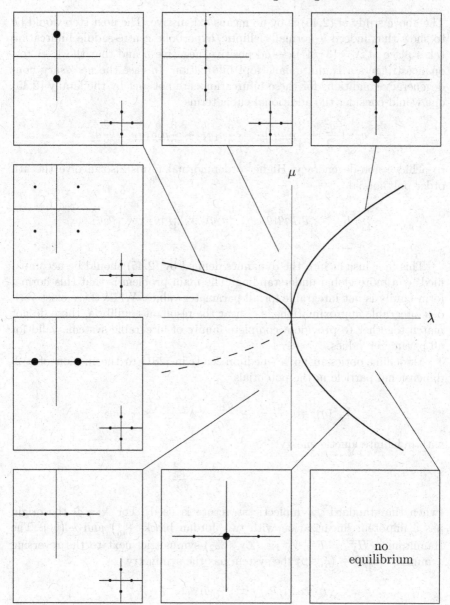

Fig. 2.15. The various eigenvalue configurations of the equilibria of the Hamiltonian system defined by (2.45).

The above study of (2.45) is by no means exhaustive. The next step would be to show that indeed a normally elliptic/hyperbolic centre-saddle bifurcation takes place at $\lambda = (3^4/4^8)\mu^2$ and positive/negative μ and that the 1:$-$1 resonance triggers a Hamiltonian Hopf bifurcation. In case the necessary non-degeneracy conditions for these bifurcations are not met by the family (2.45) one would consider the additional cubic terms

$$q_1 \cdot \left(\frac{q_2^2}{2} + \frac{3}{4}p_2 q_1\right) \quad, \quad \frac{3}{2}p_1 q_1^2 + \frac{3}{2}p_2 q_1 q_2 + \frac{2}{3}q_2^3$$

to achieve non-degeneracy. Higher order normal forms also involve the 4th order polynomial

$$\frac{3}{4}p_1^2 q_1^2 + \frac{3}{2}p_1 p_2 q_1 q_2 - \frac{1}{2}p_2^3 q_1 + \frac{2}{3}p_1 q_2^3 - \frac{1}{4}p_2^2 q_2^2 \quad,$$

see again [86].

This first insight into the dynamics defined by (2.45) should be accompanied by a more global understanding. The main problem is that this normal form family is not integrable for all parameter values. While there exist various integrable approximations, e.g. near the resonant equilibria, these do not match together to provide a complete family of integrable systems valid for all parameter values.

Modelling optics in an h²-medium leads in [281] to the motion of a 2-dimensional particle in the potentials

$$V_\lambda^\pm(q) = \frac{q_1^2 q_2}{2} \mp \frac{q_2^3}{6} + \lambda \frac{q_1^2 \pm q_2^2}{2} + \mu q_2$$

with indefinite kinetic energy

$$T(p) = \frac{p_1^2}{2} - \frac{p_2^2}{2}$$

(when the standard[26] symplectic structure is used). For $\lambda = 0$ the origin has a nilpotent linearization with two Jordan blocks $\left(\begin{smallmatrix} 0 & 1 \\ 0 & 0 \end{smallmatrix}\right)$ and $-\left(\begin{smallmatrix} 0 & 1 \\ 0 & 0 \end{smallmatrix}\right)$. The Hamiltonian $H_\lambda^\pm = T + V_\lambda^\pm$ is $(\mathbb{Z}_2 \times \mathbb{Z}_2)$-symmetric; next to the reversing symmetry $(q, p) \mapsto (q, -p)$ the system has the symmetry

$$(q_1, q_2, p_1, p_2) \mapsto (-q_1, q_2, -p_1, p_2) \ .$$

This implies that the linear co-dimension is indeed 2 and the two bifurcation diagrams of $X_{H_\lambda^\pm}$ in [281] display a "linear richness" similar to Fig. 2.15.

Where an (externally given) S^1-symmetry has to be respected the resulting conserved quantity always ensures integrability. While there is no linear S^1-action on $\mathbb{R}^4$ that leaves (2.44) invariant, the Hamiltonian

---

[26] In [281] the minus sign of the kinetic energy is transferred to a minus sign in the symplectic structure.

$$\mathcal{H}(q,p) \;=\; a\,\frac{q_1^2 + q_2^2}{2}$$

is invariant under the S^1-action 2.30 generated by $S = q_1 p_2 - q_2 p_1$. Under the genericity conditions $a \neq 0$ on the linear term and $b \neq 0$ on the higher order terms the family

$$\mathcal{H}_\lambda(q,p) \;=\; a\,\frac{q_1^2 + q_2^2}{2} \;+\; \frac{b}{2}\left(\frac{q_1^2 + q_2^2}{2}\right)^2 \;+\; \lambda\,\frac{p_1^2 + p_2^2}{2}$$

provides a versal S^1-symmetric unfolding and we recover the Hamiltonian Hopf bifurcation. Recall from Section 2.2.2 that the semi-simple part S was only needed to introduce the S^1-action 2.30 by means of normalization. When the Hamiltonian system already is S^1-symmetric to start with, this preliminary step is not necessary and the subsequent analysis still applies, see [146, 148]. In fact, as remarked in [107] a $\mathbb{Z}_\ell$-symmetry, $\ell \geq 5$, may already be sufficient.

3

Bifurcations of Periodic Orbits

We now restrict to the study of a single Hamiltonian system defined on a symplectic manifold $\mathcal{P}$, there is no initial dependence on an external parameter. Given a periodic orbit of X_H we use symplectic co-ordinates on $\mathcal{P}$ with one angle x parametrising the periodic orbit. Varying the action y conjugate to x yields a 1-parameter family of periodic orbits, certainly where no bifurcations occur. The unions of such 1-parameter families are called *orbit cylinders*. In case H is a generic Hamiltonian such orbit cylinders may contain bifurcations of co-dimension one. In the complement of these bifurcations the 1-parameter family of periodic orbits can also be parametrised by the value of H, the energy.

A standard way to study periodic orbits is to study the fixed points of a *Poincaré section*. In the present Hamiltonian context the dimension can be reduced once more since energy is preserved. The resulting *iso-energetic Poincaré sections* carry an induced symplectic structure which is preserved by the iso-energetic Poincaré mapping, see [3, 208]. When starting with two degrees of freedom, this yields a 1-parameter family of area-preserving mappings. Correspondingly, bifurcations of periodic orbits may be classified through thorough examination of area-preserving mappings, see [205, 207, 208]. However, our ultimate goal is to understand similar bifurcations of invariant tori. Therefore, we do not study symplectic mappings here, but instead work with an alternative approach that we later generalize in Chapter 4.

Floquet's theorem, cf. [208], yields symplectic co-ordinates (x, y, z) with $x \in \mathbb{T}^1$ along the periodic orbit $(y, z) = (y_0, 0)$ such that the equations of motion read

$$
\begin{aligned}
\dot{x} &= f(x, y, z) \\
\dot{y} &= g(x, y, z) \\
\dot{z} &= h(x, y, z)
\end{aligned}
$$

and for $y = y_0$ the right hand sides are of the form

$$
f(x, y_0, z) = \omega(y_0) + \mathcal{O}(z^2)
$$

$$g(x, y_0, z) = \mathcal{O}(z^3)$$
$$h(x, y_0, z) = \Omega(y_0) \cdot z + \mathcal{O}(z^2) \ .$$

Thus, the linear part $\Omega(y_0) \cdot z$ in z is independent of the angle x. A further normalization allows to succesively make quadratic, cubic, … terms x-independent as well, see Appendix C. Truncating the higher order terms then yields an integrable approximation – provided the resulting z-component defines an integrable system for[1] fixed y.

The eigenvalues α_j of $\Omega(y_0)$ are Floquet exponents of the periodic orbit $(y, z) = (y_0, 0)$. In case all Floquet exponents are different from zero a y-dependent translation in z allows to achieve

$$h(x, y, z) = \Omega(y) \cdot z + \mathcal{O}(z^2)$$

for all $y \in \mathbb{R}$ whence $z = 0$ determines a whole 1-parameter family of periodic orbits. However, one has to be careful when working with Floquet exponents. For instance, the symplectic change of co-ordinates

$$\begin{pmatrix} x \\ y \\ q \\ p \end{pmatrix} \mapsto \begin{pmatrix} x \\ y + \pi(q^2 + p^2) \\ q \cdot \cos 2\pi x - p \cdot \sin 2\pi x \\ q \cdot \sin 2\pi x + p \cdot \cos 2\pi x \end{pmatrix}$$

turns the Hamiltonian function

$$H(x, y, q, p) = \omega y + \alpha \frac{p^2 + q^2}{2}$$

into

$$H(x, \acute{y}, q, p) = \omega y + (\alpha + 2\pi\omega) \frac{p^2 + q^2}{2}$$

whence the Floquet exponents form a lattice

$$\pm \alpha i + 2\pi i \omega \, \mathbb{Z} \ .$$

Under $\alpha \mapsto \exp(\alpha/\omega)$ this whole lattice is mapped to a pair of Floquet multipliers on the unit circle. These constitute eigenvalues of the linearization of the iso-energetic Poincaré mapping and are thus the primary source of dynamical information.

Floquet multipliers off the unit circle lead to normally hyperbolic dynamical behaviour. Where the number of such hyperbolic multipliers changes, the family of periodic orbits undergoes a bifurcation. In a generic 1-parameter family this may happen in three ways:

---

[1] Independence of x introduces an S^1-symmetry that makes y a (distinguished) parameter. Since $(\partial/\partial y)H = \omega \neq 0$ the rôle of the parameter may as well be played by the value of the Hamiltonian.

1. A pair of Floquet multipliers on the unit circle meets at $+1$ and passes to the (positive) real axis.
2. A pair of Floquet multipliers on the unit circle meets at -1 and passes to the negative real axis.
3. Two pairs of Floquet multipliers on the unit circle meet in a Krein collision at a double pair and turn into a quartet $(\gamma, \bar{\gamma}, \gamma^{-1}, \bar{\gamma}^{-1})$.

Note that 0 cannot be a Floquet multiplier since the Poincaré mapping is invertible. In case 1 the periodic orbits may cease to exist as we shall see in the next section. In the other two cases the linearized iso-energetic Poincaré mapping has no eigenvalue $+1$ and the family of periodic orbits may be parametrised by the energy. Here the bifurcation leads to a different normal behaviour, see Sections 3.2 and 3.4 for more details.

3.1 The Periodic Centre-Saddle Bifurcation

The two-degree-of-freedom dynamics of the integrable Hamiltonian

$$H(x, y, q, p) = \omega y + \frac{a}{2}p^2 + \frac{b}{6}q^3 + dyq \qquad (3.1)$$

with $\omega, a, b, d \neq 0$ is readily analysed. Reducing the S^1-symmetry $x \mapsto x + \rho$ makes the action y conjugate to x a (distinguished) parameter and leads to the one-degree-of-freedom problem defined by

$$\mathcal{H}_y(q, p) = \frac{a}{2}p^2 + \frac{b}{6}q^3 + dyq .$$

This is the centre-saddle bifurcation (of equilibria) we already encountered in Section 2.1.1. At the bifurcation value $y = 0$ one reconstructs a parabolic[2] periodic orbit and under variation of y or, equivalently, the value of the Hamiltonian H a periodic centre-saddle bifurcation[3] takes place.

This qualitative description remains true for the more general integrable Hamiltonian

$$N(x, y, q, p) = \eta(y) + \mathcal{N}_y(q, p) \qquad (3.2)$$

with $\eta'(0) = \omega(0) \neq 0$ and

$$\mathcal{N}_y(q, p) = \frac{a(y)}{2}p^2 + \frac{b(y)}{6}q^3 + c(y)q + \text{higher order terms}$$

with $a(0), b(0) \neq 0$ and $c(0) = 0$, $d(0) := c'(0) \neq 0$. Indeed, the one-degree-of-freedom problem defined by $\mathcal{N}_y$ displays again a centre-saddle bifurcation (of equilibria) as y is varied.

---

[2] In the literature this periodic orbit is also called an extremal periodic orbit.
[3] In the literature this bifurcation is also called creation (if the product ωabd is negative) or annihilation (for ωabd positive).

Theorem 3.1. Let H define a Hamiltonian system in two degrees of freedom and let γ be a periodic orbit of X_H with all Floquet multipliers $+1$. Under generic conditions on H a periodic centre-saddle bifurcation takes place as the value of H passes through $H(\gamma)$.

Proof. We choose Floquet co-ordinates (x, y, q, p) around $\gamma = \{(y, q, p) = (0, 0, 0)\}$ and proceed to normalize with respect to $\dot{x} = \omega$, using the genericity condition

$$\omega := \int \frac{\partial H}{\partial y}(x, 0, 0, 0)\, dx \neq 0 .$$

This turns the Hamiltonian into the sum $H = N + P$ of (3.2) and $P(x, y, q, p)$ that contains the dependence on x and is of at least fourth order in (q, p). Further genericity conditions are $a, b, d \neq 0$ and ensure that (3.2) undergoes a periodic centre-saddle bifurcation.

This shows that γ is parabolic with extremal energy, and that there are two families γ_+ and γ_- of closed curves that consist of hyperbolic and elliptic periodic orbits of the dynamics defined by (3.2). One readily verifies that these can be continued to two families γ^+ and γ^- of hyperbolic and elliptic periodic orbits of the given Hamiltonian system X_H. $\square$

For a proof based on the (area-preserving) iso-energetic Poincaré mapping see [205, 207, 208]. Here the frequency condition $\omega \neq 0$ ensures that this mapping is defined while the genericity conditions $a, b, d \neq 0$ are formulated in terms of this mapping. Concerning lower-dimensional invariant tori there is no equivalent to the iso-energetic Poincaré section transverse to a periodic orbit. For the normalization procedure to yield results the frequency condition $\omega \neq 0$ then has to be replaced by Diophantine conditions on the frequency vector ω, see Chapter 4.

Remark 3.2. Introducing the weight $(\alpha_y, \alpha_q, \alpha_p, \alpha_\omega) = (4, 2, 3, 2)$ it would in fact be sufficient to normalize up to quasi-homogeneous order 6 – the terms of this order are assembled in (3.1). While there is no need for this in the present periodic case where one can easily normalize up to any desired order, this possibility becomes crucial in the quasi-periodic case of Chapter 4.

Not all qualitative features of the dynamics defined by (3.2) survive the perturbation by the higher order term P. Indeed, while stable and unstable manifolds of the hyperbolic periodic orbits coincide in the integrable dynamics, they are expected to split under addition of P. Also, the 2-parameter family of invariant tori enclosed by these separatrices turns under perturbation into a Cantor family of invariant tori. Interestingly, the iso-energetic non-degeneracy condition is necessarily violated for values (y, h) on a half line in the quadrant "opposite" to the singular values of $\mathcal{EM}$, see [100].

It is generic for Hamiltonian systems that the stable and unstable manifolds of orbit cylinders intersect transversely. However, there may be a countable set of parameter (i.e. energy) values for which the periodic orbit has

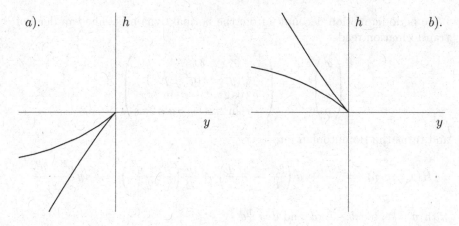

Fig. 3.1. Singular values of the energy-momentum mapping $\mathcal{EM} = (H, y)$ of (3.1) for positive a, bd and (**a**) positive ω, (**b**) negative ω.

stable and unstable manifolds that still do not coincide, but have an intersection that is not transverse, cf. [247, 248, 3]. A non-degenerate homoclinic tangency is in turn accumulated by periodic orbits with Floquet multipliers ± 1 of increasingly high period, see [274].

The dynamics of an integrable Hamiltonian system is collected in the set of singular values of the energy-momentum mapping. Around a periodic centre-saddle bifurcation this set has a characteristic cuspoidal shape, see Fig. 3.1. For (3.1) there are four cases, depending on the relative signs of ω, a and bd. Multiplying H, if necessary, by -1 makes a positive and the π-rotation $(q, p) \mapsto (-q, -p)$ allows to restrict to positive b as well. The remaining two cases with positive d are depicted in Fig. 3.1, for negative d one merely has to reflect about the vertical h-axis.

Example 3.3. We consider a family of elliptic periodic orbits in two degrees of freedom that encounters the Floquet multipliers $\exp(\pm\frac{2}{3}\pi i)$. Normalization yields in Floquet co-ordinates the integrable approximation

$$H(x, y, q, p) = y + \frac{2\pi}{3}\frac{p^2 + q^2}{2} + \frac{a}{6}\text{Re}\left((q + ip)^3 e^{2\pi i x}\right)$$
$$+ \frac{b}{2}\left(\frac{p^2 + q^2}{2}\right)^2 + dy\frac{p^2 + q^2}{2}$$

where the internal frequency of the periodic orbit is rescaled to 1. Note that the normalization is not performed with respect to the periodic motion $\dot{x} = 1$, but with respect to the periodic dynamics defined by the first two terms of H, cf. [49]. While the Hamiltonian is thus not directly made x-independent, this can now be achieved by means of a van der Pol transformation to co-rotating co-ordinates. The system is lifted to a 3-fold covering space on which the basic periodic motion decouples from the normal dynamics. The van der Pol transformation reads

$$
\begin{pmatrix} x \\ y \\ q \\ p \end{pmatrix} \longmapsto \begin{pmatrix} 3x \\ \frac{1}{3}\left(y - \pi(q^2 + p^2)\right) \\ q\cos 2\pi x + p\sin 2\pi x \\ -q\sin 2\pi x + p\cos 2\pi x \end{pmatrix}
$$

and turns the Hamiltonian into

$$
\hat{H}(x,y,q,p) = \frac{y}{3} + \hat{a}\left(\frac{q^3}{6} - \frac{p^2 q}{2}\right) + \frac{\hat{b}}{2}\left(\frac{p^2 + q^2}{2}\right)^2 + \hat{d}y\frac{p^2 + q^2}{2}
$$

with $\hat{a} = a$, $\hat{b} = b - \frac{4}{3}\pi d$ and $\hat{d} = \frac{1}{3}d$.

Note that not only the integrable approximation, but also the original Hamiltonian may be lifted to the 3-fold covering, yielding a $\mathbb{Z}_3$-equivariant system as well. Independence of x remains proper to $\hat{H}$, though, and reducing the S^1-symmetry $x \mapsto x + \rho$ yields the one-degree-of-freedom problem

$$
\mathcal{H}_y(q,p) = \hat{a}\left(\frac{q^3}{6} - \frac{p^2 q}{2}\right) + \frac{\hat{b}}{2}\left(\frac{p^2 + q^2}{2}\right)^2 + \hat{d}y\frac{p^2 + q^2}{2} \ .
$$

On the initial phase space the S^1-symmetry introduced by normalization is not free at the central family of elliptic periodic orbits; here reduction yields the Poisson space $\mathbb{R}^2/_{\mathbb{Z}_3}$. We first note that $\mathcal{H}$ undergoes at $y = 0$ the $\mathbb{Z}_3$-equivariant bifurcation (2.12) of co-dimension 1. For the original Hamiltonian system this implies that the elliptic periodic orbit momentarily loses its stability as it coincides with a nearby family of hyperbolic periodic orbits with 3 times the period. One could speak of a transcritical[4] bifurcation, although there is no lasting loss of stability.

The reason why we adopted a semi-global approach and retained the "higher order term" with coefficient $\hat{b}$ is that the transcritical bifurcation may trigger a periodic centre-saddle bifurcation nearby. Indeed, one readily checks that $\mathcal{H}$ undergoes a centre-saddle bifurcation at the three $\mathbb{Z}_3$-related points

$$
(q,p) = \left(\frac{-\hat{a}}{2\hat{b}}, 0\right) \quad \text{and} \quad \left(\frac{\hat{a}}{4\hat{b}}, \pm\frac{\sqrt{3}\hat{a}}{4\hat{b}}\right)
$$

as y passes through

$$
y_0 = \frac{\hat{a}^2}{8\hat{b}\hat{d}} \ .
$$

In two degrees of freedom all three parabolic equilibria correspond to the same 3-periodic orbit, and the hyperbolic periodic orbits born in this bifurcation then approach the central 1-periodic orbit.

---

[4] In the literature this is also called 3-bifurcation or phantom kiss.

Note that the integrable approximation H is close to the original Hamiltonian only in a neighbourhood of the initial family of elliptic periodic orbits close to the normal-internal 1:3 resonance. We thus need $\hat{a}$ to be sufficiently small for our analysis to be valid, i.e. already the original Hamiltonian should depend on x only weakly. Typical examples are forced oscillators.   □

In [102] this is discussed in the context of area-preserving mappings. The one-dimensional frequency mapping turns out to have a fold singularity precisely at the centre-saddle bifurcation. The frequency mappings of four-dimensional symplectic mappings may have fold and cusp singularities, see [101] for more details.

3.2 The Period-Doubling Bifurcation

The only real symplectic matrices that do not necessarily have a real logarithm are those with an eigenvalue -1. In Floquet theory this is remedied by squaring the matrix, i.e. by using a co-ordinate change that is not 1-periodic but has period 2. This corresponds to lifting the system to a double covering, where a Floquet multiplier -1 leads to a Floquet multiplier $+1$. Moreover, the lifted system inherits a $\mathbb{Z}_2$-symmetry. Note the similarity to the treatment of a pair of Floquet multipliers $\exp(\pm\frac{2}{3}\pi i)$ in Example 3.3.

Again the two-degree-of-freedom dynamics of the integrable Hamiltonian

$$H(x,y,q,p) \;=\; \omega y + \frac{a}{2}p^2 + \frac{b}{24}q^4 + \frac{d}{2}yq^2 \tag{3.3}$$

with $\omega, a, b, d \neq 0$ is readily analysed. Reducing the S^1-symmetry $x \mapsto x + \rho$ leads to the one-degree-of-freedom problem

$$\mathcal{H}_y(q,p) \;=\; \frac{a}{2}p^2 + \frac{b}{24}q^4 + \frac{d}{2}yq^2$$

and reducing the $\mathbb{Z}_2$-symmetry $q \mapsto -q$ as well yields a phase space with a conical singularity on which a Hamiltonian flip[5] bifurcation occurs, cf. Theorem 2.17. At the bifurcation value $y = 0$ one reconstructs a parabolic[6] periodic orbit and under variation of the energy a period-doubling bifurcation[7] takes place.

This qualitative description remains true if we replace (3.3) on the double covering by the more general integrable Hamiltonian

$$N(x,y,q,p) \;=\; \eta(y) + \mathcal{N}_y(q,p) \tag{3.4}$$

---

[5] A dual flip bifurcation in case a and b have opposite signs.
[6] In the literature this periodic orbit is also called a transitional periodic orbit.
[7] In the literature this bifurcation is also called subtle division (if ab is positive) or murder (in the dual case).

with $\eta'(0) = \omega(0) \neq 0$ and

$$\mathcal{N}_y(q,p) \;=\; \frac{a(y)}{2}p^2 \;+\; \frac{b(y)}{24}q^4 \;+\; \frac{c(y)}{2}q^2 \;+\; higher\ order\ terms$$

with $a(0), b(0) \neq 0$ and $c(0) = 0$, $d(0) := c'(0) \neq 0$. It is of course crucial that $\mathcal{N}_y$ respects the $\mathbb{Z}_2$-symmetry $q \mapsto -q$, whence Theorem 2.17 still applies.

Theorem 3.4. Let H define a Hamiltonian system in two degrees of freedom and let γ be a periodic orbit of X_H with Floquet multipliers -1. Under generic conditions on H a periodic-doubling bifurcation takes place as the value of H passes through $H(\gamma)$.

Proof. We choose Floquet co-ordinates (x, y, q, p) around the double cover $\{\,(x,0,0,0) \mid x \in \mathbb{T}^1\,\}$ of γ and proceed to normalize with respect to $\dot{x} = \omega$, using the genericity condition

$$\omega \;:=\; \int \frac{\partial H}{\partial y}(x,0,0,0)\,\mathrm{d}x \;\neq\; 0 \;.$$

This turns the Hamiltonian into the sum $H = N + P$ of (3.4) and $P(x, y, q, p)$ that contains the dependence on x and is of at least fifth order in (q, p). Further genericity conditions are $a, b, d \neq 0$ and ensure that (3.4) undergoes a periodic Hamiltonian pitchfork bifurcation, of dual type if a and b have opposite signs.

This shows that, still on the double covering $\mathcal{P}$, the periodic orbits $\{\,(x,y,0,0) \mid x \in \mathbb{T}^1\,\}$ change their normal behaviour from elliptic to hyperbolic and two additional $\mathbb{Z}_2$-related periodic orbits arise. If $ab > 0$ these are both elliptic, and hyperbolic in the dual case. When projecting from the double covering $\mathcal{P}$ to the original phase space $\mathcal{B}$, the central periodic orbits are mapped to a family of periodic orbits passing through γ and thus get their periods divided by 2. The two $\mathbb{Z}_2$-related families of periodic orbits are mapped to a single family of periodic orbits in $\mathcal{B}$ but retain the period.   □

For a proof based on the (area-preserving) iso-energetic Poincaré mapping see [205, 207, 208]. When dealing with lower-dimensional tori in Chapter 4 it is not the whole frequency vector that gets divided by 2 when passing to the appropriate covering space, but only one (well-chosen) of its components; we use again a double covering and the group of deck transformations remains $\mathbb{Z}_2$.

The singular values of the energy-momentum mapping carry no information on the normal behaviour of periodic orbits and therefore cannot distinguish between a period-doubling bifurcation and a dual period-doubling bifurcation. This brings the 8 cases of (3.3) down to four. Those with positive ω are depicted in Fig. 3.2, for negative ω one merely has to reflect about the vertical h-axis.

Remark 3.5. The swinging of the incense burners in some churches is an example for the destabilizing effect of parametric forcing with half the period. In

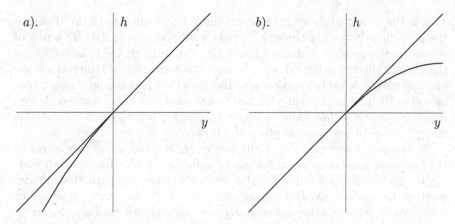

Fig. 3.2. Singular values of the energy-momentum mapping $\mathcal{EM} = (H, y)$ of (3.3) for positive ω, a, b and (**a**) positive d, (**b**) negative d.

a similar fashion the unstable equilibrium of the pendulum can be stabilized by means of oscillations of the suspension point. The mechanism is in both cases related to the period-doubling bifurcation.

The periodic orbit born in a period-doubling bifurcation inherits the stability of the initial periodic orbit and may afterwards lose its stability in another period-doubling bifurcation. For area-preserving mappings such period doubling cascades converge to "adding machines", cf. [189]. When passing to Hamiltonian systems these suspend to the simplest type Σ_2 of a solenoid (the one defined by iterated double covers), see [210] for more details.

3.3 Discrete Symmetries

We have seen in Example 3.3 how the presence of a $\mathbb{Z}_3$-symmetry, introduced by a lift to a 3-fold covering, influences the bifurcation scenario. And also the $\mathbb{Z}_2$-equivariant dynamics of (3.3) considerably differs from the generic bifurcation scenario detailed in Theorem 3.1. More generally, one obtains $\mathbb{Z}_\ell$-equivariant dynamics when lifting a system near a periodic orbit with e.g. Floquet multipliers $\exp(\pm 2\pi i/\ell)$ to an ℓ-fold covering. Throughout this section we remain in two degrees of freedom, see [38] for the generalization to higher degrees of freedom.

Let us first consider the cases $\ell \geq 5$. Mimicking the treatment in Example 3.3, but retaining only lowest order terms, leads to (lifted) reduced dynamics in one degree of freedom defined by

$$\mathcal{H}_y(q, p) = \frac{\hat{a}}{\ell!} \mathrm{Re}\,(q + \mathrm{i}p)^\ell + \frac{\hat{m}}{2}\left(\frac{p^2 + q^2}{2}\right)^2 + \hat{d}\,y\frac{p^2 + q^2}{2}.$$

This is the universal $\mathbb{Z}_\ell$-symmetric unfolding $1_{2,\ell}$ defined in (2.15). Thus, as the periodic orbit passes through Floquet multipliers $\exp(\pm 2\pi i k/\ell)$ a pair of elliptic and hyperbolic periodic orbits with ℓ times the period branch off from the initial (elliptic) orbit cylinder. In case one terms this a bifurcation[8] one ends up with a dense bifurcation set. Also note that periodic orbits with Diophantine Floquet multipliers give rise to invariant tori in the same emission-like way. On the other hand, a dense bifurcation set becomes unavoidable along families of normally elliptic tori, cf. [49].

In the remaining case $\ell = 4$ the family $\mathcal{H}_y$ derived above specializes to the universal $\mathbb{Z}_4$-symmetric unfolding 4_4 of Section 2.1.3. Here the two leading terms are of equal order 4 and it is their relative strength that decides whether the nearby dynamics resembles that of the cases $\ell \geq 5$ above or whether the periodic orbit momentarily loses its stability during a transcritical[9] bifurcation.

3.3.1 The Periodic Hamiltonian Pitchfork Bifurcation

Reversing symmetries may influence the dynamics as well. Choosing Floquet co-ordinates and normalizing with respect to $\dot{x} = \omega$ – which preserves reversibility – yields an integrable approximation. For the reflection

$$(x, y, q, p) \;\mapsto\; (-x, y, q, -p)$$

this Hamiltonian generically is (3.2) and a periodic centre-saddle bifurcation takes place. In case of the reflection

$$(x, y, q, p) \;\mapsto\; (-x, y, -q, p)$$

we are again led to the Hamiltonian (3.4), but now the corresponding dynamics is defined on the original phase space and not on a 2-fold covering.

Theorem 3.6. Let H define a reversible Hamiltonian system in two degrees of freedom and let γ be a periodic orbit of X_H with all Floquet multipliers $+1$. Generically the linearized iso-energetic Poincaré mapping has only one eigenvector to this eigenvalue, and the reversing symmetry must either multiply this eigenvector by -1 of fix it. Generically, as the energy passes through $H(\gamma)$, there is a periodic centre-saddle bifurcation in the former case, and in the latter case a periodic Hamiltonian pitchfork bifurcation takes place.

Proof. In the first case the proof of Theorem 3.1 still applies and in the second case the first part of the proof of Theorem 3.4 shows that γ is parabolic and part of an orbit cylinder that changes at γ its normal linear behaviour from elliptic to hyperbolic. Furthermore there is a second orbit cylinder passing

---

[8] In the literature it is called ℓ-bifurcation or emission.

[9] In the literature this is also called phantom kiss, while the term 4-bifurcation is used for both cases.

through γ with elliptic normal linear behaviour off γ if $ab > 0$ and hyperbolic normal linear behaviour off γ in the dual case.   □

For a proof based on the (area-preserving) iso-energetic Poincaré mapping see [243, 244]. The set of singular values of the energy-momentum mapping near a periodic Hamiltonian pitchfork bifurcation coincides with that near a Hamiltonian period-doubling bifurcation depicted in Fig. 3.2.

3.4 The Periodic Hamiltonian Hopf Bifurcation

We now turn to the generic bifurcation triggered by a Krein collision of Floquet multipliers on the unit circle. Throughout this section we restrict to three degrees of freedom. Still, introducing a mere S^1-symmetry will not lead to an integrable system as it was the case in two degrees of freedom.

In suitable Floquet co-ordinates (x, y, z) with $z = (q_1, q_2, p_1, p_2)$ the Hamiltonian reads

$$H(x, y, z) = \omega y + \Omega(q_1 p_2 - q_2 p_1) + a\frac{q_1^2 + q_2^2}{2} + \mathcal{O}(y^2, z^3) \qquad (3.5)$$

and the genericity condition $a \neq 0$ expresses that the linearized iso-energetic Poincaré mapping with double pair $\exp(\pm i\Omega/\omega)$ is not semi-simple, but has a non-diagonisable Jordan form. Following [37, 232, 227, 228] we assume that there are no resonances[10] $2\pi k\omega + l\Omega = 0$ between the internal frequency ω of the periodic orbit $(y, z) = (0, 0)$ and the normal frequency Ω. This implies that the dynamics defined by the first two terms $H_0 = \omega y + \Omega S$ of (3.5) is quasi-periodic. Here and in the following we adopt the definitions

$$S(q, p) = q_1 p_2 - q_2 p_1 , \qquad N(q, p) = \tfrac{1}{2}(q_1^2 + q_2^2) ,$$
$$M(q, p) = \tfrac{1}{2}(p_1^2 + p_2^2) , \qquad P(q, p) = q_1 p_1 + q_2 p_2$$

of Section 2.2.2.

Normalization with respect to H_0 introduces a $\mathbb{T}^2$-symmetry and thus leads to an integrable system. As in Section 2.2.2 this is followed by an additional normalization with respect to the nilpotent part N to further simplify the $\mathbb{T}^2$-symmetric higher order terms. In this way one obtains a normal form with respect to the truncated part

$$H_a(x, y, z) = \omega y + \Omega S(z) + aN(z) \qquad (3.6)$$

of (3.5) which reads

$$H(x, y, z) = H_a(x, y, z) + \frac{b}{2}M^2 + \frac{c}{2}y^2 + dyM$$
$$+ Z(y, S, M) + P(x, y, z) , \qquad (3.7)$$

---

[10] This means $\dfrac{\Omega}{\omega} \notin 2\pi\mathbb{Q}$; to achieve (3.5) we already used $\dfrac{\Omega}{\omega} \notin 2\pi\mathbb{Z}$.

see [150, 232, 227, 228] or Appendix C. Here

$$Z(y, S, M) = e_1 yS + \frac{e_2}{2}S^2 + e_3 SM + \frac{e_4}{6}M^3 + \frac{e_5}{2}yM^2 + \dots$$

contains further normalized terms and $P(x, y, z)$ is the remainder of the normal form.

The dynamics defined by the first line of (3.7) is readily analysed, provided that the genericity conditions $a, b, d \neq 0$ are fulfilled. Indeed, reducing the S^1-symmetry $x \mapsto x + \rho$ yields the 1-parameter family of integrable two-degree-of-freedom Hamiltonians

$$\mathcal{H}_y(z) = \Omega S + aN + \frac{b}{2}M^2 + dyM .$$

The dynamics defined by $\mathcal{H}_y$ have for all values of y an equilibrium at the origin which undergoes a Hamiltonian Hopf bifurcation as y passes through zero. Alternatively one could reduce the whole $\mathbb{T}^2$-symmetry generated by H_0 and end up with the one-degree-of-freedom problem (2.20) studied in Section 2.1.4. In any case one reconstructs a periodic Hamiltonian Hopf[11] bifurcation, supercritical for $ab > 0$ and subcritical if a and b have opposite signs.

Let us fix thoughts on the supercritical case. As the periodic orbit $z = 0$ loses its stability the two "normal modes" of (normally) elliptic 2-tori detach and form a single 2-parameter family. The hyperbolic periodic orbits resulting from the bifurcation lead to a thread in the set of singular values of the energy-momentum mapping $\mathcal{EM}$, cf. Fig. 2.14a. The pre-image of such a singular value (y, σ, h) is a pinched 3-torus, consisting of the periodic orbit together with its (un)stable manifold. The (maximal) 3-tori around the thread display monodromy.

This qualitative description remains true if the additional normalized terms $Z(y, S, M)$ in (3.7) are retained. Indeed, one can still reduce symmetry and thus show that a periodic Hamiltonian Hopf bifurcation takes place. Examples of three-degree-of-freedom Hamiltonian systems that already are $\mathbb{T}^2$-symmetric (and thus integrable) and display a (supercritical) periodic Hamiltonian Hopf bifurcation are the Lagrange top and the Kirchhoff top, see [84, 199, 81, 147, 20] and Example 3.8 below.

It remains to understand how the perturbation of the integrable part by the remainder term $P(x, y, z)$ influences the dynamics. In the present quasi-periodic context this question is much more problematic than it was for the normalizations with respect to periodic flows we had encountered so far, cf. Appendix C. For instance, we had to impose infinitely many non-resonance conditions $2\pi k\omega + l\Omega \neq 0$ to ensure that normalization with respect to H_0 is even possible to any order. To obtain effective bounds on P these have to be sharpened to Diophantine conditions

---

[11] In the literature this is also called a quasi-periodic Hamiltonian Andronov–Hopf bifurcation.

$$\bigwedge_{(k,l)\in\mathbb{Z}^2\setminus\{(0,0)\}} |2\pi k\omega + l\Omega| \geq \frac{\gamma}{(|k| + |l|)^\tau} \tag{3.8}$$

with $\gamma > 0$ and $\tau > 1$. While (3.8) is not a generic condition, it determines a set of frequencies (ω, Ω) that is of large measure in that the relative measure within any compact set tends to 1 as $\gamma \searrow 0$.

A Nekhoroshev-like approach is pursued in [232, 229] where the Hamiltonian is assumed to be analytic. This is used to derive an optimal order r to which the Hamiltonian is normalized, depending on the constants γ, τ in (3.8) that (ω, Ω) satisfy and on the (complex) domain of analyticity of the original Hamiltonian. In a suitable norm the remainder P_r of the normal form (3.7) of optimal order r is then bounded on a shrunk domain by a power $\frac{1}{2}r - 1$ of the quotient of the size of the shrunk domain by the size of the original domain. Denoting these two sizes by R and R_0 the precise bound is

$$|||P_r||| \leq c_8 \left(1 - \frac{R}{R_0}\right)^{-1} \left(c_5 \frac{R}{R_0}\right)^{(r/2)-1} =: \mathcal{R}$$

where c_5 and c_8 are constants not depending on R. Moreover P_r goes to zero with the distance R to the periodic orbit γ faster than any algebraic order in R, see again [232, 229]. Note that this is weaker than the exponential decay one has in the normally elliptic case, cf. [279, 163].

The splitting (3.7) into a normal form Hamiltonian and a small remainder term allows to obtain persistence of the integrable dynamics. For an approach based on the (four-dimensional) iso-energetic Poincaré mapping see [37].

Theorem 3.7. Let H define a Hamiltonian system in three degrees of freedom and let γ be a periodic orbit of X_H with a double pair of Floquet multipliers $\exp(\pm i\Omega/\omega)$ satisfying (3.8). In the normal form (3.7) of H let $ab, \omega d > 0$ and $d^2 \neq bc$. Then a supercritical periodic Hamiltonian Hopf bifurcation takes place as the value h of H passes through $H(\gamma)$:

For $h > H(\gamma)$ the periodic orbit $z = 0$ is elliptic and turns hyperbolic for $h < H(\gamma)$. Emanating from these periodic orbits for $h \geq H(\gamma)$ and detached when $h < H(\gamma)$ there is a 2-dimensional Cantor family of normally elliptic invariant 2-tori of large relative measure – the complement within a given compact manifold extending the Whitney-C^∞-smooth Cantor family has a measure $\leq C\mathcal{R}^\alpha$, $0 < \alpha < \frac{1}{2}$, where $\mathcal{R}$ is the (optimal) bound of P_r.

For a proof the reader is referred to [232]. □

The change to $\omega d < 0$ amounts to interchanging $h > H(\gamma)$ and $h < H(\gamma)$ for the values h of the Hamiltonian. We come back to the Hamiltonian Hopf bifurcation of subcritical type $ab < 0$ in Example 4.5 in Section 4.1.1. The condition $d^2 \neq bc$ ensures that the normally elliptic 2-tori satisfy the Kolmogorov condition on their internal frequencies.

The pinched tori of stable and unstable manifolds of the hyperbolic periodic orbits are expected to break up under the perturbation by P. The results in [46] indicate that the Cantor family of surviving maximal tori should again display monodromy. In fact, following [224, 246] one can use precisely this monodromy to prove persistence of maximal tori close to the supercritical periodic Hamiltonian Hopf bifurcation.

The qualitative behaviour of the integrable approximation can already be determined by truncating the normal form (3.7) at/of order $r = 4$. To compute this normal form only low order resonances $2\pi k\omega + l\Omega = 0$ with $|l| \leq 4$ have to be excluded, and to obtain bounds on the remainder term P_4 these have to be sharpened to the finitely many conditions

$$\bigwedge_{\substack{(k,l)\in\mathbb{Z}^2 \\ 0<|l|\leq 4}} |2\pi k\omega + l\Omega| \geq \gamma$$

with $\gamma > 0$. The remaining set of double Floquet multipliers $\exp(\pm i\Omega/\omega)$ consists of the unit circle with small neighbourhoods of 3rd and 4th roots of unity excluded.

The perturbation by P_4 still has no effect on the periodic orbit changing from elliptic to hyperbolic. However, the family of normally elliptic 2-tori may now fall into a resonance gap since their limiting internal frequencies upon approaching the bifurcating periodic orbit are precisely ω and $\Omega/(2\pi)$. This happens when $\Omega/(2\pi\omega)$ is (sufficiently close to) a rational number and should under perturbation by P_4 result in the 2-tori to break up into "long" elliptic and hyperbolic periodic orbits.

An accurate analysis of such a resonant periodic Hamiltonian Hopf bifurcation relies on a normal form of such high order that it includes "resonance terms" breaking the $\mathbb{T}^2$-symmetry to an S^1-symmetry, cf. [38]. For Floquet multipliers $\exp(\pm\frac{2}{3}\pi i)$ and $\pm i$ such resonance terms already enter the normal form of order 4. Krein collisions at ± 1 are expected to combine the phenomena of the present section with those of Sections 3.1, 3.2 and 2.2.4.

A Krein collision at a pre-described Floquet multiplier is a co-dimension 2 phenomenon. Correspondingly, in [38] 2-parameter unfoldings are used to study resonant periodic Hamiltonian Hopf bifurcations. The analysis takes place on the level of symplectic maps where already the first unfolding parameter is externally given whence the addition of a second external parameter looks more natural, and it is claimed that a 2-parameter unfolding provides a clearer picture of the (non-resonant) periodic Hamiltonian Hopf bifurcation as well. We come back to this in Section 4.3.1 where the actions conjugate to the bifurcating n-torus do provide $n \geq 2$ unfolding parameters.

Example 3.8. The Kirchhoff top is a symmetric rigid body with a fixed point subject to a symmetric linear force field, cf. [20, 134]. Since we want to study gyroscopic stabilization we assume the linear force field to be directed towards the origin. The system is completely integrable on its phase space $T^*SO(3)$ with canonical symplectic form. The two symmetries are a right S^1-action corresponding to rotation about the axis of symmetry of the body and a left

S^1-action corresponding to rotation about the vertical axis in space. Reduction with respect to the left S^1-action gives the Euler–Poisson equations

$$\dot{x} = x \times \frac{\partial H}{\partial y} \ , \quad \dot{y} = x \times \frac{\partial H}{\partial x} + y \times \frac{\partial H}{\partial y} \tag{3.9}$$

for the top. Here $x \in \mathbb{R}^3$ is the so-called Poisson vector, the direction of the vertical axis measured in a body set of axes, whence the Casimir $C_1 = x_1^2 + x_2^2 + x_3^2$ of (3.9) takes the value 1. Similarly $y \in \mathbb{R}^3$ is the angular momentum vector measured in the body set of axes. The Casimir $C_2 = x_1 y_1 + x_2 y_2 + x_3 y_3$ therefore takes the value a of the angular momentum along the vertical axis. This defines the reduced phase space

$$T_a S^2 \ = \ \left\{ (x,y) \in \mathbb{R}^6 \ \middle| \ C_1 = 1 \, , \ C_2 = a \right\} \ \subseteq \ \mathbb{R}^6$$

which is a 4-dimensional smooth submanifold invariant under the flow of the vector field given by (3.9) . Rescaling time and length the Hamiltonian reads

$$H(x,y) \ = \ \tfrac{1}{2}(y_1^2 + y_2^2) + \frac{\gamma}{2} y_3^2 + x_3^2$$

where $\gamma \geq \tfrac{1}{2}$ is the quotient of the principal moments $I_1 = I_2$ and I_3 of inertia. The equilibrium points $(x,y) = (0,0,\pm 1, 0, 0, a)$ are the upright standing and downward hanging positions of the top, which become gyroscopically stabilized if the top spins sufficiently fast.

The reduced system still has the integral of motion $L(x,y) = y_3$, which is the angular momentum corresponding to rotation about the symmetry axis of the top. The corresponding action is

$$\rho : \ S^1 \times \mathbb{R}^6 \ \longrightarrow \ \mathbb{R}^6$$
$$(t,x,y) \ \mapsto \ (R_t x, R_t y)$$

with

$$R_t \ = \ \begin{pmatrix} \cos t & -\sin t & 0 \\ \sin t & \cos t & 0 \\ 0 & 0 & 1 \end{pmatrix} .$$

Inspired by the treatment of the Lagrange top in [84, 81, 147] we perform a second (and singular) reduction. The invariants of ρ are

$$\pi_1 = x_1^2 + x_2^2 \ , \quad \pi_3 = x_1 y_1 + x_2 y_2 \ , \quad \pi_5 = x_3 \ ,$$
$$\pi_2 = y_1^2 + y_2^2 \ , \quad \pi_4 = x_1 y_2 - x_2 y_1 \ , \quad \pi_6 = y_3$$

with relations

$$\pi_3^2 + \pi_4^2 \ = \ \pi_1 \pi_2 \ , \quad \pi_1 \geq 0 \ , \quad \pi_2 \geq 0 \ .$$

Furthermore we have

$$C_1 = \pi_1 + \pi_5^2 = 1 \ , \quad C_2 = \pi_3 + \pi_5\pi_6 = a \ , \quad L = \pi_6 = b$$

where b is the value of the third integral of motion. Consequently, we can eliminate π_1, π_3, π_6 and the twice reduced phase space is given by

$$V_a^b = \left\{ (\pi_2, \pi_4, \pi_5) \in \mathbb{R}^3 \ \middle| \ R_a^b(\pi) = 0 \ , \ \pi_2 \geq 0 \ , \ |\pi_5| \leq 1 \right\}$$

with

$$R_a^b(\pi) = \pi_4^2 + (a - b\pi_5)^2 - (1 - \pi_5^2)\pi_2 \ .$$

After skipping constant terms the reduced Hamiltonian on V_a^b is

$$\mathcal{H}(\pi) = \frac{1}{2}\pi_2 + \pi_5^2$$

and the trajectories are given by the intersections $\{\mathcal{H} = h\} \cap V_a^b$. Note that the energy level sets $\{\mathcal{H} = h\} \subseteq \mathbb{R}^3$ are parabolic cylinders. Our aim is to use the geometry of V_a^b to check that non-degenerate Hamiltonian Hopf bifurcations take place at $|a| = |b| = \sqrt{2}$. Using the discrete symmetries $x \mapsto -x$ and $y \mapsto -y$ we may concentrate on $a = b = 2\sqrt{2}$. The twice reduced phase space is smooth for $|a| \neq |b|$ and degenerates into a cone at $a = b \neq 0$; this means that locally the invariants form a Lie algebra isomorphic to $\mathfrak{sl}_2(\mathbb{R})$.

While the actual eigenvalue movement is lost through the second reduction, the bifurcation value can also be deduced from the one-degree-of-freedom system – for this value of $a = b$ the energy level set $\{\mathcal{H} = 1\}$ enters the conical singularity with the same "slope" as V_a^a at some direction of the cone. Note that for tangency of $\{\mathcal{H} = 1\}$ to the reduced phase space V_a^a one must have $\pi_4 = 0$. Consequently we have to compute the behaviour of $\mathcal{H}$ with respect to V_a^a within the (π_5, π_2)-plane. Both on $\{\mathcal{H} = 1\}$ and on $V_a^a \cap \{\pi_4 = 0\}$ we can write π_2 as a function of π_5, which leads us to considering the difference function

$$f_a(\pi_5) = a^2\frac{1 - \pi_5}{1 + \pi_5} - 2(1 - \pi_5^2) \ .$$

The equilibrium $(\pi_2, \pi_4, \pi_5) = (0, 0, 1)$ bifurcates if and only if the derivative $f_a'(1) = -\frac{1}{2}a^2 + 4$ vanishes, that is at $a = 2\sqrt{2}$.

The function f_a contains the necessary information on the relative position of $\{\mathcal{H} = 1\}$ to V_a^a within $\{\pi_4 = 0\}$. Linear versality follows from

$$\frac{\mathrm{d}}{\mathrm{d}a}f_a'(1) = -a = -2\sqrt{2} \neq 0$$

and

$$f_{2\sqrt{2}}''(1) = 8$$

implies quadratic contact from the outside, see Fig. 3.3. Hence, the Kirchhoff top undergoes supercritical Hamiltonian Hopf bifurcations as the angular momenta of the vertically upwards and downwards spinning relative equilibria

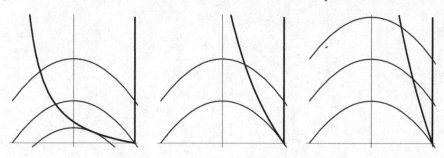

Fig. 3.3. Relative position of $\{\mathcal{H} = h\}$ (thin) and V_a^a (thick) within $\{\pi_4 = 0\}$ for $|a| < 2\sqrt{2}$, $|a| = 2\sqrt{2}$ and $|a| > 2\sqrt{2}$. The horizontal axis is π_5 and the vertical axis is $\frac{1}{3}\pi_2$.

pass through $\pm 2\sqrt{2}$. Theorem 3.7 implies that this remains true if the axial symmetries of body and force field are only approximate, though to sufficiently high precision. $\qquad\qquad\square$

The four Hamiltonian Hopf bifurcations in the Kirchhoff top are of super-critical type and therefore create monodromy where the thread parametrising periodic orbits detaches from the surface parametrising elliptic 2-tori, see Fig. 2.14a. Because of the discrete symmetry $\pi_5 \mapsto -\pi_5$ the two threads meet at $a = b = 0$ resulting in a 2-pinched torus. Correspondingly, the two monodromy matrices of the form $\left(\begin{smallmatrix}0&1\\0&0\end{smallmatrix}\right)$ add up to $\left(\begin{smallmatrix}0&2\\0&0\end{smallmatrix}\right)$, see [20, 225] for more details.

An example displaying monodromy that is not already integrable to start with is the swing-spring. Balancing the physical parameters as in [188, 53], but now in three degrees of freedom, puts the equilibrium state into 1:1:2 resonance. The initial axial symmetry together with the S^1-symmetry aquired from normalization yields the integrable approximation studied in [152, 98]. The ramified torus bundle of this system displays monodromy and the step-wise precession of the swing-spring turns out to be a manifestation of this, see [98].

In molecular physics the 1:1:2 resonance is known as the Fermi resonance and occurs when modelling bending and $\mathbb{Z}_2$-symmetric stretching vibrations of the CO_2 molecule. Here monodromy manifests itself as a lattice defect in the joint spectrum of energy and vertical angular momentum, for more details see [104, 82, 269] and references therein. The results in [245, 246, 46] allow to conclude that monodromy is not only present in the integrable approximation, but also in the original swing-spring system.

4

Bifurcations of Invariant Tori

We continue to study a single Hamiltonian system defined on a symplectic manifold $\mathcal{P}$. Given an isotropic invariant torus T of X_H we use symplectic coordinates (x, y, z) on $\mathcal{P}$ with angles $x_1, \ldots, x_n$ parametrising $\mathsf{T} = \{\, (y, z) = (0, 0) \,\}$ and conjugate actions $y_1, \ldots, y_n$ whence the equations of motion read

$$
\begin{aligned}
\dot{x} &= f(x, y, z) \\
\dot{y} &= g(x, y, z) \\
\dot{z} &= h(x, y, z) \ .
\end{aligned}
$$

For $y = 0$ we expand the right hand sides in z and obtain

$$
\begin{aligned}
f(x, 0, z) &= \omega + \mathcal{O}(z^2) \\
g(x, 0, z) &= \langle \Omega'(x) \cdot z, z \rangle + \mathcal{O}(z^3) \\
h(x, 0, z) &= \Omega(x) \cdot z + \mathcal{O}(z^2)
\end{aligned}
$$

with x-dependent higher order terms. If we can achieve $\Omega(x) \equiv \Omega$ to be x-independent the invariant torus T is said to be *reducible* to Floquet form. We mainly concentrate on this case and comment on the non-reducible case in Section 4.3.2. In the reducible case the eigenvalues of Ω are called Floquet exponents.

A normally hyperbolic invariant n-torus T has all its Floquet exponents off the imaginary axis. Note that T is not a normally hyperbolic manifold. Indeed, the normal linear behaviour involves the n zero eigenvalues in the y-direction as well; the constant matrix Ω describes the symplectic normal linear behaviour, cf. [159, 56, 28]. Instead of speaking of symplectic-normally hyperbolic invariant tori we simplify to hyperbolic[1] tori that have no purely imaginary Floquet exponents. Similarly for (normally) elliptic, parabolic, ... invariant tori.

---

[1] In the literature a torus is called hyperbolic, in a different context, if the dynamics *on* the torus is hyperbolic. In the present context the dynamics $\dot{x} = \omega$ on our invariant tori is parallel.

Near a hyperbolic torus T we can improve upon the Floquet co-ordinates and obtain

$$
\begin{aligned}
f(x,y,z) &= \omega + \mathcal{O}(y) + \mathcal{O}(z^2) \\
g(x,y,z) &= \mathcal{O}(y) + \mathcal{O}(z^3) \\
h(x,y,z) &= \Omega(y) \cdot z + \mathcal{O}(z^2)
\end{aligned}
$$

since zero is not a Floquet exponent. This makes $\{\, z = 0 \,\}$ a normally hyperbolic manifold which is the centre manifold of T. In this way looking for further hyperbolic n-tori near the given T boils down to the study of maximal tori in the n-degree-of-freedom Hamiltonian system on $\{\, z = 0 \,\}$. In particular, if we start with an integrable system, where

$$
f(x,y,0) = \omega(y) \quad \text{and} \quad g(x,y,0) = 0
$$

then any slightly perturbed system has a large Cantor family of hyperbolic tori near $(y,z) = (0,0)$ provided that the Kolmogorov condition $D\omega(0) \neq 0$ on the internal frequencies is met.

The persistence of elliptic tori under perturbation from an integrable system is more involved, see [55] and references therein. Here the necessary Diophantine conditions involve not only the internal frequencies $\omega_1, \ldots, \omega_n$, but also the normal frequencies $\alpha_1, \ldots, \alpha_m$, i.e. the (positive) imaginary parts of the (purely imaginary) Floquet exponents $\pm i\alpha_j$ and read

$$
\bigwedge_{k \in \mathbb{Z}^n \setminus \{0\}} \bigwedge_{\substack{\ell \in \mathbb{Z}^m \\ |\ell| \leq 2}} |2\pi \langle k, \omega \rangle + \langle \ell, \alpha \rangle| \geq \frac{\gamma}{|k|^\tau} . \tag{4.1}
$$

Next to the internal resonances this excludes the normal-internal resonances

$$
\begin{aligned}
2\pi \langle k, \omega \rangle &= \alpha_j & \text{(4.2a)} \\
2\pi \langle k, \omega \rangle &= 2\alpha_j & \text{(4.2b)} \\
2\pi \langle k, \omega \rangle &= \alpha_i + \alpha_j & \text{(4.2c)} \\
2\pi \langle k, \omega \rangle &= \alpha_i - \alpha_j . & \text{(4.2d)}
\end{aligned}
$$

In Chapter 3 the first three resonances led to the different bifurcations and we study the corresponding bifurcations of invariant tori in the three sections of this chapter. For the resonance (4.2d), which generalizes an equilibrium in 1:1 resonance, it has been shown [161] that the n-tori persist and remain elliptic.

The approach in [30, 31, 294] yields persistence already if (4.1) is satisfied only with $|\ell| \leq 1$ whence only internal resonances and normal-internal resonances (4.2a) are excluded. The price to pay is that one loses control on the normal linear behaviour. Thus, when passing through resonances (4.2b) and (4.2c) the lower-dimensional tori may (and as we shall see do) lose ellipticity and aquire hyperbolic Floquet exponents.

Whenever there are hyperbolic Floquet exponents a centre manifold reduction, see [211], allows to reduce to a Hamiltonian system with less degrees of freedom where all Floquet exponents lie on the imaginary axis. Persistence of elliptic tori hence shows that hypo-elliptic tori persist as Cantor families with persistent[2] normal linear behaviour. It remains to understand the dynamical consequences when (4.1) is violated.

We stay away from internally resonant tori and consider lower-dimensional tori with solely purely imaginary Floquet exponents and at least one of the ensuing normal frequencies satisfying a normal-internal resonance (4.2a), (4.2b) or (4.2c). To simplify the exposition in the subsequent sections we assume that there are no further purely imaginairy Floquet exponents. Thus, for (4.2a) and (4.2b) we study n-tori in $n+1$ degrees of freedom and for (4.2c) in $n+2$ degrees of freedom. Concerning additional ellipticity this is mainly a technical simplification as it relieves us from additional Diophantine conditions that would have to be dragged through the complete analysis.

The restriction to a single normal-internal resonance rather is dictated by our present possibilities. Indeed, in Chapter 2 we could not treat bifurcating equilibria corresponding to two resonances (4.2). Thus, a satisfactory study of such tori which already may form one-parameter families in integrable Hamiltonian systems with five degrees of freedom has to await further progress in local bifurcation theory.

4.1 Bifurcations with Vanishing Normal Frequency

We are given a perturbed integrable Hamiltonian $H = N + P$ on the phase space

$$\mathcal{P} = \mathbb{T}^n \times \mathbb{R}^n \times \mathbb{R}^2$$

such that

$$N(x, y, z) = \eta(y) + \mathcal{N}_y(z) \tag{4.3}$$

with $D\mathcal{N}_0(0) = 0$, whence $\mathsf{T} = \mathbb{T}^n \times \{0\} \times \{0\}$ is an invariant torus of the unperturbed dynamics. This integrability implies that T is already reduced to Floquet form and our assumption is that the two Floquet exponents form a purely imaginary pair $\pm i\alpha$ with normal frequency α satisfying the normal-internal resonance (4.2a), where $\omega = D\eta(0)$ is the (internal) frequency vector of T. We follow [294, 49] and choose our Floquet co-ordinates (x, y, z) in such a way that α is zero.

The integrability of the system defined by (4.3) stems from the $\mathbb{T}^n$-symmetry $x \mapsto x + \xi$ and reducing this symmetry yields the n-parameter family of one-degree-of-freedom systems with Hamiltonian $\mathcal{N}_y$ on $\mathbb{R}^2$. At $y = 0$

---

[2] Reducibility to Floquet form is only guaranteed within the centre manifold and may fail on the whole phase space, cf. [162, 55, 279, 163].

the origin is a regular equilibrium with vanishing eigenvalues which is the situation treated in Section 2.1.1. On $\mathbb{R}^2$ we use again co-ordinates $z = (q, p)$ in which the symplectic structure becomes the area element $dq \wedge dp$.

4.1.1 Parabolic Tori

In the parabolic case the one-degree-of-freedom Hamiltonian at $y = 0$ reads

$$\mathcal{N}_0(q, p) \;=\; \frac{a}{2}p^2 \,+\, \mathcal{O}(q^3, p^3)$$

with $a \neq 0$. Generically this is a singularity $A_{d-1}^{\pm}$ with $3 \leq d \leq n+2$ and there is a re-parametrisation $y \mapsto \lambda(y)$ such that the Hamiltonian system defined by $\mathcal{N}_y$ is locally (topologically) equivalent to that defined by the universal unfolding (2.3).

In the simplest case $d = 3$ a centre-saddle bifurcation takes place as y passes through $\lambda^{-1}(0)$. From this one reconstructs a quasi-periodic centre-saddle bifurcation in the $(n + 1)$-degree-of-freedom system defined by (4.3). Before generalizing to $d \geq 4$ let us show that the perturbed system with Hamiltonian $H = N + P$ displays a quasi-periodic centre-saddle bifurcation as well.

In the present quasi-periodic setting the n actions conjugate to the toral angles not only unfold the bifurcation scenario, but are also needed to control the frequency vector $\omega(y) = D\eta(y)$. Indeed, we can only expect n-tori that have a Diophantine frequency vector to persist and upon a small perturbation a single Diophantine frequency vector $\omega(0)$ may become resonant. For the quasi-periodic centre-saddle bifurcation these two goals are better compatible than for general bifurcations of parabolic tori, leading to an improved result.

The Quasi-Periodic Centre-Saddle Bifurcation

The key step is to first decouple both the unfolding parameter λ and the frequencies $\omega_1, \ldots, \omega_n$ from the actions $y_1, \ldots, y_n$ and to consider

$$N(x, y, q, p; \lambda, \omega) \;=\; \langle \omega, y \rangle \,+\, \frac{a(\omega)}{2}p^2 \,+\, \frac{b(\omega)}{6}q^3 \,+\, \lambda q \qquad (4.4)$$

with externally given parameters λ and ω. Only after the perturbation problem is solved in this parameter-dependent context, conditions are formulated that allow y to successfully play the rôle of both λ and ω. It is in this last step where the main differences between the present case $d = 3$ with 1-dimensional λ and the general case with $(d - 2)$-dimensional λ appear.

The parameter-dependent Hamiltonian (4.4) is defined on the product of phase space and parameter space

$$\mathbb{T}^n \times \mathbb{Y} \times \mathbb{S} \times \Lambda \times \Sigma$$

where $\mathbb{Y} \subseteq \mathbb{R}^n$ and $\mathbb{S} \subseteq \mathbb{R}^2$ are small neighbourhoods of the origin and $\Lambda \subseteq \mathbb{R}$ is an open interval around zero. The domain $\Sigma \subseteq \mathbb{R}^n$ of frequency vectors is compact, but not necessarily localized around a Diophantine ω_0. We obtain results only for frequencies in

$$\Sigma_\gamma' = \left\{ \omega \in \Sigma_\gamma \;\middle|\; d(\omega, \partial\Sigma) \geq \gamma \right\}$$

that are bounded away from the boundary $\partial\Sigma$ and satisfy the Diophantine conditions

$$\bigwedge_{k \in \mathbb{Z}^n \setminus \{0\}} |\langle k, \omega \rangle| \geq \frac{\gamma}{|k|^\tau} . \tag{4.5}$$

Here we fix $\tau > n - 1$ but allow $\gamma > 0$ to vary, denoting by Σ_γ the set of frequency vectors satisfying (4.5). Furthermore $|..|_A$ stands for the supremum norm on the set A.

When proving a persistence theorem the difficult part is to keep track of the most degenerate "object" in the perturbed system. Our first step is therefore to look for the bifurcating parabolic tori of X_H.

Theorem 4.1. Let the functions $a, b : \Sigma \longrightarrow \mathbb{R}$ in (4.4) satisfy the inequalities

$$|a|_\Sigma, \; |b|_\Sigma, \; \left|\frac{1}{a}\right|_\Sigma, \; \left|\frac{1}{b}\right|_\Sigma, \; |Da|_\Sigma, \; |Db|_\Sigma \; < \; \Gamma$$

for some constant $\Gamma > 0$. Then there is $\delta > 0$ with the following property. For any analytic perturbation $H = N + P$ of (4.4) with

$$|P|_{\mathbb{T}^n \times \mathbb{Y} \times \mathbb{S} \times \Lambda \times \Sigma} \; < \; \delta$$

there exists a C^∞-diffeomorphism

$$\Phi \; : \; \mathbb{T}^n \times \mathbb{R}^n \times \mathbb{R}^2 \times \mathbb{R} \times \Sigma \; \longrightarrow \; \mathbb{T}^n \times \mathbb{R}^n \times \mathbb{R}^2 \times \mathbb{R} \times \Sigma$$

such that

1). Φ is analytic for fixed ω.
2). Φ is symplectic for fixed (λ, ω).
3). Φ is C^∞-close to the identity.
4). On $\mathbb{T}^n \times \mathbb{R}^n \times \mathbb{R}^2 \times \mathbb{R} \times \Sigma_\gamma' \cap \Phi^{-1}(\mathbb{T}^n \times \mathbb{Y} \times \mathbb{S} \times \Lambda \times \Sigma)$ one can split $H \circ \Phi = N_\infty + P_\infty$ into an integrable part N_∞ and higher order terms P_∞. Here N_∞ has the same form (4.4) as N. The x-dependence is pushed into the higher order terms, i.e.

$$\frac{\partial^{|j|+k+l+|m|} P_\infty}{\partial y^j \partial p^k \partial q^l \partial \lambda^m}(x, 0, 0, 0; 0, \omega) \; = \; 0$$

for all $(x, \omega) \in \mathbb{T}^n \times \Sigma_\gamma'$ and all j, k, l, m satisfying $6|j| + 3k + 2l + 4m \leq 6$.

For the proof see [139, 50] or Appendix D.   □

In the new co-ordinates the "origin" $y = 0 = q = p = \lambda$ consists of parabolic tori. Since the new co-ordinates provide valid information on the perturbed vector field X_H only for Diophantine frequency vectors $\omega \in \Sigma'_\gamma$, this yields a Cantor family of parabolic tori. Our next step is to show that the families of elliptic and hyperbolic meeting at the parabolic tori also persist under perturbation.

Corollary 4.2. Under the conditions of Theorem 4.1 the Hamiltonian system defined by $H = N + P$ undergoes a quasi-periodic centre-saddle bifurcation as λ traverses Λ.

Proof. We work in the new co-ordinates provided by Theorem 4.1 where $H_\infty = H \circ \Phi$ is a small perturbation of N_∞. We concentrate on positive a_∞ and b_∞ whence the integrable dynamics defined by N_∞ has a family

$$\mathbb{T}^n \times \{0\} \times \left\{ (q, p; \lambda, \omega) \;\middle|\; q = -\sqrt{\frac{-2\lambda}{b_\infty(\omega)}} \, , \, p = 0, \, \lambda < 0, \, \omega \in \Sigma \right\}$$

of hyperbolic tori. These are moved to the "origin" $y = 0 = q = p$ by the parameter-dependent translation

$$(x, y, q, p; \lambda, \omega) \;\longmapsto\; \left(x, y, q + \sqrt{\frac{-2\lambda}{b_\infty(\omega)}}, p; \lambda, \omega \right)$$

after which the Hamiltonian reads

$$N_\infty \;=\; \langle \omega, y \rangle + \frac{a_\infty(\omega)}{2} p^2 - \sqrt{\frac{-\lambda b_\infty(\omega)}{2}} q^2 + \frac{b_\infty(\omega)}{6} q^3$$

where we have dropped the constant term. Since the hyperbolicity is of order $\sqrt[4]{-\lambda}$ we scale according to

$$(x, y, q, p; \lambda, \omega) \;\longmapsto\; \left(x, (-\lambda)^{-9/4} y, (-\lambda)^{-1} q, (-\lambda)^{-5/4} p; \lambda, (-\lambda)^{-1/4} \omega \right)$$

and divide the Hamiltonian by $\sqrt{-\lambda^5}$. This turns the perturbed Hamiltonian into

$$H_\infty \;=\; \langle \omega, y \rangle + \frac{a_\infty(\omega)}{2} p^2 - \sqrt{\frac{b_\infty(\omega)}{2}} q^2 + R_\infty \qquad (4.6)$$

with

$$R_\infty(x, y, q, p; \lambda, \omega) \;=\; \sqrt{-\lambda} \frac{b_\infty(\omega)}{6} q^3$$

$$+ \frac{1}{\sqrt{-\lambda^5}} P_\infty\left(x, (-\lambda)^{9/4} y, -\lambda \left(q - \sqrt{\frac{-2\lambda}{b_\infty(\omega)}} \right), (-\lambda)^{5/4} p; \lambda, (-\lambda)^{1/4} \omega \right) .$$

Using that P_∞ has no low order terms we can estimate R_∞ by $\delta \sqrt[4]{-\lambda}$ where $\delta > 0$ is the small constant in Theorem 4.1. Thus, for sufficiently small $-\lambda > 0$ the hyperbolic tori of the integrable dynamics defined by the first three terms of (4.6) persist under the perturbation by R_∞, cf. [55]. The necessary Diophantine conditions are of the same form as those we already impose since the new co-ordinates provided by Theorem 4.1 allow to obtain valid information only for $\omega \in \Sigma'_\gamma$.

For the elliptic tori of the integrable dynamics defined by N_∞ at

$$
\mathbb{T}^n \times \{0\} \times \left\{ (q, p; \lambda, \omega) \;\middle|\; q = +\sqrt{\frac{-2\lambda}{b_\infty(\omega)}} \,, p = 0, \lambda < 0, \omega \in \Sigma \right\}
$$

the corresponding translation and scaling yields

$$
H_\infty \;=\; \langle \omega, y \rangle + \frac{a_\infty(\omega)}{2} p^2 + \frac{\sqrt{2 b_\infty(\omega)}}{2} q^2 + R_\infty \;.
$$

Again we infer persistence from [55] provided that $-\lambda > 0$ is sufficiently small. However, the necessary Diophantine conditions now also involve the normal frequency and read

$$
\bigwedge_{k \in \mathbb{Z}^n \setminus \{0\}} \; \bigwedge_{\ell \in \{0, \pm 1, \pm 2\}} \left| 2\pi \langle k, \omega \rangle + \ell \sqrt{a_\infty(\omega)} \sqrt{-2\lambda b_\infty(\omega)} \right| \;\geq\; \frac{\gamma}{|k|^\tau} \quad (4.7)
$$

whence the parametrising Cantor subset of $\Sigma \times \Lambda$ has gaps in the λ-direction as well.   □

Remark 4.3. The normal form (4.4) has many homoclinic orbits to hyperbolic n-tori, where stable and unstable manifolds coincide. In the present Hamiltonian context homoclinic orbits are a typical phenomenon. However, one expects the stable and unstable manifolds to split. For a generic perturbation P this leads to transverse homoclinic orbits. Similar observations apply to homoclinic orbits of parabolic n-tori, cf. [18, 19].

To apply Theorem 4.1 and Corollary 4.2 to a family $H_\mu = N_\mu + P_\mu$ of nearly integrable Hamiltonians it is sufficient that the mapping $\mu \mapsto (\lambda(\mu), \omega(\mu))$ be a submersion whence the external parameter μ accounts for λ and ω. In our last step we want to let the actions $y_1, \ldots, y_n$ conjugate to the toral angles play the rôle of both the bifurcation parameter λ and the frequency vector ω. One difficulty is that $\mathbb{Y}$ is only n-dimensional whence $y \mapsto (\lambda(y), \omega(y))$ cannot be submersive.

For the present co-dimension 1 bifurcation we lack only one parameter and this difficulty can be overcome as follows. Already in the periodic case treated in Section 3.1 the correspondence between perturbed and unperturbed periodic orbits is not a conjugacy, but only an equivalence. We now again allow for a re-parametrisation of time and refrain from controlling the frequency vector ω itself, but only aim to control the mutual ratios $[\omega_1 : \omega_2 : \ldots : \omega_n]$. In [54] the iso-energetic KAM theorem is in a similar fashion derived from the "ordinary" KAM theorem. Correspondingly, the Cantor set of lines defined by (4.5) is replaced by *Cantor dust* parametrising the parabolic n-tori. While the hyperbolic tori born in the quasi-periodic centre-saddle bifurcation are parametrised by the Cantor set of half lines obtained by the Cartesian product with $]\lambda_0, 0[$ the additional inequalities in (4.7) lead again to the elliptic tori being parametrised by Cantor dust.

Theorem 4.4. Let $\mathbb{Y} \subseteq \mathbb{R}^n$ and $\mathbb{S} \subseteq \mathbb{R}^2$ be small neighbourhoods of the origin and supply $\mathcal{P} = \mathbb{T}^n \times \mathbb{Y} \times \mathbb{S}$ with the symplectic structure $\sum \mathrm{d}x_i \wedge \mathrm{d}y_i + \mathrm{d}q \wedge \mathrm{d}p$. Consider a small analytic perturbation $H = N + P$ of an x-independent Hamiltonian function N on $\mathcal{P}$ that starts as a Taylor series in q and p as

$$N(x, y, q, p) = \eta(y) + \frac{a(y)}{2}p^2 + \frac{b(y)}{6}q^3 + c(y)q + \ldots$$

with $c(0) = 0$, but $a(0) \neq 0$ and $b(0) \neq 0$. Furthermore the mapping

$$
\begin{aligned}
\mathbb{Y} &\longrightarrow \qquad\qquad \mathbb{RP}^n \\
y &\mapsto [c(y) : \omega_1(y) : \ldots : \omega_n(y)]
\end{aligned}
\tag{4.8}
$$

is a submersion, where $\omega_i(y) = \partial\eta/\partial y_i(y)$. Then the Hamiltonian system defined by H displays a quasi-periodic centre-saddle bifurcation.

Proof. Define $\hat{N} := \xi \cdot N$ with an additional parameter ξ varying in a small interval around 1 and $\lambda := \hat{c}(y, \xi) := \xi \cdot c(y)$. Choose co-ordinates $\nu_1, \ldots, \nu_n$ on $\Sigma := \ker D\hat{c}(0, 1)$ and apply Corollary 4.2.   □

The bifurcation occurs at the hyperplane $c^{-1}(0) \subseteq \mathbb{Y}$ as $c(y)$ changes sign. This makes (4.8) a submersion if and only if $Dc(0) \neq 0$ and

$$[\omega_1 : \ldots : \omega_n] \ : \ c^{-1}(0) \ \longrightarrow \ \mathbb{RP}^{n-1}$$

is a submersion. If we use the former condition to change co-ordinates on $\mathbb{Y}$ and let y_1 play the rôle of the bifurcation parameter, then the latter condition can be rephrased as

$$< \left. \frac{\partial^{|\ell|}\omega}{\partial y^\ell} \ \right| \ |\ell| \leq 1, \ell \neq (1, 0, \ldots, 0) > \ = \ \mathbb{R}^n \ . \tag{4.9}$$

Thus, the "remaining" directions in $\mathbb{Y}$ are used to control ω in the best possible way. Since the additional parameter ξ is close to 1 the frequency vector of a surviving torus is still close to the unperturbed frequency vector, but only the frequency ratios can be perfectly matched. Where more than one direction is needed to unfold the bifurcation scenario one has to allow for $|\ell| \leq 2$ in (4.9) to obtain a Rüssmann-like condition on the curvature of the frequency mapping $y \mapsto \omega(y)$, see below.

Example 4.5. We return to the periodic Hamiltonian Hopf bifurcation of Section 3.4 and consider the subcritical type. Thus, the coefficients a and b in the normal form (3.7) have opposite signs. Then the integrable dynamics defined by the lowest order terms

$$H(x, y, q, p) = \omega y + \Omega(q_1 p_2 - q_2 p_1) + a \frac{q_1^2 + q_2^2}{2}$$
$$+ \frac{b}{2} \frac{(p_1^2 + p_2^2)^2}{4} + \frac{c}{2} y^2 + d y \frac{p_1^2 + p_2^2}{2} \qquad (4.10)$$

displays a quasi-periodic centre-saddle bifurcation, cf. Theorem 2.20 or Section 2.2.2. We claim that this bifurcation persists under perturbation by the higher order terms in (3.7).

For the normalized part Z_r in (3.7) this follows immediately from Theorem 2.25. We therefore concentrate on the remainder term in (3.7), but neglect possible contributions from Z_r to the formulas, cf. [232, 228]. Then the integrable quasi-periodic centre-saddle bifurcation takes place at

$$16 d^3 y^3 = 27 a b^2 S^2$$
$$2 d^2 y^2 = -9 a b N$$
$$2 d y = -3 b M$$
$$P = 0 ,$$

adopting the definitions

$$S = q_1 p_2 - q_2 p_1 , \qquad N = \tfrac{1}{2}(q_1^2 + q_2^2) ,$$
$$M = \tfrac{1}{2}(p_1^2 + p_2^2) , \qquad P = q_1 p_1 + q_2 p_2$$

of Section 2.2.2 and 3.4. Introducing for $M > 0$ the variable $Q = -\tfrac{1}{2} \ln M$ conjugate to P and the angle ρ conjugate to S by means of

$$q_1 = \frac{S}{\sqrt{2}} \exp Q \sin \rho + \frac{P}{\sqrt{2}} \exp Q \cos \rho , \qquad p_1 = \frac{\sqrt{2} \cos \rho}{\exp Q} ,$$
$$q_2 = -\frac{S}{\sqrt{2}} \exp Q \cos \rho + \frac{P}{\sqrt{2}} \exp Q \sin \rho , \qquad p_2 = \frac{\sqrt{2} \sin \rho}{\exp Q}$$

the symplectic structure becomes

$$dx \wedge dy + d\rho \wedge dS + dQ \wedge dP ,$$

cf. [232, 228]. In these variables the integrable Hamiltonian (4.10) reads

$$H(x, y, \rho, S, Q, P) = \omega y + \Omega S + \frac{c}{2} y^2$$
$$+ \frac{a}{4} P^2 e^{2Q} + \frac{a}{4} S^2 e^{2Q} + \frac{b}{2} e^{-4Q} + d y e^{-2Q}$$

and the parabolic 2-tori located at

$$16d^3y^3 \;=\; 27ab^2S^2 \tag{4.11a}$$

$$Q \;=\; -\frac{1}{2}\ln\frac{2dy}{-3b} \tag{4.11b}$$

$$P \;=\; 0 \tag{4.11c}$$

have parallel dynamics

$$\dot{x} \;=\; \nu_1 \;=\; \omega + (c - \frac{8d^2}{9b})y + \frac{3ab}{8d}\frac{S^2}{y^2}$$

$$\dot{\rho} \;=\; \nu_2 \;=\; \Omega - \frac{3ab}{4d}\frac{S}{y}\;.$$

Expanding $\exp(2Q)$ around $-3b/(2dy)$ yields at the parabolic tori the coefficients

$$\frac{-3b}{dy} \quad \text{of } P^2, \qquad -\frac{32d^2y^2}{3b} \quad \text{of } Q^3$$

and the coefficient

$$\Gamma(y,S) \;:=\; \frac{4d^2y^2}{9b} - \frac{3abS^2}{4dy} \quad \text{of } Q \;.$$

(which vanishes at the parabolic tori). While the partial derivatives of Γ at the bifurcation are given by

$$\frac{\partial \Gamma}{\partial y} \;=\; \frac{4d^2y}{3b} \quad \text{and} \quad \frac{\partial \Gamma}{\partial S} \;=\; -\frac{2}{\sqrt{3}}\sqrt{ady}\;,$$

the corresponding expressions for the frequency ratio $\Delta = \nu_1/\nu_2$ are unwieldy expressions. However, it turns out that the limit

$$\lim_{y\to 0}\det\begin{pmatrix}\frac{\partial \Gamma}{\partial y} & \frac{\partial \Gamma}{\partial S}\\ \frac{\partial \Delta}{\partial y} & \frac{\partial \Delta}{\partial S}\end{pmatrix} \;=\; \frac{ad\omega}{3\Omega^2}$$

is finite. This shows that (4.10) satisfies for (fixed) small nonzero (y,S) at (4.11a) the assumptions of Theorem 4.4.

To understand the behaviour as $(y,S) \to (0,0)$ we restrict (y,S) to a small wedge

$$|16d^3y^3 - 27ab^2S^2| \;<\; |dy|^3$$

and use $\mu = dy$ as a small parameter. This makes ν_1, ν_2 and $\lambda := \mu^{-2}\Gamma$ of order 1 as $\mu \to 0$ in our truncated expansion

$$H \;=\; \nu_1 y + \nu_2 S - \frac{3b}{\mu}P^2 - \frac{32}{3b}\mu^2Q^3 + \lambda\mu^2 Q$$

around (4.11) and allows us to consider these as external parameters. Then the scaling

$$(x, y, \rho, S, Q, P; \lambda, \nu) \;\mapsto\; \left(x, |\mu|^{-3/2} y, \rho, |\mu|^{-3/2} S, Q, |\mu|^{-3/2} P; \lambda, |\mu|^{-1/2} \nu \right)$$

together with dividing the Hamiltonian by μ^2 turns (3.7) into

$$\nu_1 y + \nu_2 S - 3b P^2 - \frac{32}{3b} Q^3 + \lambda Q + R$$

with a sufficiently small remainder term R to allow for application of Theorem 4.1.   □

Higher Co-Dimensions

In the n-parameter family of invariant n-tori parametrised by the actions y conjugate to the toral angles one may encounter parabolic tori up to co-dimension n. However, under non-integrable perturbations these latter may fall into a resonance gap, so for persistence results we are mainly interested in the case that the singularities $A_{d-1}^{\pm}$ satisfy $d \leq n + 1$. Again we decouple unfolding parameters and frequencies from y and consider

$$N(x, y, q, p; \lambda, \omega) \;=\; \langle \omega, y \rangle + \frac{a(\omega)}{2} p^2 + \frac{b(\omega)}{d!} q^d + \sum_{j=1}^{d-2} \frac{\lambda_j}{j!} q^j \qquad (4.12)$$

on

$$\mathbb{T}^n \times \mathbb{Y} \times \mathbb{S} \times \Lambda \times \Sigma \qquad (4.13)$$

where $\mathbb{Y} \subseteq \mathbb{R}^n$, $\mathbb{S} \subseteq \mathbb{R}^2$ and $\Lambda \subseteq \mathbb{R}^{d-2}$ are neighbourhoods of the respective origins.

We also need $\Sigma_\gamma' := \left\{ \omega \in \Sigma_\gamma \mid d(\omega, \partial\Sigma) \geq \gamma \right\}$ where the subset Σ_γ of the compact $\Sigma \subseteq \mathbb{R}^n$ consists of those frequency vectors $\omega \in \Sigma$ that satisfy the Diophantine conditions (4.5).

Theorem 4.6. Let the functions $a, b : \Sigma \longrightarrow \mathbb{R}$ in the normal form (4.12) satisfy $|a|_\Sigma, |b|_\Sigma, |1/a|_\Sigma, |1/b|_\Sigma, |Da|_\Sigma, |Db|_\Sigma < \Gamma$ for some constant $\Gamma > 0$. Then there exists a small positive constant δ with the following property. For any analytic perturbation $H = N + P$ of (4.12) with

$$|P|_{\mathbb{T}^n \times \mathbb{Y} \times \mathbb{S} \times \Lambda \times \Sigma} \;<\; \delta$$

there exists a C^∞-diffeomorphism Φ on $\mathbb{T}^n \times \mathbb{R}^n \times \mathbb{R}^2 \times \mathbb{R}^{d-2} \times \Sigma$ such that

1). *Φ is analytic for fixed ω.*
2). *Φ is symplectic for fixed (λ, ω).*
3). *Φ is C^∞-close to the identity.*
4). *On $\mathbb{T}^n \times \mathbb{R}^n \times \mathbb{R}^2 \times \mathbb{R}^{d-2} \times \Sigma_\gamma' \cap \Phi^{-1}(\mathbb{T}^n \times \mathbb{Y} \times \mathbb{S} \times \Lambda \times \Sigma)$ one can split $H \circ \Phi = N_\infty + P_\infty$ into an integrable part N_∞ and higher order terms P_∞. Here N_∞ has the same form (4.12) as N. The x-dependence is pushed into the higher order terms, i.e.*

$$\frac{\partial^{|j|+k+l+|m|} P_\infty}{\partial y^j \partial p^k \partial q^l \partial \lambda^m}(x, 0, 0, 0; 0, \omega) = 0$$

for all $(x, \omega) \in \mathbb{T}^n \times \Sigma'_\gamma$ and all j, k, l, m satisfying $2d|j| + dk + 2l + (2d - 2)m_1 + \ldots + 4m_{d-2} \leq 2d$.

For the proof see [50] or Appendix D. □

In the new co-ordinates the parabolic tori $(y, q, p; \lambda) = 0$ inherit the normal one-degree-of-freedom dynamics governed by the singularity $A^\pm_{d-1}$. The subordinate parabolic tori governed by adjacent singularities $A^\pm_{k-1}$, $k < d$ persist in the same way that we scaled from A_2 to A_1 in Corollary 4.2. For a detailed account see [50]. This yields a *Cantor stratification* of the product (4.13) of phase space and parameter space, cf. Appendix B.

Remark 4.7. A scaling $\omega \mapsto \mu^{-1}\omega$ of the frequencies leads in the limit $\mu \to 0$ to an unbounded frequency domain. The compactness of Σ in Theorems 4.1 and 4.6 ensures that the analytic perturbation P defined on (4.13) has a "uniform" holomorphic extension to a complex neighbourhood of (4.13) of product structure. Where there are other means to prevent that a radius of the complex neighbourhood approaches zero as ω varies in Σ the compactness condition is unnecessary. We already used this refinement of Theorem 4.1 in Example 4.5.

Example 4.8. A direct application of Theorem 4.6 is given in [50] concerning the quasi-periodically forced oscillator

$$\ddot{q} = -V'_\lambda(q) + \delta f(t) \ .$$

The small perturbation is given by $f(t) = F(t\omega)$ where $F : \mathbb{T}^n \longrightarrow \mathbb{R}$ is an analytic function and $\omega \in \mathbb{R}^n$ is a (fixed) Diophantine frequency vector. For $\delta = 0$ the dynamics describes the motion of a 1-dimensional particle in the family of potentials

$$V_\lambda(q) = \frac{b}{d!}q^d + \sum_{j=1}^{d-2} \frac{\lambda_j}{j!}q^j \ .$$

Since the perturbation leaves ω fixed the "Cantorisation" is restricted to the subfamilies of surrounding n-tori (and to the $(n+1)$-tori). In particular, d may exceed $n + 2$ in this slightly theoretical class of examples.

Figure 4.1 depicts the "Cantorised" surface parametrising the Diophantine invariant 2-tori of the 3-degree-of-freedom Hamiltonian

$$y_1 + \omega_2 y_2 + \frac{y_2^2}{2} + \frac{p^2}{2} + \frac{q^4}{24} + y_1 q + y_2 \frac{q^2}{2} \tag{4.14}$$

where $\omega_2 = \frac{1}{2}(\sqrt{5}-1)$ is the golden ratio. While the hyperbolic tori existing in the small "wegde" between the two quasi-periodic centre-saddle bifurcations

Fig. 4.1. "Cantorisation" of the surface in (y_1, y_2, q)-space parametrising Diophantine 2-tori of (4.14). Lighter grey indicates higher values of q.

(at the folds) are parametrised by a Cantor set of lines, the elliptic tori are parametrised by Cantor dust. The central torus $y = 0$ with normal behaviour governed by the singularity A_3^+ is isolated and visible only because of the special choice of ω_2; under perturbation it may easily fall into a resonance gap. $\square$

It is for this reason that we restrict to $d \leq n+1$ from now on. Using Diophantine approximation of dependent quantities, cf. [55] and references therein, we then can still achieve that the n actions y compensate for both the $d-2$ unfolding parameters λ and the n frequencies ω of (4.12). Let the unperturbed integrable Hamiltonian have the form

$$N(x, y, q, p) \;=\; \eta(y) \;+\; \frac{a(y)}{2}p^2 \;+\; \frac{b(y)}{d!}q^d \;+\; \sum_{j=1}^{d-2} \frac{c_j(y)}{j!}q^j \qquad (4.15)$$

with $a(0), b(0) \neq 0$ and $c(0) = 0$. In order that the bifurcation diagram of (2.3) be faithfully represented by means of y we require the mapping

$$\begin{aligned} c \;:\; \mathbb{R}^n &\longrightarrow \mathbb{R}^{d-2} \\ y &\longmapsto c(y) \end{aligned}$$

to be a submersion at the origin, i.e. $Dc(0)$ is surjective. Recall that $d - 2 < n$, so this is generically fulfilled. For the quasi-periodic centre-saddle bifurcation this left $n - 1$ directions in the action space which we could use to control the frequency ratio. For bifurcating parabolic tori of higher co-dimension there are not enough directions left for a direct control of ω. Instead, we allow for $|\ell| \leq 2$ in (4.9) and use a Rüssmann-like condition on the curvature of the frequency mapping

$$\begin{aligned} \omega \;:\; \mathbb{R}^n &\longrightarrow \mathbb{R}^n \\ y &\longmapsto D\eta(y) \end{aligned} \;.$$

The Diophantine conditions (4.5) "disallow" part of the frequency space and a condition like (4.9) reformulates that the image $\operatorname{im}\omega$ crosses these gaps transversely. This ensures that only a small portion of the action space disappears in these gaps and the Diophantine tori are of large relative measure. Since (4.5) excludes small neighbourhoods of (linear) hyperplanes, this goal can already be achieved by forcing $\operatorname{im}\omega$ to be sufficiently curved. Note that this approach requires $\tau > 2n - 1$ for the Diophantine constant in (4.5), cf. [55].

Theorem 4.9. Let $\mathbb{Y} \subseteq \mathbb{R}^n$ and $\mathbb{S} \subseteq \mathbb{R}^2$ be small neighbourhoods of the origin and supply $\mathcal{P} = \mathbb{T}^n \times \mathbb{Y} \times \mathbb{S}$ with the symplectic structure $\sum \mathrm{d}x_i \wedge \mathrm{d}y_i + \mathrm{d}q \wedge \mathrm{d}p$. Consider a small analytic perturbation $H = N + P$ of (4.15) with $a(0), b(0) \neq 0 = c(0)$ and $d \leq n + 1$. Assume that the mapping $c : \mathbb{Y} \longrightarrow \mathbb{R}^{d-2}$ is a submersion and the $\binom{n+2}{2}$ vectors

$$\frac{\partial^{|\ell|}}{\partial y^\ell}\binom{c}{\omega} \;, \quad |\ell| \leq 2$$

span $\mathbb{R}^{d-2} \times \mathbb{R}^n$, where $\omega(y) = D\eta(y)$. Then the semi-local stratification of $\mathcal{P}$ into maximal tori, hyperbolic and elliptic tori and parabolic tori of various co-dimensions of X_N induces a similar Cantor stratification of $\mathcal{P}$ into invariant tori of X_H.

Proof. Use the submersion $c : \mathbb{Y} \longrightarrow \mathbb{R}^{d-2}$ to pull back the bifurcation diagram to the space of actions. The remaining first derivatives together with the higher derivatives then ensure that most frequencies perturbed from the $\omega(y)$ are Diophantine and, hence, yield invariant lower-dimensional tori in the perturbed system. For the maximal tori the additional frequency $\omega_{n+1}(y)$ is an

analytic function that tends to zero as the maximal tori approach the separatrices of hyperbolic tori and the partial derivatives with respect to the
additional action y_{n+1} ensures that $\mathbb{R}^{n+1}$ is again spanned by first and second
derivatives of the frequency mapping, cf. [55].   □

Note that we did not address the subordinate global bifurcations. Whenever
two unstable n-tori have the same energy they may be connected by heteroclinic orbits. For the integrable X_N this leads to lower-dimensional strata
where connection bifurcations occur as the energy difference changes from a
positive to a negative value. These circumstances of the formation of heteroclinic orbits change drastically under non-integrable perturbation. In the
generic[3] case the stable and unstable manifolds that coincide for X_N have
transverse intersections, which, however, are expected to be exponentially
small. As a result the n-tori connected by heteroclinic orbits form a subset of
their respective Cantor families of non-zero relative measure.

4.1.2 Vanishing Normal Linear Behaviour

Let now the one-degree-of-freedom Hamiltonian $\mathcal{N}_0$ in (4.3) have a vanishing
2-jet. Generically $\mathcal{N}_0$ is a planar singularity of co-dimension $d \leq n$ and $\mathcal{N}_y$
provides a versal unfolding. Thus, the integrable $(n+1)$-degree-of-freedom dynamics defined by (4.3) displays a torus bifurcation mimicking the equilibrium
bifurcation governed by $\mathcal{N}_y$. Again we are mainly interested in co-dimensions
$d \leq n - 1$ since isolated tori may fall under non-integrable perturbation into
a resonance gap, but leave n and d for the parameter-dependent result unre-
lated.

   When proving Theorem 4.6, the weight $(\alpha_q, \alpha_p) = (2, d)$ of the quasi-
homogeneous polynomial $(a/2)p^2 + (b/d!)q^d$ of order $2d$ is extended by means
of $\alpha_y = 2d$ and $\alpha_{\lambda_j} = 2d - 2j$ to a weight that makes (4.12) a quasi-
homogeneous polynomial of order $2d$ in (y, q, p, λ). This allowed for a generalization to all torus bifurcations governed by simple singularities in [51, 141].
Obviously, such an extension to

$$N(y, q, p; \lambda) = \langle \omega, y \rangle + \mathcal{N}_\lambda(q, p) \tag{4.16}$$

can be defined whenever $\mathcal{N}_0$ is a quasi-homogeneous[4] polynomial, suggesting
a further generalization to the unimodal singularities $X_9^\pm$ and J_{10}.

   For the final result a modulus is not problematic. Indeed, we expect a Cantor family of, say, J_{10}-tori where the parametrising Cantor set has dimension
$n - d \geq 1$. This Cantor set results from taking gaps out of the continuum of
values of the modulus. However, from Theorem 2.14 we know that such moduli are an artefact of the C^∞-(left)-right classification and carry no dynamical

---

[3] The genericity conditions are conditions on the perturbation P and not easily
   checked in applications.

[4] Since the co-dimension d is finite $\mathcal{N}_0$ is non-degenerate and thus also semi-quasi-
   homogeneous, cf. Definition 2.11.

information – provided that exceptional values are avoided. The latter secures e.g. the non-degeneracy conditions $200m^3 + ab^2 \neq 0$ for the modulus m of J_{10} and $9\mu^2 \neq ab$ for the modulus μ of $X_9^{\pm}$, cf. Section 2.1.1.

In the KAM iteration scheme used in [50] the terms of weighted order $\leq 2d$ are made x-independent and then transformed into the form (4.12), while terms of strictly greater weighted order are sufficiently small to join the perturbation on the next iteration step. In this way, one loses control on these terms and a necessary inequality like $m \neq 0$ involving moduli in such terms cannot be preserved during the iteration procedure.

We therefore work with a more general versal unfolding $\mathcal{N}_\lambda$ of $\mathcal{N}_0$ in (4.16). For semi-quasi-homogeneous $\mathcal{N}_0$ in standard (polynomial) form we can use the weighted order $\alpha_p k + \alpha_q l$ on monomials $p^k q^l$ to determine the maximal order α_y of monomials in $\mathcal{N}_0$ and extend the gradation to $\mathbb{R}[y, q, p]$, making sure that the linear term $\langle \omega, y \rangle$ in (4.16) is of maximal order. Let furthermore α_x denote the minimal order in (4.16) – this is merely a notation, we do not extend the gradation to x as well. The versal unfolding $\mathcal{N}_\lambda$ of $\mathcal{N}_0$ is then chosen in a trivial way, simply containing all monomials $p^k q^l$ of the universal unfolding together with those of order α satisfying $\alpha_x \leq \alpha \leq \alpha_y$ and we still label the parameters $\lambda_1, \ldots, \lambda_d$ (so that d no longer denotes the co-dimension). The gradation is then further extended to $\mathbb{R}[y, q, p; \lambda]$ by distributing weights α_{λ_j} in such a way that the order $\alpha_p k + \alpha_q l + \alpha_{\lambda_j}$ of the unfolding terms $\lambda_j p^k q^l$ is at least α_x and at most $\alpha_y + 1$.

In case $\mathcal{N}_0$ is not semi-quasi-homogeneous its stratum of constant multiplicity is defined by a *filtration* that is not induced by a gradation, cf. [11, 15]. We nevertheless put $\mathcal{N}_0$ into the Procrustes bed of semi-quasi-homogeneous singularities by choosing suitable integer weights α_q and α_p and proceed as above. The important point turns out to be that the resulting order of unfolding terms $\lambda_j p^k q^l$ with $k + l \leq 2$ is raised to $\alpha_x > \alpha_q + \alpha_p$ while it does not matter that deformations in certain directions $p^k q^l$ with order $\alpha_p k + \alpha_q l \geq \alpha_x$ now may lower the multiplicity.

Letting the coefficients and moduli of $\mathcal{N}_0$ depend on the freqency ω we consider N as a function on

$$\mathbb{T}^n \times \mathbb{Y} \times \mathbb{S} \times \Lambda \times \Sigma$$

where $\mathbb{Y} \subseteq \mathbb{R}^n$, $\mathbb{S} \subseteq \mathbb{R}^2$, $\Lambda \subseteq \mathbb{R}^d$ are neighbourhoods of the respective origins. To prevent that the coefficient functions $\Sigma \longrightarrow \mathbb{R}$ attain the exceptional values at which $\mathcal{N}_0(\omega_0)$ ceases to be C^0-equivalent to $\mathcal{N}_0(\omega)$ we choose a compact set $\Sigma \subseteq \mathbb{R}^n$ of frequencies that does not include such frequencies ω_0. In this way the coefficient functions in (4.16) are bounded away from such exceptional values. Again $\Sigma'_\gamma := \{ \omega \in \Sigma_\gamma \mid \mathrm{d}(\omega, \partial \Sigma) \geq \gamma \}$ where Σ_γ is defined by (4.5) with fixed $\tau > 2n - 1$.

Theorem 4.10. *Under the above conditions there exists a small constant δ with the following property. For any analytic perturbation $H = N + P$ of (4.16) with*

$$|P|_{\mathbb{T}^n \times \mathbb{Y} \times \mathbb{S} \times \Lambda \times \Sigma} \; < \; \delta$$

there exists a C^∞-diffeomorphism Φ on $\mathbb{T}^n \times \mathbb{R}^n \times \mathbb{R}^2 \times \mathbb{R}^d \times \Sigma$ such that

1). Φ is analytic for fixed ω.
2). Φ is symplectic for fixed (λ, ω).
3). Φ is C^∞-close to the identity.
4). On $\mathbb{T}^n \times \mathbb{R}^n \times \mathbb{R}^2 \times \mathbb{R}^d \times \Sigma'_\gamma \cap \Phi^{-1}(\mathbb{T}^n \times \mathbb{Y} \times \mathbb{S} \times \Lambda \times \Sigma)$ one can split $H \circ \Phi = N_\infty + P_\infty$ into an integrable part N_∞ and higher order terms P_∞. Here N_∞ has the same form (4.16) as N. The x-dependence is pushed into the higher order terms, i.e.

$$\frac{\partial^{|j|+k+l+|m|} P_\infty}{\partial y^j \partial p^k \partial q^l \partial \lambda^m}(x, 0, 0, 0; 0, \omega) \; = \; 0$$

for all $(x, \omega) \in \mathbb{T}^n \times \Sigma'_\gamma$ and all j, k, l, m satisfying $\alpha_y |j| + \alpha_p k + \alpha_q l + \alpha_{\lambda_1} m_1 + \ldots + \alpha_{\lambda_d} m_d \leq \alpha_y$.

For the proof see Appendix D. □

How can we reduce the number of parameters from d to the co-dimension ∂ of $\mathcal{N}_0$? The monomials $p^k q^l$ deforming $\mathcal{N}_0$ along its stratum of constant multiplicity are of three types. Those that do not figure in the universal unfolding can be transformed away by means of a C^∞-equivalence on $\mathbb{S}$. While this does in general change the symplectic structure, the stratification into families of invariant tori is preserved. In this way we are left with the moduli, and we single out those along which the universal unfolding is topologically trivial as these can be transformed away by means of a C^0-equivalence on $\mathbb{S}$. In case one of the remaining monomials $p^k q^l$ has weighted order α_y we replace[5] the maximal order α_y in (4.16) by $\alpha_y + 1$. The remaining moduli together with the unfolding parameters that deform off from the stratum of constant multiplicity then add up to the co-dimension.

Again the requirement that Σ be compact can be dropped where there are other means to ensure that the coefficient functions $\Sigma \longrightarrow \mathbb{R}$ in (4.16) are bounded, bounded away from exceptional values, have bounded derivatives, and the analytic perturbation P has a holomorphic extension to a complex neighbourhood of $\mathbb{T}^n \times \mathbb{Y} \times \mathbb{S} \times \Lambda \times \Sigma$ of product structure. This allows to invoke Theorem 4.10 in an inductive manner to recover the subordinate tori governed by adjacent singularities, yielding a Cantor stratification into invariant tori of the perturbed system. From this result we finally derive the parameter-independent version.

Theorem 4.11. Let $\mathbb{Y} \subseteq \mathbb{R}^n$, $\mathbb{S} \subseteq \mathbb{R}^2$ *be small neighbourhoods of the origin and supply* $\mathcal{P} = \mathbb{T}^n \times \mathbb{Y} \times \mathbb{S}$ *with the symplectic structure* $\sum \mathrm{d}x_i \wedge \mathrm{d}y_i + \mathrm{d}q \wedge \mathrm{d}p$. *Consider an* x-*independent Hamiltonian function* N *on* $\mathcal{P}$ *such that* $\mathcal{N}_0$ *is a*

---

[5] With an efficient choice of α_q, α_p this even does not result in additional terms.

non-degenerate planar singularity of co-dimension $\partial \leq n - 1$ for the one-degree-of-freedom part $\mathcal{N}_y(q, p)$ in (4.3) and assume that N is generic within the set of all such Hamiltonians. Then the semi-local stratification of $\mathcal{P}$ defined by X_N induces a similar Cantor stratification of $\mathcal{P}$ defined by a small analytic perturbation $H = N + P$.

Proof. Genericity yields a submersive re-parametrisation

$$c : \mathbb{R}^n \longrightarrow \mathbb{R}^\partial$$

such that the one-degree-of-freedom Hamiltonian systems defined by $\mathcal{N}_y$ and by $\mathcal{M}_{c(y)}$ (where the ∂-parameter family $\mathcal{M}_\lambda$ unfolds $\mathcal{M}_0 := \mathcal{N}_0$ universally) are (topologically) equivalent, after both systems are restricted to suitable neighbourhoods of the origin. This allows to pull back the bifurcation diagram to the space of actions. Letting $\omega(y) := D\eta(y)$ for the (q, p)-independent part η in (4.3) it is furthermore generic that the $\binom{n+2}{2}$ vectors

$$\frac{\partial^{|\ell|}}{\partial y^\ell} \begin{pmatrix} c \\ \omega \end{pmatrix} , \quad |\ell| \leq 2$$

span $\mathbb{R}^\partial \times \mathbb{R}^n$. This ensures that most frequencies perturbed from the $\omega(y)$ are Diophantine. $\qquad\square$

4.1.3 Reversibility

The Hamiltonians (4.4), (4.12) and (4.15) are reversible with respect to the reflection

$$(x, y, q, p) \mapsto (-x, y, q, -p)$$

and correspondingly one also has persistence of parabolic tori and their bifurcations in reversible Hamiltonian systems. More interesting in this context is reversibility with respect to the reflection

$$(x, y, q, p) \mapsto (-x, y, -q, p) \tag{4.17}$$

since it reduces the co-dimension of the parabolic torus. In this way the parameter dependent Hamiltonian

$$N(x, y, q, p; \lambda, \omega) = \langle \omega, y \rangle + \frac{a(\omega)}{2}p^2 + \frac{b(\omega)}{24}q^4 + \frac{\lambda}{2}q^2 \tag{4.18}$$

defined on

$$\mathbb{T}^n \times \mathbb{Y} \times \mathbb{S} \times \Lambda \times \Sigma$$

undergoes a quasi-periodic Hamiltonian pitchfork bifurcation as $\lambda \in \Lambda \subseteq \mathbb{R}$ passes through 0 and we claim persistence under non-integrable perturbations.

Theorem 4.12. Let the functions $a, b : \Sigma \longrightarrow \mathbb{R}$ in (4.18) satisfy the inequalities

$$|a|_\Sigma, \ |b|_\Sigma, \ \left|\frac{1}{a}\right|_\Sigma, \ \left|\frac{1}{b}\right|_\Sigma, \ |Da|_\Sigma, |Db|_\Sigma \ < \ \Gamma$$

for some constant $\Gamma > 0$. Then there is $\delta > 0$ with the following property. For any reversible analytic perturbation $H = N + P$ of (4.18) with

$$|P|_{\mathbb{T}^n \times \mathbb{Y} \times \mathbb{S} \times \Lambda \times \Sigma} \ < \ \delta$$

there exists a C^∞-diffeomorphism

$$\Phi : \ \mathbb{T}^n \times \mathbb{R}^n \times \mathbb{R}^2 \times \mathbb{R} \times \Sigma \ \longrightarrow \ \mathbb{T}^n \times \mathbb{R}^n \times \mathbb{R}^2 \times \mathbb{R} \times \Sigma$$

respecting (4.17) such that

1). Φ is analytic for fixed ω.
2). Φ is symplectic for fixed (λ, ω).
3). Φ is C^∞-close to the identity.
4). On $\mathbb{T}^n \times \mathbb{R}^n \times \mathbb{R}^2 \times \mathbb{R} \times \Sigma'_\gamma \cap \Phi^{-1}(\mathbb{T}^n \times \mathbb{Y} \times \mathbb{S} \times \Lambda \times \Sigma)$ one can split $H \circ \Phi = N_\infty + P_\infty$ into an integrable part N_∞ and higher order terms P_∞. Here N_∞ has the same form (4.18) as N. The x-dependence is pushed into the higher order terms, i.e.

$$\frac{\partial^{|j|+k+l+|m|} P_\infty}{\partial y^j \partial p^k \partial q^l \partial \lambda^m}(x, 0, 0, 0; 0, \omega) \ = \ 0$$

for all $(x, \omega) \in \mathbb{T}^n \times \Sigma'_\gamma$ and all j, k, l, m satisfying $4|j| + 2k + l + 2m \leq 4$.

For the proof see [50] or Appendix D. □

As always persistence of the elliptic tori (or hyperbolic tori in the dual case) born in the bifurcation is obtained by means of a scaling argument.

Theorem 4.13. Let $\mathbb{Y} \subseteq \mathbb{R}^n$ and $\mathbb{S} \subseteq \mathbb{R}^2$ be small neighbourhoods of the origin and supply $\mathcal{P} = \mathbb{T}^n \times \mathbb{Y} \times \mathbb{S}$ with the symplectic structure $\sum \mathrm{d}x_i \wedge \mathrm{d}y_i + \mathrm{d}q \wedge \mathrm{d}p$. Consider a small reversible analytic perturbation $H = N + P$ of the x-independent Hamiltonian function

$$N(x, y, q, p) \ = \ \eta(y) + \frac{a(y)}{2}p^2 + \frac{b(y)}{24}q^4 + \frac{c(y)}{2}q^2$$

with $c(0) = 0$, but $a(0) \neq 0$ and $b(0) \neq 0$. Furthermore the mapping

$$\begin{array}{ccc} \mathbb{Y} & \longrightarrow & \mathbb{RP}^n \\ y & \mapsto & [c(y) : \omega_1(y) : \ldots : \omega_n(y)] \end{array}$$

is a submersion, where $\omega_i(y) = \partial \eta/\partial y_i(y)$. Then the Hamiltonian system defined by H displays a quasi-periodic Hamiltonian pitchfork bifurcation.

The proof follows almost verbatim that of Theorem 4.4.   □

The reflection (4.17) can be dragged through the whole proof in Appendix D thus generalizing Theorems 4.6 and 4.10 to the reversible context. Then the arguments proving Theorem 4.11 also yield the following result.

Theorem 4.14. Let $\mathbb{Y} \subseteq \mathbb{R}^n$, $\mathbb{S} \subseteq \mathbb{R}^2$ *be small neighbourhoods of the origin and supply* $\mathcal{P} = \mathbb{T}^n \times \mathbb{Y} \times \mathbb{S}$ *with the symplectic structure* $\sum \mathrm{d}x_i \wedge \mathrm{d}y_i + \mathrm{d}q \wedge \mathrm{d}p$. *Consider an x-independent Hamiltonian function N on $\mathcal{P}$ that is reversible under (4.17) such that $\mathcal{N}_0$ is a non-degenerate planar singularity of co-dimension $\leq n - 1$ for the one-degree-of-freedom part $\mathcal{N}_y(q, p)$ in (4.3) and assume that N is generic within the set of all such Hamiltonians. Then the semi-local stratification of $\mathcal{P}$ defined by X_N induces a similar Cantor stratification of $\mathcal{P}$ defined by a small reversible analytic perturbation $H = N + P$.*   □

Example 4.15. The quasi-periodic Hamiltonian pitchfork bifurcation is easier to handle than the quasi-periodic centre-saddle bifurcation since the reversing symmetry forces the initial torus to remain invariant. Consequently, it has been in this context that the fate of resonant parabolic tori has first been studied. In [178, 181] the phenomenological model Hamiltonian

$$\frac{1 + 2\alpha_1}{4} y_1^2 + \omega_2 y_2 + \frac{1}{2} y_2^2 + \alpha_3 y_1 y_2 + \frac{1}{2} p^2 + \frac{1}{4} q^4 - \frac{y_1}{2} q^2 \qquad (4.19)$$

is derived, with perturbation

$$\varepsilon \left[\left(1 - \frac{q^2}{2} \right) \cos(k_1 x_1) + \cos(k_2 x_2) \right] , \quad k \in \mathbb{N}^2$$

respecting the reversing symmetry

$$(x, y, q, p) \mapsto (-x, y, -q, p)$$

of (4.19). For the unperturbed system the two families of elliptic tori $(y, q, p) = (y, \pm\sqrt{y_1}, 0)$ born in the bifurcation have energy

$$\frac{\alpha_1}{2} y_1^2 + \omega_2 y_2 + \frac{1}{2} y_2^2 + \alpha_3 y_1 y_2$$

which explains the form of the first coefficient in (4.19).

Figure 4.2 depicts the "Cantorised" surface parametrising the Diophantine invariant 2-tori. The numerical results in [178, 181] clearly show how the perturbation (with $\varepsilon = 10^{-3}$) leads to a large fluctuation (values ranging between -0.05 and 0.05) of y_1 whence the projection of the dynamics to the (q, p)-plane oscillates between motion near the origin and motion near both points $\pm \left(\sqrt{y_1}, 0 \right)$ with the largest deviations as y_1 passes through zero. The variation of y_2 is of order ε and the projection to the (x_2, y_2)-plane closely

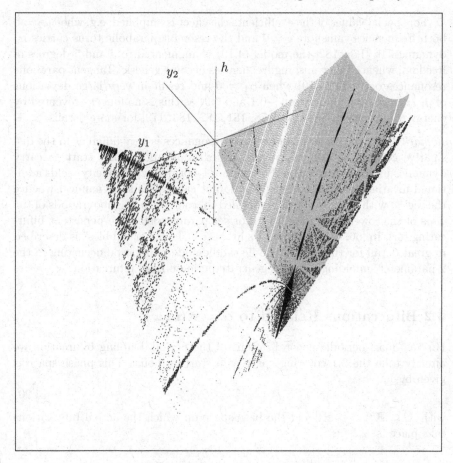

Fig. 4.2. "Cantorisation" of the singular values of the energy-momentum mapping parametrising Diophantine 2-tori of (4.19) for $y \in [-0.9, 0.9] \times [-1.2, 0]$ and $(\alpha_1, \omega_2, \alpha_3) = (0.3, 1, 1)$. The grey Cantor set of half lines parametrises the hyperbolic tori while its end points form the 1-dimensional Cantor dust parametrising the parabolic tori.

follows an oscillatory motion, while the projection to the (x_1, y_1)-plane is captured in a librational motion.

Initial conditions $(y_1, y_2, q, p) = (0, -\omega_2, 0, 0)$ form an exception to the above pattern, as it is y_1 that remains almost constant with the dynamics projected to the (x_1, y_1)-plane closely following an oscillatory motion, while the projection to the (x_2, y_2)-plane is captured in a librational motion with large extension in the y_2-direction (values ranging for $\omega_2 = 0.8$ between -0.85 and -0.75). The projection of the motion to the (q, p)-plane is narrowly confined and creates an intricating structure.

For special values of the coefficients the effect is amplified, e.g. when $\omega_2 = 0$ both frequencies vanish at $y = 0$ and the resonant parabolic torus carries no dynamics. In [178, 182] the model (4.19) is augmented to 4 and 5 degrees of freedom, whence two vanishing frequencies become generic. Tangent parabolic resonances occur in (4.19) when $\alpha_1 = 0$ and result in very large deviations of y_1 (values ranging between -0.1 and 0.9) as this is no longer prevented by energy conservation. See [180, 178, 181, 182, 183, 179] for more details.   □

Remark 4.16. The situation concerning resonances is very different in the dissipative context as the parameters are external and do not start to carry dynamics under perturbation. Furthermore, normal hyperbolicity yields additional invariant tori (although these do not carry parallel dynamics) whence the region with destroyed tori is confined to small open neighbourhoods of the gaps of the lower dimensional Cantor set parametrising the persistent bifurcating tori. In [66, 67] the dynamics in these "Chenciner bubbles" is described in great detail for the quasi-periodic saddle-node bifurcation occurring in the 2-parameter unfolding of the degenerate periodic Hopf bifurcation.

4.2 Bifurcations Related to $\alpha = \pi \langle k, \omega \rangle$

For the quasi-periodic generalizations of the period-doubling bifurcation we already take the 2:1 covering space as a starting point. This phase space is given by

$$\mathcal{P} = \mathbb{T}^n \times \mathbb{Y} \times \mathbb{S} \tag{4.20}$$

with $\mathbb{Y} \subseteq \mathbb{R}^n$, $\mathbb{S} \subseteq \mathbb{R}^2$ and the base space on which the actual bifurcations take place is

$$\mathcal{B} = \mathcal{P}/\mathbb{Z}_2$$

where the deck group $\mathbb{Z}_2 = \{\mathrm{id}, \sigma\}$ is generated by the involution

$$\sigma : \quad \begin{array}{ccc} \mathcal{P} & \longrightarrow & \mathcal{P} \\ (x_1, x_2, \ldots, x_n, y, q, p) & \mapsto & (x_1 + \tfrac{1}{2}, x_2, \ldots, x_n, y, -q, -p) \end{array} \quad . \tag{4.21}$$

In this setting $\mathsf{T} = \mathbb{T}^n \times \{0\} \times \{0\}$ is an invariant torus of the unperturbed dynamics defined by

$$N(x, y, q, p) = \eta(y) + \mathcal{N}_y(q, p) \tag{4.22}$$

with $D\mathcal{N}_0(0, 0) = 0$. The $\mathbb{T}^n$-symmetry $x \mapsto x + \xi$ of N on $\mathcal{P}$ implies that the lift from $\mathcal{B}$ to $\mathcal{P}$ has already reduced the system to Floquet form and we followed [49] in choosing the co-ordinates (x, y, q, p) in such a way that the lifted system has vanishing Floquet exponents.

On the base space $\mathcal{B}$ the $\mathbb{T}^n$-symmetry $x \mapsto x + \xi$ is not free and reduction yields a 2-dimensional Poisson space $\mathcal{C}$ with a conical singularity. Correspondingly, the one-degree-of-freedom system on the regularly reduced phase space is equivariant with respect to the π-rotation

$$(q, p) \;\longmapsto\; (-q, -p) \tag{4.23}$$

inherited from the involution σ.

4.2.1 Parabolic Tori

In the parabolic case the lifted one-degree-of-freedom Hamiltonian at $y = 0$ reads

$$\mathcal{N}_0(q, p) \;=\; \frac{a}{2}p^2 + \mathcal{O}(q^4, p^4)$$

with $a \neq 0$. Generically this is a singularity $A^{\pm}_{2d-1}$ with $2 \leq d \leq \frac{1}{2}n + 1$ and correspondingly the family $\mathcal{N}_y$ undergoes on the one-degree-of-freedom base space $\mathcal{C}$ a Hamiltonian flip bifurcation, degenerate if $d \geq 3$. In this way the parameter dependent Hamiltonian

$$N(x, y, q, p; \lambda, \omega) \;=\; \langle \omega, y \rangle + \frac{a(\omega)}{2}p^2 + \frac{b(\omega)}{(2d)!}q^{2d} + \sum_{j=1}^{d-1} \frac{\lambda_j}{(2j)!}q^{2j} \tag{4.24}$$

defined on $\mathcal{P} \times \Lambda \times \Sigma$ induces on $\mathcal{B} \times \Lambda \times \Sigma$ a (degenerate) *frequency-halving bifurcation* and we claim persistence under non-integrable perturbations. To this end $\mathbb{Y} \subseteq \mathbb{R}^n$, $\Lambda \subseteq \mathbb{R}^{d-1}$ are small neighbourhoods of the origin and $\mathbb{S} \subseteq \mathbb{R}^2$ is a small neighbourhood of the origin that is invariant under the π-rotation (4.23), while $\Sigma \subseteq \mathbb{R}^n$ is compact, $\Sigma_\gamma \subseteq \Sigma$ consists of frequencies satisfying the Diophantine conditions (4.5) and $\Sigma'_\gamma = \{\omega \in \Sigma_\gamma \mid d(\omega, \partial\Sigma) \geq \gamma\}$.

Theorem 4.17. Let the functions $a, b : \Sigma \longrightarrow \mathbb{R}$ in the normal form (4.24) satisfy $|a|_\Sigma, |b|_\Sigma, |1/a|_\Sigma, |1/b|_\Sigma, |Da|_\Sigma, |Db|_\Sigma < \Gamma$ for some constant $\Gamma > 0$. Then there exists a small positive constant δ with the following property. For any σ-invariant analytic perturbation $H = N + P$ of (4.24) with

$$|P|_{\mathbb{T}^n \times \mathbb{Y} \times \mathbb{S} \times \Lambda \times \Sigma} \;<\; \delta$$

there exists a σ-equivariant C^∞-diffeomorphism Φ on $\mathbb{T}^n \times \mathbb{R}^n \times \mathbb{R}^2 \times \mathbb{R}^{d-1} \times \Sigma$ such that

1). Φ is analytic for fixed ω.
2). Φ is symplectic for fixed (λ, ω).
3). Φ is C^∞-close to the identity.
4). On $\mathbb{T}^n \times \mathbb{R}^n \times \mathbb{R}^2 \times \mathbb{R}^{d-1} \times \Sigma'_\gamma \cap \Phi^{-1}(\mathbb{T}^n \times \mathbb{Y} \times \mathbb{S} \times \Lambda \times \Sigma)$ one can split $H \circ \Phi = N_\infty + P_\infty$ into an integrable part N_∞ and higher order terms P_∞. Here N_∞ has the same form (4.24) as N. The x-dependence is pushed into the higher order terms, i.e.

$$\frac{\partial^{|j|+k+l+|m|} P_\infty}{\partial y^j \partial p^k \partial q^l \partial \lambda^m}(x, 0, 0, 0; 0, \omega) \;=\; 0$$

for all $(x, \omega) \in \mathbb{T}^n \times \Sigma'_\gamma$ and all j, k, l, m satisfying $2d|j| + dk + l + (2d - 2)m_1 + \ldots + 2m_{d-1} \leq 2d$.

For the proof see [50] or Appendix D.   □

In the new co-ordinates provided by Theorem 4.17, the parabolic tori $(y, q, p; \lambda) = 0$ inherit the normal one-degree-of-freedom dynamics governed by the singularity $A^{\pm}_{2d-1}$. Persistence of the subordinate tori governed by adjacent singularities $A^{\pm}_{2k-1}, k < d$ follows from a scaling argument. Let us formulate the parameter-independent version for $d = 2$ as a separate result.

Theorem 4.18. Let $\mathbb{Y} \subseteq \mathbb{R}^n$ be a neighbourhood of the origin and $\mathbb{S} \subseteq \mathbb{R}^2$ be a neighbourhood of the origin that is invariant under the π-rotation (4.23). Supply $\mathcal{P} = \mathbb{T}^n \times \mathbb{Y} \times \mathbb{S}$ with the symplectic structure $\sum \mathrm{d}x_i \wedge \mathrm{d}y_i + \mathrm{d}q \wedge \mathrm{d}p$ and consider a (real analytic) integrable Hamiltonian function H on the base space $\mathcal{B} = \mathcal{P}/_{\mathbb{Z}_2}$ defined by the deck transformation (4.21). Let the lift of H to $\mathcal{P}$ be a small perturbation of

$$N(x, y, q, p) = \eta(y) + \frac{a(y)}{2}p^2 + \frac{b(y)}{24}q^4 + \frac{c(y)}{2}q^2$$

with $c(0) = 0$, but $a(0) \neq 0$ and $b(0) \neq 0$. Furthermore assume that the mapping

$$\begin{aligned} \mathbb{Y} &\longrightarrow & \mathbb{RP}^n \\ y &\mapsto & [c(y) : \omega_1(y) : \ldots : \omega_n(y)] \end{aligned}$$

is a submersion, where $\omega_i(y) = \partial \eta / \partial y_i(y)$. Then the dynamics defined by H undergoes a frequency-halving bifurcation.   □

Remark 4.19. The normal-internal resonance $\pi \langle k, \omega \rangle = \alpha$ on the base space $\mathcal{B}$ violates (4.1) with $\ell = 2$. Correspondingly, a frequency-halving bifurcation does not lead to the destruction of the parabolic torus and only the normal linear behaviour changes from elliptic to hyperbolic. This scenario can already be captured in the $\mathbb{Z}_2$-symmetric linear centraliser unfolding

$$N(x, y, q, p; \lambda, \omega) = \langle \omega, y \rangle + \frac{a(\omega)}{2}p^2 + \frac{\lambda}{2}q^2 \qquad (4.25)$$

on the 2:1 cover $\mathcal{P}$. The way in which additional n-tori bifurcate off from $y = 0 = q = p$ as λ passes through zero is not captured in (4.25) since this is governed by the higher order terms of the frequency-halving bifurcation (and also the higher order unfolding terms in the degenerate case).

The other side of this medal is that a persistence theorem like Theorem 4.17 on the basis of (4.25) instead of (4.24) would not need any genericity conditions on the higher order terms. The results in [30, 31, 294] in turn yield no control on the perturbed normal linear behaviour but also require no versality of the unperturbed normal linear behaviour. To prove persistence of (4.25) under non-integrable perturbations the $\mathbb{Z}_2$-symmetry σ becomes essential – already the quasi-periodic centre-saddle bifurcation shows how symmetry breaking of (4.25) leads to the destruction of tori.

A similar result should be attainable for perturbations of (4.25) that do not necessarily respect the involution σ, but are reversible with respect to (4.17), see also [297]. The reason that we neither explicitly formulate nor prove these two claims is that the proof of Theorem 4.6 allows for a direct generalization to these two contexts by dragging σ or (4.17) through every step of that proof. This is different where resonances (4.2c) are encountered whence Theorem 4.24 is the first result addressed in Section 4.3.1.

Theorem 4.20. Consider an analytic Hamiltonian function H on the base space $\mathcal{B}$ such that the lift of H to the 2:1 cover $\mathcal{P}$ of $\mathcal{B}$ is a small perturbation of

$$N(x, y, q, p) = \eta(y) + \frac{a(y)}{2}p^2 + \frac{b(y)}{(2d)!}q^{2d} + \sum_{j=1}^{d-1} \frac{c_j(y)}{(2j)!}q^{2j}$$

with $d \geq 3$ and coefficient functions satisfying $a(0), b(0) \neq 0$ and $c(0) = 0$. Furthermore the mapping

$$c : \mathbb{R}^n \longrightarrow \mathbb{R}^{d-1}$$
$$y \longmapsto c(y)$$

is a submersion and the vectors

$$\frac{\partial^{|\ell|}}{\partial y^\ell}\begin{pmatrix} c \\ \omega \end{pmatrix}, \quad |\ell| \leq 2$$

span $\mathbb{R}^{d-1} \times \mathbb{R}^n$, where $\omega(y) = D\eta(y)$. Then the dynamics defined by H on $\mathcal{B}$ undergoes a degenerate frequency-halving bifurcation. □

Thus in case $d = 2$ we are again able to control the frequency ratios of the perturbed tori, while for $d \geq 3$ only closeness of the perturbed to the unperturbed frequency vectors is guaranteed and $\tau > 2n - 1$ in (4.5) is needed.

4.2.2 Vanishing Normal Linear Behaviour

On the 2:1 cover $\mathcal{P} \times \Lambda \times \Sigma$ we consider a parameter dependent integrable Hamiltonian

$$N(x, y, q, p; \lambda, \omega) = \langle \omega, y \rangle + \mathcal{N}_\lambda(q, p) \tag{4.26}$$

where $\mathcal{N}_0$ has a vanishing 2-jet and co-dimension $d \leq n-1$. We require that $\mathcal{N}_\lambda$ is invariant under (4.23) and provides a $\mathbb{Z}_2$-versal unfolding of $\mathcal{N}_0$ similar to the one constructed for (4.16) with weights $(\alpha_y, \alpha_q, \alpha_p, \alpha_{\lambda_1}, \ldots, \alpha_{\lambda_d})$.

Theorem 4.21. Let the coefficient functions $\Sigma \longrightarrow \mathbb{R}$ of $\mathcal{N}_0$ in (4.26) be bounded, bounded away from exceptional values and have bounded derivatives. Then there exists a small positive constant $\delta > 0$ with the following property. For any analytic perturbation $H = N + P$ of (4.26) with

$$|P|_{\mathbb{T}^n \times \mathbb{Y} \times \mathbb{S} \times \Lambda \times \Sigma} \; < \; \delta$$

that is invariant under the deck transformation σ there exists a σ-equivariant C^∞-diffeomorphism Φ on $\mathbb{T}^n \times \mathbb{R}^n \times \mathbb{R}^2 \times \mathbb{R}^d \times \Sigma$ such that

1). Φ is analytic for fixed ω.
2). Φ is symplectic for fixed (λ, ω).
3). Φ is C^∞-close to the identity.
4). On $\mathbb{T} \times \mathbb{R}^n \times \mathbb{R}^2 \times \mathbb{R}^d \times \Sigma'_\gamma \cap \Phi^{-1}(\mathbb{T} \times \mathbb{Y} \times \mathbb{S} \times \Lambda \times \Sigma)$ one can split $H \circ \Phi = N_\infty + P_\infty$ into an integrable part N_∞ and higher order terms P_∞. Here N_∞ has the same form (4.26) as N. The x-dependence is pushed into the higher order terms, i.e.

$$\frac{\partial^{|j|+k+l+|m|} P_\infty}{\partial y^j \partial p^k \partial q^l \partial \lambda^m}(x, 0, 0, 0; 0, \omega) \;=\; 0$$

for all $(x, \omega) \in \mathbb{T} \times \Sigma'_\gamma$ and all j, k, l, m satisfying $\alpha_y |j| + \alpha_p k + \alpha_q l + \alpha_{\lambda_1} m_1 + \ldots + \alpha_{\lambda_d} m_d \leq \alpha_y$.

For the proof see Appendix D.   □

In the same way that Theorem 4.10 leads to Theorem 4.11 we obtain from Theorem 4.21 the parameter-independent version.

Corollary 4.22. Consider a $\mathbb{Z}_2$-invariant x-independent Hamiltonian function N on the 2:1 cover $\mathcal{P}$ of $\mathcal{B}$ such that N_0 is a non-degenerate singularity of co-dimension $\leq n-1$ for the one-degree-of-freedom part $N_y(q, p)$ in (4.22). Let N be the lift of a Hamiltonian M on $\mathcal{B}$ and assume that M is generic within the set of all such Hamiltonians. Then the semi-local stratification of $\mathcal{B}$ defined by X_M induces a similar Cantor stratification of $\mathcal{B}$ defined by a small analytic perturbation $H = M + P$.

Proof. The genericity conditions on the lift N of M formulated in the proof of Theorem 4.11 define genericity conditions on M.   □

All $\mathbb{Z}_2$-symmetric planar singularities with vanishing 2-jet have moduli. The simplest case is X_9 which has co-dimension three in the $\mathbb{Z}_2$-symmetric context and may thus persistently appear in Hamiltonian systems with (at least) five degrees of freedom. An additional reversing symmetry lowers this number to four degrees of freedom, cf. [144].

Remark 4.23. As explained in Remarks D.7 and D.8 the KAM-iteration respects not only the $\mathbb{Z}_2$-symmetries of a reversible system or coming from the lift to a 2:1 covering space, but in fact any discrete symmetry. In particular, both Theorems 4.10 and 4.21 extend to symmetries beyond reversibility and yield an obvious generalization of Theorem 4.14. In fact, for the symmetric singularities $1_{2,\ell}$ defined in (2.15) one even obtains the sharper formulation of Theorem 4.13 with the (topologically irrelevant) modulus of $1_{2,\ell}$ varying along the Cantor set parametrising the bifurcating tori.

4.3 Bifurcations Involving Two Normal Frequencies

We now consider Hamiltonian dynamics around an invariant n-torus $\mathsf{T} = \mathbb{T}^n \times \{0\} \times \{0\}$ in $n+2$ degrees of freedom when the phase space is

$$\mathcal{P} = \mathbb{T}^n \times \mathbb{R}^n \times \mathbb{R}^4 .$$

Here the $\mathbb{T}^n$-symmetry $x \mapsto x + \xi$ no longer suffices to render the system integrable as it did in the previous sections. Let T be elliptic whence there are two normal frequencies α_1 and α_2 next to the internal frequency vector ω. We bound these away from the resonances (4.2a) and (4.2b) by assuming the Diophantine conditions (4.1) to hold for all $|\ell| \leq 2$ with the exception of $\ell = (\pm 1, \pm 1)$.

Resonances $2\pi\langle k, \omega\rangle = \langle \ell, \alpha\rangle$ with $|\ell_1 \ell_2| \geq 2$ have a "semi-simple" Hamiltonian

$$H(x,y,z) = \langle \omega, y\rangle + \alpha_1 \frac{p_1^2 + q_1^2}{2} + \alpha_2 \frac{p_2^2 + q_2^2}{2} + \mathcal{O}\left(|y|^2, |z|^3\right)$$

where $z = (q,p) = (q_1, q_2, p_1, p_2)$. We have seen in Section 2.2.3 how this may lead to complicated dynamics, especially if ℓ is $(\pm 1, \pm 2)$ or $(\pm 1, \pm 3)$. However, under sufficiently small perturbations the normal linear behaviour remains elliptic, cf. [55], and as shown in [161] this is also true for the 1:1 resonance (4.2d). Therefore we restrict from now on to two normal frequencies in 1:-1 resonance (4.2c), follow [294] and choose Floquet co-ordinates (x,y,z) that turn (4.2c) into $\alpha_1 + \alpha_2 = 0$. As in Sections 2.2.2 and 3.4 we use the letter Ω for this double frequency.

4.3.1 The Quasi-Periodic Hamiltonian Hopf Bifurcation

The analysis of 1:-1 resonant equilibria and periodic orbits in Sections 2.2.2 and 3.4 shows that a crucial rôle is played by the generators

$$\begin{aligned} S &= q_1 p_2 - q_2 p_1 , & N &= \tfrac{1}{2}(q_1^2 + q_2^2) , \\ M &= \tfrac{1}{2}(p_1^2 + p_2^2) , & P &= q_1 p_1 + q_2 p_2 \end{aligned}$$

of the ring $C^\infty(\mathbb{R}^4)^\varrho$ of functions that are invariant with respect to the S^1-action (2.30) generated by S. We generalize (3.6) to the present quasi-periodic setting and take the Hamiltonian

$$H_2 = \langle \omega, y\rangle + \Omega_0 S + aN \tag{4.27}$$

as starting point. Again we leave the semi-simple case $a = 0$ aside and require $a \neq 0$ to hold true. Then the linear centraliser unfolding (2.31) leads to

$$H_2(x,y,z; \lambda, \omega, \Omega) = \langle \omega, y\rangle + \Omega S + a(\omega)N + \lambda M \tag{4.28}$$

where $z = (q, p) \in \mathbb{R}^4$ and the parameter Ω unfolds Ω_0. As λ passes through zero there is a Krein collision of Floquet exponents at the purely imaginary pair of double exponents $\pm i\Omega$.

Our parameter spaces are an open interval $\Lambda \subseteq \mathbb{R}$ around zero for λ, but instead of restricting Ω to an interval around Ω_0 we work with a compact domain $\Sigma \subseteq \mathbb{R}^{n+1}$ for (ω, Ω). Furthermore $\mathbb{Y} \subseteq \mathbb{R}^n$ and $\mathbb{S} \subseteq \mathbb{R}^4$ are small neighbourhoods of the respective origins and the x-independent Hamiltonian H with "quadratic part" (4.28) is defined on the product

$$\mathbb{T}^n \times \mathbb{Y} \times \mathbb{S} \times \Lambda \times \Sigma \tag{4.29}$$

of phase space and parameter space. The Diophantine conditions

$$\bigwedge_{k \in \mathbb{Z}^n \setminus \{0\}} \bigwedge_{\ell \in \{0, \pm 1, \pm 2\}} |2\pi \langle k, \omega \rangle + \ell \Omega| \geq \frac{\gamma}{|k|^\tau} \tag{4.30}$$

bound (ω, Ω) away from the resonances (4.2a) and (4.2b). These turn out to be sufficient for $a\lambda \geq 0$ where the invariant torus $\mathsf{T} = \mathbb{T}^n \times \{0\} \times \{0\} \times \{\lambda\} \times \{(\omega, \Omega)\}$ of X_H is hyperbolic or in (normal) 1:-1 resonance. When $a\lambda < 0$ the torus T is elliptic with normal frequencies $\alpha_1 = \Omega + \sqrt{-a\lambda}$ and $\alpha_2 = \Omega - \sqrt{-a\lambda}$ whence the Diophantine conditions (4.1) become

$$\bigwedge_{k \in \mathbb{Z}^n \setminus \{0\}} \bigwedge_{\ell \in L} \left| 2\pi \langle k, \omega \rangle + \ell_1 \Omega + \ell_2 \sqrt{-a\lambda} \right| \geq \frac{\gamma}{|k|^\tau} \tag{4.31}$$

with $L = \{(0, 0), (\pm 1, \pm 1), (\pm 2, \pm 2), (\pm 2, 0), (0, \pm 2)\}$. We therefore define

$$\Upsilon_\gamma' = \left\{ (\lambda, \omega, \Omega) \in \Upsilon_\gamma^+ \cup \Upsilon_\gamma^- \;\middle|\; \mathrm{d}((\lambda, \omega, \Omega), \partial(\Lambda \times \Sigma)) \geq \gamma \right\}$$

with

$$\Upsilon_\gamma^+ = \left\{ \lambda \in \Lambda \;\middle|\; a\lambda \geq 0 \right\} \times \left\{ (\omega, \Omega) \in \Sigma \;\middle|\; (\omega, \Omega) \text{ satisfies (4.30)} \right\}$$

$$\Upsilon_\gamma^- = \left\{ (\lambda, \omega, \Omega) \in \Lambda \times \Sigma \;\middle|\; a\lambda < 0, \, (\omega, \Omega) \text{ satisfies (4.31)} \right\}$$

where $\gamma > 0$ is allowed to vary, while $\tau > 2n - 1$ is fixed.

Theorem 4.24. Let $H = H_2 + \mathcal{O}(|y|, |z|^2)$ be an x-independent Hamiltonian on (4.29) with $a : \Sigma \longrightarrow \mathbb{R}$ in (4.28) satisfying $|a|_\Sigma, |1/a|_\Sigma, |Da|_\Sigma < \Gamma$ for some constant $\Gamma > 0$. Then there exists $\delta > 0$ with the following property. For any analytic perturbation K of H with

$$|K - H|_{\mathbb{T}^n \times \mathbb{Y} \times \mathbb{S} \times \Lambda \times \Sigma} < \delta$$

there exists a C^∞-diffeomorphism Φ on $\mathbb{T}^n \times \mathbb{R}^n \times \mathbb{R}^4 \times \mathbb{R} \times \Sigma$ such that

1). Φ is an analytic symplectomorphism for fixed $(\lambda, \omega, \Omega)$.

2). Φ is C^∞-close to the identity.

3). On $\mathbb{T}^n \times \mathbb{R}^n \times \mathbb{R}^4 \times \Upsilon'_\gamma \cap \Phi^{-1}(\mathbb{T}^n \times \mathbb{Y} \times \mathbb{S} \times \Lambda \times \Sigma)$ one can split $K \circ \Phi = K_2 + K_\infty$ into an integrable part K_2 and higher order terms K_∞. Here K_2 has the same form (4.28) as H_2. The x-dependence is pushed into the higher order terms, i.e.

$$\frac{\partial^{|j|+|k|} K_\infty}{\partial y^j \partial z^k}(x, 0, 0; \lambda, \omega, \Omega) = 0$$

for all $(x, \lambda, \omega, \Omega) \in \mathbb{T}^n \times \Upsilon'_\gamma$ and all j, k satisfying $2|j| + |k| \leq 2$.

For a proof the reader is referred to [52, 153]. □

This result is the analogue in the present context to the claim presented in Remark 4.19. Note that there are no genericity conditions on higher order terms, in particular one may have $H = H_2$.

Remark 4.25. The result derived in [52, 153] is much more general and addresses multiple resonant Floquet exponents as well. For instance, the conclusions remain true if $a(\omega) \equiv 0$ and (4.28) is replaced by the linear centraliser unfolding of the semi-simple 1:−1 resonance.

The dynamics near the torus in (normal) 1:−1 resonance is not completely determined by the "quadratic terms" (4.28). Motivated by Sections 2.2.2 and 3.4 we normalize with respect to (4.27) and obtain an integrable approximation

$$Z(x, y, S, N, M, P; \lambda, \omega, \Omega) = \tilde{Z} + \frac{b(\omega, \Omega)}{2} M^2 + \ldots \qquad (4.32)$$

with $\tilde{Z}$ of the form (4.28), cf. [47, 153] or Appendix C. Reducing the $\mathbb{T}^{n+1}$-symmetry introduced by normalization yields a one-degree-of-freedom problem on $\mathfrak{sl}_2(\mathbb{R})$, see Section 2.1.4. In particular, as λ passes through zero, X_Z undergoes for $b > 0$ a supercritical quasi-periodic Hamiltonian Hopf bifurcation, and for $b < 0$ one of subcritical type, cf. Sections 2.2.2 and 3.4. Note that we cannot make $\tilde{Z} + M^2 \in \mathbb{R}[y, S, N, M, P; \lambda](\omega, \Omega)$ a quasi-homogeneous polynomial since a weight α with $\alpha_y = \alpha_S = \alpha_N = 2\alpha_M = 2\alpha_\lambda$ contradicts the syzygy $2NM = \frac{1}{2}S^2 + \frac{1}{2}P^2$. In the degenerate case $b(\omega) \equiv 0$ this analogously applies to $\tilde{Z} + M^3 + SM$ and more generally to $\tilde{Z} + M^d + f(S, M)$ with $f \in \mathbb{R}[S, M]$ in higher degenerate cases.

On the other hand, we do not have to prove persistence of the dynamics defined by Z from scratch. Indeed, instead of normalizing the given x-independent Hamiltonian H (whence $K - Z$ would be merely small) we use the co-ordinates provided by Theorem 4.24 and directly normalize K, whence $R := K - Z$ consists of (x-dependent) higher order terms in (y, q, p). We formulate the resulting conclusions only for the non-degenerate case and sharpen (4.30) to

$$\bigwedge_{k \in \mathbb{Z}^{n+1} \setminus \{0\}} |2\pi k_1 \omega_1 + \ldots + 2\pi k_n \omega_n + k_{n+1}\Omega| \geq \frac{\gamma}{|k|^\tau} . \qquad (4.33)$$

Furthermore $\Sigma'_\gamma = \{ (\omega, \Omega) \in \Sigma_\gamma \mid d((\omega, \Omega), \partial\Sigma) \geq \gamma \}$ with $\Sigma_\gamma \subseteq \Sigma$ defined by the Diophantine conditions (4.33).

Corollary 4.26. Under the conditions of Theorem 4.24 and given $r \in \mathbb{N}$ the co-ordinate transformation Φ can be chosen to split $K \circ \Phi = Z_r + R_r$ into an integrable part Z_r and higher order terms R_r. Here Z_r has the form (4.32) and depends on (x, y, q, p) only as a function in (y, S, N, M). The remainder term R_r satisfies

$$\frac{\partial^{|j|+|k|+|l|+m} R_r}{\partial y^j \partial p^k \partial q^l \partial \lambda^m}(x, 0, 0, 0; 0, \omega, \Omega) = 0$$

for all $(x, \omega, \Omega) \in \mathbb{T}^n \times \Sigma'_\gamma$ and all j, k, l, m satisfying $4|j| + |k| + |l| + 2m \leq r$.   □

As in Section 3.4 the dynamics defined by the integrable Hamiltonian Z can be reduced to the dynamics studied in Sections 2.2.2 and 2.1.4 and is thus readily analysed. We claim that the subordinate $(n + 1)$-tori survive the perturbation by R as Cantor families. To fix thoughts we restrict to positive a (multiplying Z by -1 if necessary) whence the sign of b distinguishes between the supercritical and the subcritical case.

Since $a > 0$ does not vanish the invariant $(n+1)$-tori of X_Z are determined by

$$aS^2 = 4M^2 \frac{\partial Z}{\partial M} = 4\lambda M^2 + 4bM^3 + \ldots \qquad (4.34)$$

and satisfy $M > 0$. As in Example 4.5 this allows us to introduce the variable $Q = -\frac{1}{2}\ln M$ conjugate to P and the angle ρ conjugate to S by means of

$$q_1 = \frac{S}{\sqrt{2}} \exp Q \sin\rho + \frac{P}{\sqrt{2}} \exp Q \cos\rho , \qquad p_1 = \frac{\sqrt{2}\cos\rho}{\exp Q} ,$$

$$q_2 = -\frac{S}{\sqrt{2}} \exp Q \cos\rho + \frac{P}{\sqrt{2}} \exp Q \sin\rho , \qquad p_2 = \frac{\sqrt{2}\sin\rho}{\exp Q} ,$$

cf. [232, 228]. In these variables the symplectic structure reads

$$\sum_{i=1}^n dx_i \wedge dy_i + d\rho \wedge dS + dQ \wedge dP$$

and the Hamiltonian (4.32) becomes

$$Z = \langle \omega, y \rangle + \Omega S + \frac{a}{4} P^2 e^{2Q} + \frac{a}{4} S^2 e^{2Q} + \lambda e^{-2Q} + \frac{b}{2} e^{-4Q} + \ldots .$$

Near an invariant torus $(Q, P) = (Q_0, 0)$ we translate co-ordinates to $(q, p) = (Q - Q_0, P)$. This is completed to a symplectic transformation by means of

$\rho \mapsto \rho - (\partial P/\partial S)Q_0$ and $x \mapsto x - (\partial P/\partial y)Q_0$. Furthermore we localise to $s = S - S_0$ and write $S_0 = \lambda^{3/2}\sigma$ whence λ controls the approach to the origin in the (S_0, λ)-plane and can be used as a scaling parameter below. The coefficient of q^2 is given by

$$\frac{a}{2}\lambda^3 e^{2Q_0} + 2\lambda e^{-2Q_0} + 4be^{-4Q_0} + \dots$$

and can be put into the form

$$\frac{\lambda^2\beta}{2}$$

using (4.34) which expresses that the coefficient of q vanishes. Similarly we obtain

$$0 < \frac{a}{4}e^{2Q_0} =: \frac{\alpha}{2\lambda}$$

for the coefficient of p^2. Here α and β are of order one as λ approaches zero.

In the supercritical case $b > 0$ there is locally a single solution of (4.34) and the corresponding $(n+1)$-torus is elliptic, cf. Sections 2.1.4, 2.2.2 and 3.4. The normal frequency is of order $\sqrt{\lambda}$ whence the scaling

$$(x, y, \rho, s, q, p; \lambda, \omega, \Omega) \mapsto \left(x, \lambda^{-\frac{9}{2}}y, \rho, \lambda^{-\frac{9}{2}}s, \lambda^{-\frac{3}{2}}q, \lambda^{-3}p; \lambda, \lambda^{-\frac{1}{2}}\omega, \lambda^{-\frac{1}{2}}\Omega\right)$$

together with a division of Z by λ^5 yields

$$Z(x, y, \rho, s, q, p) = \langle \omega, y \rangle + \Omega s + \frac{\alpha}{2}p^2 + \frac{\beta}{2}q^2 + \dots .$$

As $\lambda \to 0$ the perturbation by both the higher order terms of Z and the non-integrable remainder term R is sufficiently small to yield persistence of elliptic tori $(q, p) = 0$ that satisfy the Diophantine condition (4.1) on the internal frequencies $\omega_1, \dots, \omega_n, \omega_{n+1} = \Omega/(2\pi)$ and the normal frequency $\sqrt{\alpha\beta}$.

In the subcritical case $b < 0$ there are no positive solutions of (4.34) for $\lambda < 0$ and for $\lambda > 0$ two families of elliptic and hyperbolic tori meet in a subordinate quasi-periodic centre-saddle bifurcation. The considerations above again yield persistence of elliptic tori satisfying (4.1), while hyperbolic tori only need to satisfy the Diophantine conditions (4.33). This latter condition also determines the Cantor set parametrising the persistent parabolic tori involved in the subordinate quasi-periodic centre-saddle bifurcation, cf. Example 4.5.

Theorem 4.27. Let $\mathbb{Y} \subseteq \mathbb{R}^n$ and $\mathbb{S} \subseteq \mathbb{R}^4$ be small neighbourhoods of the origin and supply $\mathcal{P} = \mathbb{T}^n \times \mathbb{Y} \times \mathbb{S}$ with the symplectic structure $\sum \mathrm{d}x_i \wedge \mathrm{d}y_i + \sum \mathrm{d}q_j \wedge \mathrm{d}p_j$. Consider a small analytic perturbation $H = N + P$ of the x-independent Hamiltonian function

$$N(x, y, q, p) = \eta(y) + \Omega(y)(q_1 p_2 - q_2 p_1) + a(y)\frac{q_1^2 + q_2^2}{2}$$
$$+ \frac{b(y)}{2}\frac{(p_1^2 + p_2^2)^2}{4} + c(y)\frac{p_1^2 + p_2^2}{2}$$

with $c(0) = 0$, but $a(0) \neq 0$ and $b(0) \neq 0$. Furthermore $Dc(0) \neq 0$ and the vectors

$$\frac{\partial^{|\ell|}}{\partial y^\ell} \begin{pmatrix} c \\ \Omega \\ \omega \end{pmatrix}, \quad |\ell| \leq 2$$

span $\mathbb{R} \times \mathbb{R} \times \mathbb{R}^n$, where $\omega(y) = D\eta(y)$. Then X_H displays a quasi-periodic Hamiltonian Hopf bifurcation.

Proof. Since $c : \mathbb{Y} \longrightarrow \mathbb{R}$ is a submersion one of the n actions $y_1, \ldots, y_n$ plays the rôle of the bifurcation parameter λ, while the "remaining" $n-1$ directions in $\mathbb{Y}$ are used to control the $n+1$ frequencies Ω and ω in the best possible way. $\qquad\Box$

An alternative proof can be obtained by generalizing the considerations in [232, 228] to the present quasi-periodic setting.

In the supercritical case monodromy is created in the family of (maximal) $(n+2)$-tori as the bifurcating n-tori become hyperbolic. This is used in [153, 48, 47] to obtain persistence of maximal tori by means of Kolmogorov's condition. Furthermore this result is applied to the forced Lagrange top. Depending on how many symmetries are broken by the (quasi)-periodic forcing, the ramified torus bundle is "Cantorised" in different ways, with parametrising action spaces ranging from Cantor dust to smooth manifolds. See [48, 47, 143] for more details.

4.3.2 Bifurcations of Non-Reducible Tori

Linear (non-Hamiltonian) systems

$$\begin{aligned} \dot{x} &= \omega \\ \dot{z} &= \Omega^\lambda(x)\,z \end{aligned}$$

with z taking values in some compact semi-simple Lie group G are considered in [167]. Denoting by $\mathfrak{g}$ the Lie algebra of G, and by $\Lambda \subseteq \mathbb{R}$ a bounded interval, the family $\Omega^\lambda : \mathbb{T}^n \longrightarrow \mathfrak{g}$ is given by $\Omega^\lambda(x) = \lambda\Omega_0 + F(x)$, $\lambda \in \Lambda$ with F analytic and small. The system is then shown to be reducible to Floquet form for almost every λ.

One may speculate what can be proven for families $\Omega_\mu^\lambda(x) = \lambda\Omega_\mu + F_\mu(x)$, where the movement of the eigenvalues of Ω_μ initiates a bifurcation (which in turn requires well-determined higher order terms in z, to have finite codimension). The obstruction to reducibility may also be of topological nature, see [68, 282, 60, 62, 273] for the dissipative analogue.

Example 4.28. On the phase space $\mathcal{P} = \mathbb{R}^2 \times \mathbb{R}^2 \times \mathbb{R}^4$ with symplectic structure $dx \wedge dy + dq \wedge dp$ we consider the Hamiltonian

$$H(x, y, q, p) = y_1 + \omega y_2 + 2\pi k(x_2 - \omega x_1)(q_1 p_2 - q_2 p_1) \qquad (4.35)$$

with frequency $\omega \in \mathbb{R}$ and $k \in \mathbb{Z}$ a fixed integer. Note that neither H nor the equations of motion

$$
\begin{aligned}
\dot{x}_1 &= 1 \\
\dot{x}_2 &= \omega \\
\dot{y}_1 &= 2\pi k \omega \, (q_1 p_2 - q_2 p_1) \\
\dot{y}_2 &= -2\pi k \, (q_1 p_2 - q_2 p_1) \\
\dot{q}_1 &= -2\pi k \, (x_2 - \omega x_1) \, q_2 \\
\dot{q}_2 &= 2\pi k \, (x_2 - \omega x_1) \, q_1 \\
\dot{p}_1 &= -2\pi k \, (x_2 - \omega x_1) \, p_2 \\
\dot{p}_2 &= 2\pi k \, (x_2 - \omega x_1) \, p_1
\end{aligned}
$$

are defined on the quotient space $\mathcal{P}/\mathbb{Z}^2 = \mathbb{T}^2 \times \mathbb{R}^2 \times \mathbb{R}^4$ (except for $(q,p) = 0$ defining quasi-periodic motion on a 2-parameter family of invariant 2-tori). However, the time-1-map, for simplicity defined on $\{x_1 = 0\}$, does define a symplectic mapping

$$
\begin{aligned}
x_2 &\mapsto x_2 + \omega \\
y_2 &\mapsto y_2 - 2\pi k (q_1 p_2 - q_2 p_1) \\
q_1 &\mapsto \cos(2\pi k x_2)\, q_1 - \sin(2\pi k x_2)\, q_2 \\
q_2 &\mapsto \sin(2\pi k x_2)\, q_1 + \cos(2\pi k x_2)\, q_2 \\
p_1 &\mapsto \cos(2\pi k x_2)\, p_1 - \sin(2\pi k x_2)\, p_2 \\
p_2 &\mapsto \sin(2\pi k x_2)\, p_1 + \cos(2\pi k x_2)\, p_2
\end{aligned}
\tag{4.36}
$$

from $\mathcal{B} = \mathbb{T} \times \mathbb{R} \times \mathbb{R}^4$ to itself, cf. [140]. Here $\mathbb{T} \times \mathbb{R} \times \{0\}$ is a 1-parameter family of invariant 1-tori that cannot be transformed to Floquet form. Indeed, the integer $k \in \mathbb{Z}$ turns out to be a topological invariant, cf. [68, 282, 60] for the counterpart in the dissipative context. In the present symplectic context the x-dependence of the normal linear part implies that the y-directions cannot be kept completely fixed. This is the reason why we work with two normal degrees of freedom, with only (q_1, p_1) at our disposal the only recurrent dynamics would be that on the invariant tori $(q_1, p_1) = 0$. A possible generalization of the factor $2\pi k(q_1 p_2 - q_2 p_1)$ in (4.35) would have been $\pi k(p_1^2 + q_1^2) - \pi l(p_2^2 + q_2^2)$ with $k, l \in \mathbb{N}$.

The recurrent dynamics of (4.36) is restricted to $\{q_1 p_2 = q_2 p_1\}$ and we expect a certain robustness with respect to perturbations since the sign of

$$
S = q_1 p_2 - q_2 p_1
$$

decides whether y_2 increases or decreases. Note that S is an integral of (4.35) and since three further commuting integrals are easily identified as

$$
y_1 + \omega y_2 \,, \quad x_2 - \omega x_1 \quad \text{and} \quad P = q_1 p_1 + q_2 p_2
$$

the Hamiltonian system X_H is in fact integrable. For $S = 0$ the variables y_1 and y_2 remain independently constant. Fixing the values of $N = \frac{1}{2}(q_1^2 + q_2^2)$ and $M = \frac{1}{2}(p_1^2 + p_2^2)$ as well defines a ramified fibration of $\{S = 0\}$ into 3-dimensional invariant submanifolds with singular fibres

$$\mathbb{R}^2 \times \{(y_1, y_2)\} \times \{(0,0,0,0)\} \ .$$

For (4.36) this yields a 3-parameter family of isotropic invariant 2-tori shrinking down to $\mathbb{T} \times \mathbb{R} \times \{0\}$. Since the dynamics around this family of 1-tori is not reducible to Floquet form the dynamics on the invariant 2-tori is not quasi-periodic.

The non-reccurent dynamics of (4.36) is best explained fixing the value $\sigma \neq 0$ of S. Here the isotropic 2-tori parametrised by the values of N, M and y_2 are not invariant since y_2 is no longer fixed, but moves according to $y_2 \mapsto y_2 - 2\pi\sigma$. Note that $\{S \neq 0\}$ is a regular bundle of (non-invariant) 2-tori. Keeping the motions in y_2 aside, the base space of the ramified 2-torus bundle defined by (4.36) is the portion $2NM - \frac{1}{2}P^2 = \frac{1}{2}S^2 \geq 0$ of $\mathfrak{sl}_2(\mathbb{R})$ we already encountered for the Hamiltonian Hopf bifurcation in Section 2.2.2.   □

This first study of (4.35) is by no means complete. For instance, the linear centraliser unfolding of the semi-simple $1:-1$ resonance suggests to experiment with unfolding terms

$$2\pi k \,(x_2 - \omega x_1)\, N \ , \quad 2\pi k \,(x_2 - \omega x_1)\, M \quad \text{and} \quad 2\pi k \,(x_2 - \omega x_1)\, P \ .$$

In the dissipative analogue the number of necessary unfolding terms increases linearly with $|k|$. Furthermore one should find appropriate higher order terms to obtain an initial dynamic on $\mathfrak{sl}_2(\mathbb{R})$, compare again with the Hamiltonian Hopf bifurcation in Section 2.2.2. The ultimate goal is to identify structures in the dynamics that are persistent under non-integrable perturbations.

5

Perturbations of Ramified Torus Bundles

Our starting point is an integrable Hamiltonian system with compact energy levels $H^{-1}(h)$. This makes the phase space $\mathcal{P}$ a ramified[1] torus bundle. The regular fibres are invariant isotropic tori that have trivial normal dynamics. Singular fibres are invariant tori of lower dimension, together with their stable and unstable manifolds.

Granted the indeed restrictive condition of integrability, this is still a quite general situation where many different constellations are possible. Best known and most prominent in applications are phase spaces $\mathcal{P}$ that are $2d$-dimensional symplectic manifolds. Where $G_1 = H, G_2, \ldots, G_d$ are d independent integrals in involution the invariant tori $G^{-1}(g)$ are Lagrangean, i.e. isotropic and of maximal dimension d. Critical values g of G yield singular fibres of the ramified torus bundle

$$\mathcal{P} = \bigcup_{g \in \text{im}\,G} \bigcup_{\iota} G^{-1}(g)_{\iota}$$

where the $G^{-1}(g)_{\iota}$ are the connected components of $G^{-1}(g)$. In case of non-commutative integrability, cf. [212, 16, 112], the regular fibres are isotropic tori of dimension

$$n = \dim \mathcal{P} - \dim(\text{im}\,G) < d .$$

Again singular fibres $G^{-1}(g)_{\iota}$ consist of invariant tori of dimension $< n$, together with their stable and unstable manifolds. Further complications arise where $\mathcal{P}$ is a Poisson space, already divided into symplectic leaves before any dynamics is defined.

---

[1] This fibration is also called a singular foliation in the literature.

5.1 Non-Degenerate Integrable Systems

The flow on the Lagrangean tori of a Liouville integrable system is conditionally periodic. Locally around such a torus there are action angle variables $(x, y) \in \mathbb{T}^d \times \mathbb{R}^d$ in which the symplectic structure becomes $dx \wedge dy$ and the Hamiltonian function $H = H(y)$ does not depend on the angles. The equations of motion read

$$\dot{x} = \omega(y) := DH(y)$$
$$\dot{y} = 0$$

and where the frequency vector ω is non-resonant the quasi-periodic flow on $\mathbb{T}^n$ is dense, excluding the existence of further integrals of motion. We speak of a *non-degenerate* Liouville integrable system if almost all Lagrangean tori have dense orbits. Sufficient conditions are the Kolmogorov non-degeneracy condition $\det D^2 H(y) \neq 0$ for almost all y or iso-energetic non-degeneracy.

The Lagrangean tori form d-parameter families and the singular fibres of the ramified d-torus bundle determine how these families fit together. At the $(d - 1)$-parameter families of elliptic $(d - 1)$-tori the Lagrangean tori shrink down in the same way as periodic orbits shrink down to centres in one degree of freedom. Different families of Lagrangean tori are separated by $(d - 1)$-parameter families of hyperbolic $(d - 1)$-tori and their (un)stable manifolds.

This picture is repeated in how the $(d - 1)$-tori shrink down to $(d - 2)$-parameter families of (partially) elliptic $(d - 2)$-tori and are separated by $(d - 2)$-parameter families of (partially) hyperbolic $(d - 2)$-tori and (part of) their (un)stable manifolds. Furthermore there are $(d - 2)$-parameter families of hyperbolic $(d - 2)$-tori with Floquet exponents $\pm \Re \pm i \Im$, together with their (un)stable manifolds these form "pinched" d-tori. In these three ways we are led to invariant tori of smaller and smaller dimension until we end up with 1-parameter families of periodic orbits and isolated equilibria.

Within the family of all $(d - 1)$-tori we encounter quasi-periodic centre-saddle and frequency halving bifurcations along $(d - 2)$-parameter subfamilies and more generally bifurcations of co-dimension $k \leq d - 1$ along $(d - k - 1)$-parameter subfamilies. Similarly invariant $(d - 2)$-tori undergoing a quasi-periodic Hamiltonian Hopf bifurcation form $(d - 3)$-parameter families and the n-parameter families of invariant n-tori have $(n - k)$-parameter subfamilies where bifurcations of co-dimension $k \leq n$ occur. Such bifurcations are not restricted to those of semi-local type, but may also involve coinciding stable and unstable manifolds of different invariant tori. For instance, heteroclinic orbits between hyperbolic $(d - 1)$-tori form $(2d - 2)$-dimensional submanifolds of the phase space $\mathcal{P}$.

5.1.1 Persistence under Perturbation

Our aim is to answer the question: *"What happens to the ramified d-torus bundle under small perturbations of the Hamiltonian ?"* Let us collect the partial answers that are already known and indicate possible directions of future research.

Persistence of Lagrangean tori is addressed by classical KAM theory. Most tori survive a small perturbation if the Kolmogorov condition $\det D^2 H(y) \neq 0$ is satisfied – near such y the relative measure of surviving tori tends to 1 as the perturbation strength tends to zero. These tori form a (Whitney)-smooth Cantor family, being parametrised over a Cantor set that has the local structure

$$\mathbb{R} \times \textit{Cantor dust} \; .$$

Where the energy level sets are transverse to the continuous direction one has persistence of most Lagrangean tori on each energy shell, parametrised by Cantor dust. The same result is obtained under the condition of iso-energetic non-degeneracy, which is independent of Kolmogorov's condition. Note that it is generic for an integrable system to satisfy both conditions almost everywhere. However, in applications it is a non-trivial task to actually check this and to determine the hypersurfaces in action space where these determinants vanish.

The Cantor set structure defined by Diophantine conditions can be used to weaken the necessary non-degeneracy condition. Indeed, since the gaps are defined by linear inequalities the conditions on the first derivatives of the frequency mapping $y \mapsto \omega(y) = DH(y)$ can be replaced by conditions on the curvature or even higher derivatives. Such Rüssmann-like conditions still guarantee that the relative measure of surviving tori tends to 1 as the perturbation strength tends to zero, but at a price. For instance, the highest derivative $L \in \mathbb{N}$ needed in

$$< \frac{\partial^{|\ell|}\omega}{\partial y^\ell} \; \Big| \; |\ell| \leq L > \; = \; \mathbb{R}^d \tag{5.1}$$

enters the Diophantine conditions on the frequency vector by means of the inequality $\tau > dL - 1$ on the Diophantine constant τ. For more details the reader is referred to [55, 251] and references therein.

For hyperbolic n-tori the above criteria remain valid almost verbatim; the key step is to pass to a centre manifold (and to replace d by n in the formulas). A technical difficulty is that even for analytic Hamiltonians centre manifolds may only be of finite differentiability. KAM-theorems remain true in this context, during the proof one has to intersperse an analytic approximation at each iteration step. Still, the analytic context has its advantages – for instance (5.1) is satisfied for some $L \in \mathbb{N}$ for an analytic frequency mapping ω if and only if imω does not lie within a linear hyperplane. An alternative is therefore to prove persistence of hyperbolic tori directly, see [251, 249] and references therein. This also gives a more direct hold on their stable and unstable manifolds.

Elliptic $(d-1)$-tori need one extra parameter to control the normal frequency as well. Similar to the iso-energetic case one can use time re-paramet-

risation and obtain Cantor families of persistent elliptic $(d-1)$-tori parametrised by Cantor dust without the use of an external parameter. Where there
are more than one normal frequency to control, this can no longer be done
in a linear way; a problem solved by Rüssmann-like conditions on the higher
derivatives of the frequency vector, see [55, 251] and references therein. In
case the mapping of internal frequencies satisfies Kolmogorov's condition, the
higher order derivatives are only needed of normal frequencies. Now normal
frequencies α_j enter the Diophantine conditions (4.1) only as combinations
$\langle \ell, \alpha \rangle$ with $|\ell| \leq 2$. This allows to extend the result to finite-dimensional elliptic tori in infinitely many degrees of freedom, cf. [241, 168]. For hypo-elliptic
tori one may deal with the hyperbolic part by means of a centre manifold or
use a direct approach, cf. [159, 56, 251].

Where (lower-dimensional) n-tori undergo a semi-local bifurcation the
n actions y conjugate to the toral angles x first of all have to versally unfold the bifurcation scenario. It is generic for the integrable Hamiltonian H
that the n-parameter families of n-tori, $1 \leq n \leq d-1$, do not encounter bifurcations of co-dimension higher than n, so this is possible. The curvature of the
frequency mapping is then used to ensure Diophanticity of most bifurcating
tori, i.e. a Rüssmann-like condition with $L=2$ is sufficient, cf. Chapter 4.

While we have kept the proof in Appendix D as simple as possible, restricting to $n = d-1$, it should be feasible to include additional elliptic and
hyperbolic normal directions. On the other hand, additional vanishing Floquet exponents pose a much harder problem, as we have seen in Section 2.2.4.
Thus, if we explicitly require that the bifurcation is one of those considered in
Chapter 4, the quasi-periodic bifurcation scenario should persist for all n-tori
with $2 \leq n \leq d-1$ and in fact also in infinite-dimensional Hamiltonian systems. Recall that the maximal co-dimension of occurring bifurcations is the
dimension n of the bifurcating torus and not related to the number of degrees
of freedom. For instance, the above curvature requirement is not necessary for
2-tori; these may undergo the quasi-periodic analogues of the co-dimension
one bifurcations of Chapter 3. Indeed, co-dimension two bifurcations are isolated within these 2-parameter families and cannot be prevented to disappear
in resonance gaps.

Let an $(n-k)$-parameter family of n-tori that undergo a bifurcation of
co-dimension k have m additional pairs of purely imaginary Floquet exponents. Then excitation of normal modes, cf. [164, 261], leads for $l = 1, \ldots, m$
to $(n+l-k)$-parameter families of $(n+l)$-tori undergoing that co-dimension k
bifurcation in the integrable system. This whole structure should persist under a (sufficiently small) non-integrable perturbation on pertinent Cantor sets.
Additional hyperbolic directions augment the dimension of stable and unstable manifolds.

5.1.2 Resonant Maximal Tori

Up to now the reported changes of the ramified d-torus bundle under a small perturbation of the Hamiltonian were of the form *"Diophantine tori persist"* leading to a Cantor-like ramified d-torus bundle – the stratification of the action space into various subfamilies parametrising the tori is replaced by a Cantor stratification.

Completely different phenomena occur where invariant tori with resonant frequency vector are perturbed. We concentrate on Lagrangean tori and furthermore restrict to a single resonance. In this way the following example shows that the phenomena of Section 4.1.1 emerge in virtually every perturbed integrable system.

Example 5.1. Our starting point is "Poincaré's fundamental problem of dynamics", the real analytic perturbed $(n + 1)$-degree-of-freedom Hamiltonian

$$H_\varepsilon(\varphi, I) = H_0(I) + \varepsilon H_1(\varphi, I; \varepsilon)$$

defined on $\mathbb{T}^{n+1} \times \mathbb{R}^{n+1}$ with perturbation parameter $0 < \varepsilon \ll 1$. We assume H_0 to be quasi-convex; this implies that Kolmogorov's non-degeneracy condition $\det D^2 H_0(I) \neq 0$ holds true for every $I \in \mathbb{R}^{n+1}$ whence invariant Lagrangean tori $\mathbb{T}^{n+1} \times \{I\}$ with Diophantine frequency vector $\varpi = DH_0(I)$ persist by classical KAM theory.

The resonance $\langle \ell, \varpi \rangle = 0$ (with $0 \neq \ell \in \mathbb{Z}^{n+1}$ fixed) determines a hypersurface $\mathbb{Y} \subset \mathbb{R}^{n+1}$, for $I \in \mathbb{Y}$ the unperturbed Lagrangean torus $\mathbb{T}^{n+1} \times \{I\}$ is foliated into invariant n-tori. Let us consider $\mathbb{Y}$ as an open and bounded subset of $\mathbb{R}^n$ and choose co-ordinates $I = (y, p)$ with $y \in \mathbb{Y}$ and $\varphi = (x, q)$ with $x \in \mathbb{T}^n$ conjugate to y in which the single resonance reads $(\partial/\partial p)H_0(y, 0) = 0$, cf. [69, 70, 262]. The remaining frequencies form the vector $\omega = \nabla_y H_0(y, 0)$ which is non-resonant at a single resonance $(y, p) = (y^*, 0)$. We expect results only for Diophantine ω (the gaps of the resulting Cantor set correspond to multiple resonances) and treat ω as external parameter. Following [240, 159, 56], we localise writing $y = y^* + \mu$ and restrict to the lowest order terms

$$H_0(y^*, p; \mu, \omega) = \langle \omega, y^* \rangle + \frac{a(y^*)}{2} p^2 .$$

From quasi-convexity we infer

$$\bigwedge_{y^* \in \mathbb{Y}} a(y^*) \neq 0$$

whence we find a lower bound of the function $|a|$ by shrinking the co-ordinate domain $\mathbb{Y}$ a bit if necessary.

We now apply a normalizing transformation that turns the perturbed Hamiltonian into

$$H_\varepsilon(x, y^*, q, p; \mu, \omega) = H_0(y^*, p; \mu, \omega) + \varepsilon \bar{H}_1(y^*, q, p; \mu, \omega) + \mathcal{O}(\varepsilon^2)$$

where $\bar{H}_1$ is the $\mathbb{T}^n$-average along x of H_1 at $\varepsilon = 0$. In the expansion

$$\bar{H}_1 = \eta(y^*; \mu, \omega) + \alpha(y^*; \mu, \omega)p + \beta(y^*; \mu, \omega)q$$
$$+ \frac{A(y^*)}{2}p^2 + \frac{B(y^*)}{2}q^2 + C(y^*)pq + \cdots$$

we may have $A(y^*) \equiv 0$, but more importantly $|a(y^*) + \varepsilon A(y^*)|$ is still bounded from below on $\mathbb{Y}$. Re-parametrising $\omega \mapsto \omega + \varepsilon \nabla_{y^*} \eta(y^*; \mu, \omega)$ the Hamiltonian H_ε still starts with $\langle \omega, y^* \rangle$. By means of an ε-small shear transformation in p we get rid of terms that are linear in p, and scaling p by $\sqrt{\varepsilon}$ we achieve

$$H_\varepsilon(x, y^*, q, p; \mu, \omega) = \langle \omega, y^* \rangle + \varepsilon \mathcal{H}(y^*, q, p; \mu, \omega) + \mathcal{O}(\varepsilon^2)$$

with

$$\mathcal{H}(y^*, q, p; \mu, \omega) = \frac{\tilde{a}(y^*)}{2}p^2 + V_\mu(q) .$$

Here V_μ can be interpreted as a family of 1-dimensional potentials, and critical points $q^* = q^*(\mu)$ of V_μ correspond to invariant n-tori $\mathbb{T}^n \times \{(y^*, q^*, 0; \mu, \omega)\}$ of the "intermediate" integrable system with Hamiltonian $\mathcal{H}_\varepsilon = \omega + \varepsilon \mathcal{H}$.

While critical points of a single potential are generically non-degenerate, it is generic for the n-parameter family V_μ of potentials to encounter critical points up to co-dimension n. More precisely, the μ-values parametrising a potential V_μ with a degenerate critical point of co-dimension k form an $(n-k)$-dimensional submanifold Λ_k in μ-space. Let $\mu^* \in \Lambda_k$ and put $d = k + 2$. Then

$$V_{\mu^*}(q) = \frac{b(y^*)}{d!}(q - q^*)^d + \mathcal{O}\left((q - q^*)^{d+1}\right) \tag{5.2}$$

with $b(y^*) \neq 0$ and $2 \leq d \leq n + 2$ near the critical point q^* of V_{μ^*}. When $d = 2$ the critical point q^* is non-degenerate and $(q, p) = (q^*, 0)$ is a non-degenerate equilibrium of the one-degree-of-freedom system – a saddle if $\tilde{a}b < 0$ and a centre if both coefficients have the same sign. In case $d \geq 3$ the equilibrium is parabolic and it is furthermore generic for the family V_μ to provide a universal unfolding of the degenerate critical point. Both genericity conditions are conditions on the perturbation H_1 of H_0.

Let us concentrate on a degenerate critical point with $d \leq n + 1$ and translate co-ordinates to $q^* = 0$. Then

$$\mathcal{H}(y^*, q, p; \mu, \omega) = \frac{\tilde{a}(y^*)}{2}p^2 + \frac{\tilde{b}(y^*)}{d!}q^d + \sum_{j=1}^{d-2} \frac{c_j(\mu)}{j!}q^j + \cdots$$

and $\mathcal{H}_\varepsilon = \omega + \varepsilon \mathcal{H}$ has the form (4.12) with $\lambda_j = \varepsilon c_j(\mu)$. We conclude that the resonant Lagrangean torus leads under generic small perturbation to the whole bifurcation scenario detailed in Theorem 4.6.

This allows to recover the results in [69, 70] for generic perturbations εH_1 of a quasi-convex Hamiltonian H_0. Indeed, the angular variable q takes values in $\mathbb{T}$ and on this compact set each potential V_μ, μ fixed, assumes minimum and maximum. It is a genericity condition on H_1 for these to be different from each other, *i.e.* $V_\mu \neq 0$ for all μ, and furthermore for V_μ to assume the form (5.2) with d even and $b(y^*) < 0$ at a maximum q^* while $b(y^*) > 0$ at a minimum; let us denote the latter by q_*. Furthermore we concentrate on $\tilde{a}(y^*) > 0$ (which may always be achieved by reversing time if necessary). Then the maximum q^* corresponds to a hyperbolic-type torus $\mathbb{T}^n \times \{(y^*, q^*, 0)\}$ and we have persistence for ω satisfying the Diophantine conditions (4.5).

The persistence of at least one n-torus from the H_0-resonant $\mathbb{T}^n \times \{y^*\} \times \mathbb{T} \times \{0\}$ for Diophantine ω had already been established in [69], and without any genericity condition on the perturbation. What we add to this is a precise description how occurring degenerate maxima of the potential, called "of weaker persistency" in [69], lead to a "Cantorised" bifurcation scenario of the corresponding n-tori. This latter result cannot be obtained without genericity conditions.

The minima q_* are treated in [70]. In the non-degenerate case these correspond to normally elliptic n-tori, whence normal-internal resonances have to be avoided as well. Therefore, persistence of a second n-torus is obtained in [70] only on a smaller (though still measure-theoretically large) subset S. For generic perturbations we can now explain the fine structure. In particular the tori coming from degenerate minima q_* do not have to be excluded. In fact, these give the opportunity to enlarge S a bit. For instance, when $d = 4$ in (5.2) we recover the bifurcation diagram Fig. 4.1 and next to at least one centre for all nonzero $\lambda \in \mathbb{R}^2$ we have an additional saddle in the small wedge of λ satisfying $9\lambda_1^2 < -8\lambda_2^3$; hence, in this wedge only Diophanticity of internal frequencies is needed to obtain two invariant n-tori. □

5.1.3 Dynamics Induced by the Perturbation

Next to the destruction of resonant tori there are many more changes that make sure that the non-integrable perturbed dynamics is indeed qualitatively different from the integrable unperturbed dynamics. While the persistence results of Section 5.1.1 are obtained upon genericity conditions on the unperturbed system, such changes require the perturbation to be generic.

One of the effects of a small generic perturbation is that stable and unstable manifolds of hyperbolic tori no longer coincide, but split and intersect transversely, cf. [247, 248, 90, 91]. Where this concerns heteroclinic orbits between two different families of hyperbolic tori this leads to drastic changes of the connection bifurcation scenario. Indeed, heteroclinic orbits exist in the integrable system only at $\mu = 0$ for an appropriately chosen transverse parameter μ. For a sufficiently small generic perturbation there is a whole interval of μ-values containing a Cantor subset of relative measure near 1 for which there are heteroclinic orbits between surviving hyperbolic tori. Similar observations apply to stable and unstable manifolds of parabolic and other bifurcating tori.

Completely new phenomena are also to be expected in the gaps of the Cantor sets parametrising persistent tori. Disintegrating Lagrangean tori lead to invariant n-tori, where $d - n$ is the number of independent resonances $\langle k, \omega \rangle = 0$ of the (internal) frequencies. Most of these lower dimensional tori will be elliptic or hyperbolic, cf. [275]. The new hyperbolic tori lie at the basis of the example in [7] of dynamical instablility. This approach to Arnol'd diffusion relies on the splitting of separatrices which also leads to transverse intersections of stable and unstable manifolds of neighbouring hyperbolic tori in the same energy shell. These hyperbolic tori form a Cantor family, and one of the main problems is to make sure that the transition chain of hyperbolic tori and their heteroclinic connections bridges the occuring gaps, cf. [90, 91] and references therein.

The dynamics in the gaps of Cantor families of hyperbolic tori can already be studied in the perturbation near resonant singular fibres of the ramified d-torus bundle. On the centre manifold these become again (resonant) regular fibres, but the full perturbed motion is superposed by the hyperbolic dynamics in the symplectic normal directions. In particular, secondary hyperbolic tori – maximal tori on the centre manifold that appear in the resonance gap – are used in [90, 91] together with hyperbolic tori of even lower dimension to continue a transition chain through the resonance gap.

A Lagrangean torus with $d - 1$ independent resonances consists of periodic orbits. When the torus breaks up under the perturbation, only finitely many of these are expected to survive. At the same time the trivial normal behaviour of these periodic orbits changes, resulting in hyperbolic and elliptic periodic orbits. The latter can serve as starting points for the construction of solenoids, cf. [25, 192, 203, 189]. This construction should carry over to elliptic tori, where the "encircling" tori emerge from the normal-internal resonances studied in [49]. This might also result in solenoids that are limits of tori with varying dimension.

The nature of the gaps defined by (4.2) for elliptic tori is twofold. Internal resonances $\langle k, \omega \rangle = 0$ lead again to the destruction of the torus. The study [49] of normal-internal resonances relates boundary points of the resulting gaps to quasi-periodic bifurcations. In particular resonance gaps

$$| 2\pi \langle k, \omega \rangle + 2\alpha | < \frac{\gamma}{|k|^\tau}$$

are completely filled by hyperbolic tori (in accordance with [30, 31, 294]) that terminate in frequency halving bifurcations. One may speculate that resonance gaps

$$| 2\pi \langle k, \omega \rangle + \alpha_1 + \alpha_2 | < \frac{\gamma}{|k|^\tau}$$

are similarly filled by hyperbolic tori obtained in quasi-periodic Hamiltonian Hopf bifurcations generated by the perturbation.

The results in Chapter 4 address persistence of Diophantine tori involved in a bifurcation and the corresponding gaps trigger again new phenomena. A first step has been made in [178, 181, 182, 183] where (internally) resonant parabolic tori involved in a quasi-periodic Hamiltonian pitchfork bifurcation are considered. This may result in large dynamical instabilities, especially where multiple parabolic resonances are encountered. The effect is further amplified for tangent (or flat) parabolic resonances, which fail to satisfy the iso-energetic non-degeneracy condition.

Lower-dimensional tori involved in a bifurcation can also have additional elliptic normal directions. This leads to normal-internal resonances that may interact with the bifurcation scenario. For instance, a family of para-elliptic tori encountering a resonance (4.2a) is expected to contain tori with the normal linear behaviour of Section 2.2.4.

5.2 Perturbations of Superintegrable Systems

We speak of a superintegrable system if the regular fibres[2] of the ramified torus bundle are isotropic tori of dimension $< d \,(= \frac{1}{2} \dim \mathcal{P})$. Determined by the dimension of these "maximal" tori this defines a whole hierarchy, starting with the minimally superintegrable systems where the regular fibres are $(d-1)$-tori (and almost all of them have dense quasi-periodic orbits) up to maximally superintegrable systems where almost all orbits are periodic. In the case of $d = 2$ degrees of freedom all these notions coincide. According to [219] the Hamiltonian of a superintegrable system only depends on the r actions conjugate to the toral angles, so the $2(d-r)$ "extra integrals" are mute parameters and a family of n-tori still encounters only bifurcations up to co-dimension n – although these are no longer isolated but form $2(d-r)$-parameter families.

Our aim remains to answer the question: *"What happens to the ramified r-torus bundle under small perturbations of the Hamiltonian?"* The strategy is to find an "intermediate" system that is also integrable, but non-degenerately so.

Definition 5.2. The perturbation εP of a superintegrable Hamiltonian N removes the degeneracy if the perturbed Hamiltonian $H = N + \varepsilon P$ can be written in the form

$$H \;=\; N \,+\, \varepsilon S \,+\, \varepsilon^2 R$$

where $N + \varepsilon S$ is a non-degenerate integrable Hamiltonian.

Let (x, q, y, p) complete the action angle variables (x, y) of $N = N(y)$ to action angle variables of $N(y) + \varepsilon S(y, p)$. If N satisfies "its" Kolmogorov condition $\det D^2 N(y) \neq 0$ for almost all y then

---

[2] Furthermore the regular fibration is symplectically complete, admitting a co-isotropic polar foliation, see [111] and references therein.

$$\det D_2^2 S(y,p) \;=\; \det \begin{pmatrix} \dfrac{\partial^2 S}{\partial p_1^2} & \cdots & \dfrac{\partial^2 S}{\partial p_1 \partial p_{d-r}} \\ \vdots & \ddots & \vdots \\ \dfrac{\partial^2 S}{\partial p_1 \partial p_{d-r}} & \cdots & \dfrac{\partial^2 S}{\partial p_{d-r}^2} \end{pmatrix} \;\neq\; 0$$

ensures that the integrable Hamiltonian $N + \varepsilon S$ is non-degenerate. Under this condition most invariant tori

$$\mathbb{T}^r \times \mathbb{T}^{d-r} \times \{(y,p)\}$$

persist under the perturbation of the intermediate system by $\varepsilon^2 R(x,q,y,p)$, cf. [6]. If N is iso-energetically non-degenerate in y then this holds true on every energy shell.

5.2.1 Minimally Superintegrable Systems

One way to put $\varepsilon P(x,q,y,p)$ into the form $\varepsilon S(y,p) + \varepsilon^2 R(x,q,y,p)$ is to compute a normal form of εP with respect to N. This results in an intermediate Hamiltonian $\bar{H} = N + \varepsilon \bar{P}$ where $\bar{P}$ is the average of P along the fibres of the ramified torus bundle defined by N. On the regular part of this bundle this defines a $\mathbb{T}^r$-symmetry and regular reduction makes $\bar{P}$ a Hamiltonian in $d - r$ degrees of freedom.

In the minimally superintegrable case $r = d - 1$ this is a one-degree-of-freedom system and always integrable. Furthermore, it is generic for $\bar{P}$ to have non-trivial dynamics in one degree of freedom, so P removes the degeneracy with $S = \bar{P}$. The remainder term $\varepsilon^2 R$ is given by

$$R \;=\; \frac{1}{\varepsilon}\left(P \circ \Psi - \bar{P}\right)$$

where Ψ is the normalizing transformation. Note that the dynamics defined by N is fast with respect to the ε-slow one-degree-of-freedom dynamics defined by $\varepsilon \bar{P}$.

From the ramified $(d-1)$-torus bundle and the one-degree-of-freedom dynamics we now construct the ramified d-torus bundle defined by the intermediate system. The Lagrangean tori consist of the regular $(d-1)$-tori superposed with the slow periodic one-degree-of-freedom dynamics. To obtain singular fibres we can proceed in two different ways.

One the one hand, the (relative) equilibria of the one-degree-of-freedom system lead to singular fibres. This is already true for the $(d-1)$-tori of the "fast" ramified torus bundle and even more so for its singular fibres. A further hierarchical structure is imposed by the co-dimensions of the various equilibria of the $(d-1)$-parameter family $S_y = \bar{P}(y, ..)$, starting at saddles and centres of co-dimension zero and generically ranging to bifurcations up to co-dimension $d - 1$.

On the other hand, superposing singular fibres of the (fast) ramified $(d-1)$-torus bundle with the slow periodic one-degree-of-freedom dynamics leads to singular fibres of the ramified d-torus bundle as well. In this way the intermediate system has four kinds of motion:

(i) The regular fibres correspond to conditionally periodic motions with $d-1$ fast frequencies and one slow frequency.

(ii) Singular fibres with periodic slow motion correspond on resulting $(n+1)$-tori, $0 \leq n \leq d-2$, to conditionally periodic motions with n fast and one slow frequencies. The symplectic normal behaviour is fast as well, this is in particular true for the asymptotic motion on existing (un)stable manifolds.

(iii) Singular fibres constructed from regular $(d-1)$-tori have fast conditionally periodic motion and slow symplectic normal behaviour. In particular, the motion on existing (un)stable manifolds combines a fast rotational motion with a slow approximation of the invariant $(d-1)$-torus.

(iv) The superposition of one-degree-of-freedom equilibria and singular fast fibres leads to conditionally periodic motion with n fast frequencies, $0 \leq n \leq d-2$, while the symplectic normal behaviour is a combination of $d-n-1$ fast degrees of freedom and one slow degree of freedom.

It remains to understand what happens to the ramified d-torus bundle defined by $\bar{H} = N + \varepsilon \bar{P}$ under perturbation by $\varepsilon^2 R$. For the regular fibres (i) the result in [6] yields persistence of a Cantor family of Lagrangean tori. The proof relies on initial normalizing transformations, using the ultraviolet cut-off introduced in [5]. The lower-dimensional tori (ii) also have the slow dynamics encoded in one of the internal frequencies, the symplectic normal behaviour is of the same magnitude as the fast frequencies. This should allow to obtain their persistence along the same lines.

For the $(d-1)$-tori (iii) the two time scales distinguish the internal from the normal dynamics. In the hyperbolic case the normal hyperbolicity of the centre manifold is of order ε and thus sufficiently large with respect to the perturbation strength ε^2 to yield persistence. During local bifurcations in the slow dynamics one has the alternative between a scaling argument [50] and a direct incorporation in the KAM-iteration [139]. In the elliptic case (treated in [176, 177]) the Diophantine conditions again involve both the ε-small normal and the "large" internal frequencies.

The fast-slow dynamics is contained in the symplectic normal behaviour for lower-dimensional tori (iv). Again the hypo-elliptic case does not pose new problems. When incorporating additional elliptic and hyperbolic directions into the proof in Appendix D one should furthermore be able to let one of them be slow – where the bifurcation scenario is developing in the slow dynamics the above alternatives still apply. More interesting is the combination of two bifurcations in both the fast and the slow (symplectic normal) dynamics. Indeed, with two different time scales e.g. the dynamics of Section 2.2.4 appears to be of $(1+1)$-degree-of-freedom rather than having truly 2 degrees of freedom. This might help to obtain more detailed results.

5.2.2 Systems in Three Degrees of Freedom

For superintegrable systems with regular r-tori with $r \leq d-2$ it is no longer automatic for the average $\varepsilon \bar{P}$ of the perturbation εP to reduce to an r-parameter family of integrable systems, the remaining number of degrees of freedom being $d - r \geq 2$. It seems therefore unlikely that a given perturbation removes the degeneracy, but see [6, 116] for a treatment of the planetary system as a perturbation of the superintegrable superposition of 9 Keplerian systems that does remove the degeneracy. The ensuing problems can already be illustrated in three degrees of freedom.

The rigid body with a fixed point is a mechanical system with three degrees of freedom, the phase space $\mathcal{P} = T^*SO(3)$ being the cotangent bundle of the group $SO(3)$ of three-dimensional rotations. An example of a non-degenerate integrable system on $\mathcal{P}$ is the Lagrange top, an axially symmetric rigid body subject to a constant vertical force field. The two commuting integrals next to the energy N are the vertical component μ_3 of the angular momentum and the component ℓ_3 of the angular momentum with respect to the symmetry or figure axis of the body. Correspondingly, fast motions are a superposition of a rotation about the figure axis, a precession of the figure axis about the vertical axis and an "up and down" nutation of the figure axis. For a detailed treatment of the integrable dynamics see [81]; the singular fibres of the resulting ramified 3-torus bundle consist of elliptic 2-tori, of elliptic periodic orbits, of "pinched 3-tori" formed by hyperbolic periodic orbits and their (un)stable manifolds and of two periodic orbits that each undergo a periodic Hamiltonian Hopf bifurcation.

The Euler case is a "free" rigid body not subject to any external force or torque. This makes the spatial components μ_1, μ_2, μ_3 of the angular momentum three non-commuting integrals of motion next to the (kinetic) energy

$$N = \frac{\ell_1}{2I_1} + \frac{\ell_2}{2I_2} + \frac{\ell_3}{2I_3}$$

and replacing one of them by the sum $\mu^2 = \mu_1^2 + \mu_2^2 + \mu_3^2$ of their squares yields a second integral (next to the energy) that commutes with all other integrals. In this way the phase space $\mathcal{P}$ becomes a ramified 2-torus bundle with a complicated singular set at $\mu^2 = 0$. Redefining

$$\mathcal{P} = T^*SO(3) \backslash SO(3)$$

by taking out the zero section (where no dynamics takes place, $SO(3)$ consists of equilibria) simplifies this situation and also allows to replace μ^2 by $|\mu| = \sqrt{\mu^2}$.

In the dynamically symmetric case $I_1 = I_2$ of two equal moments of inertia the conditionally periodic motion along the regular 2-tori becomes particularly transparent. Indeed, for such an Euler top the precession of the figure axis about the angular momentum is superposed by a rotation of the body about the figure axis. At $\ell_3 = 0$ the 2-tori become (internally) 1:0 resonant as the precession consists of the body rotating about any axis perpendicular to the figure axis. Note that the Hamiltonian

$$N = \frac{|\mu|^2}{2I_1} - \frac{I_3 - I_1}{I_1 I_3} \frac{\ell_3^2}{2}$$

can be expressed as a function of $|\mu|$ and ℓ_3 whence the additional integral ℓ_3 does not lead to topological changes of the regular fibres of the ramified 2-torus bundle. This will change below when we replace the additional S^1-symmetry by an additional $SO(3)$-symmetry.

For the general free rigid body with three different moments of inertia the above 1:0 resonant 2-tori break up and the (un)stable manifolds of the rotation about the "middle" axis of inertia separate four families of regular 2-tori. Depending on which family a 2-torus belongs to, the rôle of the figure axis is played by the "longest" or the "shortest" axis of inertia, which still precesses regularly about the direction of the angular momentum. However, the rotational motion of the body about this figure axis looks more complicated as the sines and cosines of the dynamically symmetric case have to be replaced by elliptic functions. The free rigid body is a minimally superintegrable system.

The torque exerted by a perturbing external force field causes the angular momentum to slowly move in space. For the intermediate system this motion is periodic and superposed to the fast precessional-rotational motion of the free rigid body. In [196] persistence of the resulting 3-tori is explicitly proven. The structure of the ramified 3-torus bundle defined by the intermediate system depends on the precise form of (the average of) the perturbation. The case of an affine[3] force field is detailed in [135, 136, 137] for the dynamically symmetric case and persistence of parabolic 2-tori together with the whole bifurcation scenario is discussed in [139, 50].

The same phenomenon occurs in the planar three body problem, the symmetries of which allow for reduction to three degrees of freedom where the uncoupled Keplerian motion defines a minimally superintegrable system. In well-chosen domains in the product of phase space and parameter space the secular system removes the degeneracy and relative equilibria of elliptic, hyperbolic and parabolic type give rise to Cantor families of corresponding 2-tori organizing the slow-fast invariant 3-tori obtained in [6]. See [176, 113, 114, 115] for more details.

A free rigid body with three equal moments $I_1 = I_2 = I_3$ of inertia has only periodic motions (we still exclude the zero section $SO(3)$ from the phase space) since every axis through the fixed point is a principal axis of inertia. Correspondingly, one has five independent integrals of motion by choosing next to the energy

---

[3] The linear part is needed to break the rotational symmetry of the constant part of the force field.

$$N = \frac{|\mu|^2}{2I_1}$$

two of the three components ℓ_1, ℓ_2, ℓ_3 of the angular momentum about a body set of axes and two out of the μ_1, μ_2, μ_3 in a spatial frame. The free rigid body with trivial tensor of inertia is a maximally superintegrable system.

The effect of the torque of the average of a perturbing external force field is now that the direction of the angular momentum moves both in the spatial and the body frame. Fixing $|\mu|$, regular reduction of the S^1-symmetry generated by $|\mu|$ yields a two-degree-of-freedom system on $S^2_{|\mu|} \times S^2_{|\mu|}$. This system may of course be integrable, e.g. because the force field is S^1-symmetric. Note that the external force field has to "detect" the asymmetries of the rigid body, whence an affine force field is no longer sufficiently general, leading to an S^1-symmetric system. But already a generic quadratic force field has an average that cannot be used to remove the degeneracy.

Other maximally superintegrable systems in three degrees of freedom are those defined by fully resonant quadratic Hamiltonians (2.34). For generic higher order terms of the 1:1:2 resonance the cubic normal form cannot be used to remove the degeneracy, see [95], and we expect the same to be true for all the genuine first order resonances in the upper part of Table 2.2. The cubic normal form does not contribute new terms to (2.34) for the genuine second order resonances in the middle part of Table 2.2 whence the first non-trivial normal form is obtained only after the second normalization. Again we expect this fourth order normal form to be generically non-integrable and hence not able to remove the degeneracy.

In case the resonances are of order ≥ 5 the higher order terms of (2.34) do remove the degeneracy, taking for S the quartic terms of the Birkhoff normal form. Here the ramified 3-torus bundle defined by the intermediate system is quite simple: the single family of Lagrangean tori shrinks down to three families of elliptic 2-tori and these extend between three families of elliptic periodic orbits, the normal modes of the resonant equilibrium. Normalizing the terms up to order 4 also removes the degeneracy if there is one resonance of order 2 or 4 and the second resonance is of order ≥ 5, but now the structure of the ensuing ramified 3-torus bundle depends on the form of that 2nd or 4th order resonance. A single normalization suffices if there is exactly one resonance of order three whence the higher order terms of (2.34) remove the degeneracy by taking for S the resulting cubic terms. This holds in particular true for the genuine first and second order resonances in the lower part of Table 2.2.

Finally assume that there are no resonances of order ≤ 6 and that the Birkhoff normal form of order 4 reads

$$\bar{H}(I_1, I_2, I_3) = N + \varepsilon S(N, I_1)$$

where $N = \alpha_1 I_1 + \alpha_2 I_2 + \alpha_3 I_3$ is the quadratic part (2.34) and ε is introduced by means of a $\sqrt{\varepsilon}$-scaling. We already know that the intermediate

Hamiltonian $\bar{H}$ is integrable, but the non-generic assumption that S does not depend in a general way on I_1, I_2, I_3 enforces $\bar{H}$ to be superintegrable. Further assuming genericity within the "universum" of such perturbations by higher order terms, the condition

$$\det D^2 S \neq 0 \tag{5.3}$$

ensures that $\bar{H}$ is minimally superintegrable and the condition

$$\partial_3^2 T \neq 0 \tag{5.4}$$

on the Birkhoff normal form

$$\overline{\overline{H}} = N + \varepsilon S(N, I_1) + \varepsilon^2 T(N, I_1, I_2)$$

of order 6 ensures that $\overline{\overline{H}}$ is a non-degenerate integrable system. While this example is a bit constructed, the same phenomenon appears in some applications involving the Kepler system. Indeed, the regularized spatial Kepler problem is a maximally superintegrable three-degree-of-freedom system with Hamiltonian $N(K) = K$ and has a "first" normal form

$$\bar{H} = K + \varepsilon S(K, L)$$

for the lunar problem, the Rydberg (or hydrogen) atom in crossed fields and the problem of orbiting dust, cf. [201, 202, 80, 85, 268, 104]. Here L is the third component of the angular momentum and in the two former cases $S(K, L)$ is a multiple of $K \cdot L$, while it can be brought into this form by an additional transformation in the latter case as well. A second normalization yields

$$\overline{\overline{H}} = K + \varepsilon S(K, L) + \varepsilon^2 T(K, L, I)$$

with an appropriately chosen third action I.

In all these cases the conditionally periodic motion defined by $\overline{\overline{H}}$ has three time scales, while the rates of change of these frequencies are only of the two orders ε and ε^2 of magnitude. This is not a coincidence since for a maximally superintegrable system with nowhere vanishing periodic flow the first action can always be chosen to be the unperturbed Hamiltonian. While a perturbation

$$H_\varepsilon(J, \phi) = J_1 + \varepsilon S(J_1, J_2) + \varepsilon^2 T(J, \varepsilon) + \text{h.o.t.}$$

does not fulfil Definition 5.2, it could nonetheless be shown in [268] that most regular fibres of the ramified 3-torus bundle defined by $\overline{\overline{H}}$ persist as a Cantor family, provided that the non-degeneracy conditions (5.3) and (5.4) hold true.

Three-degree-of-freedom systems are still a bit special in that a superintegrable system that is not minimally superintegrable is automatically maximally superintegrable. Possible generalizations of the above considerations include integrable intermediate systems

$$H(J) = J_1 + \varepsilon S(J_1, J_2) + \ldots + \varepsilon^{d-1} T(J)$$

in d degrees of freedom or

$$H(I, J, K) = N(I) + \varepsilon S(I, J) + \varepsilon^2 T(I, J, K)$$

with three groups I, J, K of actions. In both cases the rates of frequency changes take place on (at least) three time scales.

5.3 Perturbed Integrable Systems on Poisson Spaces

The generalization from phase spaces that are symplectic manifolds to a Poisson space $\mathcal{P}$ is twofold. On the one hand, the rank of the Poisson structure need not be constant but may drop from $\dim \mathcal{P} = 2d$ to $2d - 2$ to $2d - 4$ and so on. This stratifies $\mathcal{P}$ into different symplectic leaves where the $(2k - 2)$-dimensional strata serve as singular sets of the $2k$-dimensional strata. We have encountered one-degree-of-freedom examples in Section 2.1.3, more involved cascades of singular sets are e.g. generated if one reduces the normal form Hamiltonians at fully resonant equilibria (2.34) to $\ell - 1$ degrees of freedom, cf. [156].

On the other hand, even if the rank of the Poisson structure is equal to $2d$ everywhere it need not be equal to the dimension of $\mathcal{P}$ which becomes foliated into symplectic leaves. In general these two phenomena are combined and even if starting with a Poisson manifold $\mathcal{P}$ one may end up with a disjoint union

$$\mathcal{P} = \bigcup_{\alpha \in A} \mathcal{P}_\alpha \tag{5.5}$$

of Poisson spaces $\mathcal{P}_\alpha$ that are closed subsets of $\mathcal{P}$ and whose dimension equals the maximal rank $2d$ of the Poisson structure. The $\mathcal{P}_\alpha$ are in turn stratified by their symplectic leaves, and an integrable Hamiltonian system on $\mathcal{P}$ defines on each symplectic leaf of each $\mathcal{P}_\alpha$ a ramified torus bundle.

To simplify the discussion let us assume that the index set $A \subseteq \mathbb{R}^\ell$ and that the $\mathcal{P}_\alpha$ are common level sets of $\ell = \dim \mathcal{P} - \dim \mathcal{P}_\alpha$ Casimir functions. The values α can be treated as parameters the Hamiltonian system depends upon. In the integrable case invariant n-tori form $(n + \ell)$-parameter families. Note that the actions y conjugate to the toral angles are parameters that are distinguished with respect to α, cf. Section 1.1.2.

When perturbing a Hamiltonian system, the perturbation may address both the Hamiltonian function and the phase space. In case of a symplectic manifold like $\mathcal{P} = \mathbb{T}^d \times \mathbb{R}^d$ a perturbation of $\mathcal{P}$, in particular of the symplectic structure, can easily be transformed away leading to a perturbation of the Hamiltonian function. In case $\mathcal{P}$ is a Poisson space e.g. singular points of $\mathcal{P}$ could be perturbed away. A possible remedy would be to consider a whole family $\mathcal{P}_\alpha$ of phase spaces that already contains all possible perturbations. The union (5.5) would then be robust with respect to perturbations.

The perturbation may very well preserve the phase space and its singular subsets. For instance, if the Poisson space is obtained by symmetry reduction and the original perturbation preserves that symmetry, then the reduced perturbed system has the same phase space. In case the perturbation breaks the symmetry, the perturbation analysis should take place on the original phase space.

In [175] persistence of n-tori on $\mathcal{P} = \mathbb{T}^n \times \mathbb{R}^l$ is considered. The Poisson structure is required to be invariant under the $\mathbb{T}^n$-action $(x, y) \mapsto (x + \xi, y)$ and to satisfy the involution conditions $\{y_i, y_j\} = 0$ for $i, j = 1, \ldots, l$ whence the unperturbed system associated to

$$H(x, y) \;=\; N(y) + \varepsilon P(x, y)$$

is completely integrable and the structure matrix is of the form

$$\begin{pmatrix} C(y) & -B^{\mathrm{T}}(y) \\ B(y) & 0 \end{pmatrix}$$

with $C(y) \in M_{n \times n}(\mathbb{R})$ antisymmetric. In this situation persistence of most tori $\mathbb{T}^n \times \{y\}$ is proved in [175] under Rüssmann-like non-degeneracy conditions on the frequency mapping

$$\begin{aligned} \omega \;:\; \mathbb{R}^l &\longrightarrow & \mathbb{R}^n \\ y &\mapsto & -B^{\mathrm{T}}(y) DN(y) \end{aligned}$$

and also under the Kolmogorov condition $\det D^2 N(y) \neq 0$ if B and C are constant. The standard symplectic case is recovered for $l = n$ and $B \equiv \mathrm{id}$, $C \equiv 0$; whenever $C \equiv 0$ the n-tori are isotropic. If $l > n$ and $B \equiv (\mathrm{id}, 0)$, $C \equiv 0$ then the unperturbed equations of motion are those of a superintegrable system. Writing $z_i := y_{i+n}$, $i = 1, \ldots, l - n$ this makes $\{z_i, z_j\} = \varepsilon a_{ij}^k z^k$ a natural assumption, leading to the generalization $\{y_i, y_j\} = \mathcal{O}(\varepsilon)$ in the above Poisson structure.

A

Planar Singularities

A function $\mathcal{H} : \mathcal{P} \longrightarrow \mathbb{R}$ has a critical point $z \in \mathcal{P}$ where $D\mathcal{H}(z)$ vanishes. In local co-ordinates we may arrange $z = 0$ (and similarly that it is mapped to 0 as well). Two germs $\mathcal{K} : (\mathbb{R}^n, 0) \longrightarrow (\mathbb{R}, 0)$ and $\mathcal{N} : (\mathbb{R}^n, 0) \longrightarrow (\mathbb{R}, 0)$ represent the same function $\mathcal{H}$ locally around z if and only if there is a diffeomorphism η on $\mathbb{R}^n$ satisfying

$$\mathcal{N} = \mathcal{K} \circ \eta . \tag{A.1}$$

The corresponding equivalence class is called a singularity.

In these notes we abuse language a bit and call every representing germ a singularity; the equivalence relation (A.1) is called right equivalence. Above we used translations in the target space $\mathbb{R}$ to move the critical value to 0 as well. If we allow for more general transformations in the target space we are led to left-right equivalence. A further coarsening is contact equivalence for which the "left" transformation on $\mathbb{R}$ is allowed to depend on the point $z \in \mathbb{R}$ and the graph of $\mathcal{N}$ is mapped to the graph of $\mathcal{K}$. In this appendix we content ourselves with right equivalences unless explicitly stated otherwise.

A.1 Singularities with Non-Vanishing 3-Jet

When classifying singularities one soon notices that these are organized in series, cf. [11, 15, 17, 259, 237] and references therein. For instance, the planar singularities with non-vanishing 2-jet[1] form the series A_μ, labelled by the multiplicity $\mu = d - 1$ in (2.3). One may successively obtain these singularities from their Newton diagrams by fixing a point of a ruler at p^2 and rotating the ruler about that point, drawing a straight line each time a monomial q^d is encountered. This defines a quasi-homogeneous singularity, where for even $d = 2\ell$ the monomial pq^ℓ can be transformed away by means of a translation

$$(q, p) \mapsto (q, p - \gamma q^\ell) . \tag{A.2}$$

---

[1] In the literature these are also called 1-dimensional or co-rank one singularities.

The head of the series is the non-isolated singularity $A_\infty : \frac{1}{2}p^2$ where the ruler is fixed.

In a similar way one obtains the series D_μ from the non-isolated singularity $D_\infty : \frac{1}{2}p^2q$ by fixing one point of a ruler. For odd $k = 2\ell + 1$ the monomial $pq^{\ell+1}$ on the straight line through q^k can again be transformed away by means of (A.2) to achieve the form unfolded in (2.5).

The remaining non-zero 3-jet has a triple zero. Fixing one point of a ruler at p^3 leads to a more complicated situation. The first three straight lines yield the quasi-homogeneous singularities E_6, E_7, E_8 which are still simple. The next straight line passes not only through q^6, but simultaneously through pq^4 and p^2q^2 as well. Only one of these (or a linear combination) can be transformed away by means of (A.2), leading to a modulus m. The non-isolated singularity $J_{2,\infty} : (1/6)p^3 + (m/4)p^2q^2$ is the head of a series[2] of unimodal singularities

$$J_{2,i} \quad : \qquad \frac{1}{6}p^3 + \frac{m}{4}p^2q^2 + \frac{1}{(6+i)!}q^{6+i}$$

with local algebra $\mathcal{Q}(J_{2,i})$ generated by $\mathbb{1}, p^2q^2$ and the unfolding monomials $p, pq, pq^2, pq^3, q, \ldots, q^{4+i}$. Occurring monomials pq^ℓ on the straight segment between p^2q^2 and q^{6+i}, $i = 2\ell - 6$ can be transfromed away with (A.2).

$$
\begin{array}{ccccccccc}
 & E_{6k} & \leftarrow & E_{6k+1} & \leftarrow & E_{6k+2} & \leftarrow & J_{k+1,0} & \leftarrow & J_{k+1,1} \\
 & \swarrow & & \swarrow & & \swarrow & & \swarrow & & \swarrow \\
J_{k,0} & \leftarrow \; J_{k,1} & \leftarrow & J_{k,2} & \leftarrow & J_{k,3} & \leftarrow & J_{k,4} & \leftarrow \; J_{k,5} & \leftarrow \; J_{k,6}
\end{array}
$$

Fig. A.1. Adjacency diagram of k-modal and $(k + 1)$-modal planar singularities with non-vanishing 3-jet.

This sets up the following inductive procedure. Starting point is the head $J_{k,0}$ of the series

$$J_{k,i} \quad : \qquad \frac{1}{6}p^3 + \frac{m}{2(k!)}p^2q^k + \frac{1}{(3k+i)!}q^{3k+i} \; .$$

Adjacent to $J_{k,1}$ are the three singularities

$$E_{6k} \quad : \qquad \mathcal{H}_y(q,p) = \frac{1}{6}p^3 + \frac{1}{(3k+1)!}q^{3k+1} + \frac{m}{(2k+1)!}pq^{2k+1}$$

$$E_{6k+1} \quad : \qquad \mathcal{H}_y(q,p) = \frac{1}{6}p^3 + \frac{1}{(2k+1)!}pq^{2k+1} + \frac{m}{(3k+2)!}q^{3k+2}$$

$$E_{6k+2} \quad : \qquad \mathcal{H}_y(q,p) = \frac{1}{6}p^3 + \frac{1}{(3k+2)!}q^{3k+2} + \frac{m}{(2k+2)!}pq^{2k+2}$$

---

[2] In Section 2.1.1 we labelled these by their multiplicities $\mu = 10 + i$ as in [10].

which are semi-quasi-homogeneous. These form a "prelude" to the next[3] series started by the semi-quasi-homogeneous singularity

$$J_{k+1,0} : \quad \frac{1}{6}p^3 + \frac{1}{(3k+3)!}q^{3k+3} + \frac{\mathbf{m} + m_{k-1}q^{k-1}}{2((k+1)!)}p^2q^{k+1}$$

which is furthermore adjacent to $J_{k,4}$. Here the new modulus m_{k-1} in front of p^2q^{2k} is introduced, and correspondingly

$$\mathbf{m} = m_0 + m_1 q + \ldots + m_{k-2}q^{k-2}$$

already contains $k-1$ moduli. For the singularities $J_{k,i}$ one may complete $\mathbb{1}, p^2q^k, \ldots, p^2q^{2k-2}$ by the unfolding monomials $q, \ldots, q^{3k+i-2}, p, pq, \ldots,$ pq^{2k-1} to obtain a basis of the local algebra $\mathcal{Q}(J_{k,i})$, while bases of $\mathcal{Q}(E_{2k+i})$ can be compiled from Table 2.1. From Fig. A.1 one can assemble the adjacency diagram of all singularities with 3-jet $\frac{1}{6}p^3$.

A.2 Singularities with Vanishing 3-Jet

We have seen in the previous section how the monomial p^2q^2 leads to a modulus of the singularities $J_{2,i}$. From this monomial we furthermore obtain the series

$$Y_{k,l} : \quad \frac{1}{4}p^2q^2 \pm \frac{1}{k!}p^k + \frac{m}{l!}q^l , \quad l \geq k \geq 4$$

where the sign in front of p^k can be transformed away if k is odd. For $k=4$ the labelling[4] $X_{1,l-4} := Y_{k,l}$ is used in [11, 15]. The local algebra $\mathcal{Q}(Y_{k,l})$ is generated by $\mathbb{1}, p^2q^2$ and the unfolding monomials $q, \ldots, q^{l-2}, pq, pq^2, p^2q, p, \ldots,$ p^{k-2}. The monomial p^2q^2 is also present in the minima

$$\widetilde{Y}_k : \quad \frac{1}{2}\left(\frac{p^2+q^2}{2}\right)^2 + \frac{m}{k!}q^k , \quad k \geq 5$$

for which $\mathcal{Q}(\widetilde{Y}_k) = <\mathbb{1}, p^2, p, pq, \ldots, pq^{k-2}, q, \ldots, q^k>$.

The remaining planar singularities with co-dimension up to 12 are given in Table A.1, detailing in particular the remaining unimodal planar singularities. Since N_{16} has three moduli, the classification of all planar singularities with non-vanishing 4-jet in [11, 15] yields all bimodal planar singularities. Note that the singularities in Table A.1 are semi-quasi-homogeneous, whence bases of the local algebras can be compiled from Table 2.1.

Figure A.2 shows the adjacencies between the singularities of this section, cf. [39, 15, 237]. The adjacencies to singularities of Section A.1 are $E_{12} \leftarrow Z_{13}$, $J_{2,1} \leftarrow Z_{12}$, $J_{2,2} \leftarrow W_{13}$ and $J_{2,i} \leftarrow X_{1,i+2}$. The two figures 2.3 and A.2 contain all planar singularities up to co-dimension 12.

---

[3] We adhere to [11, 15]. The labelling $E_{(i)}^k := E_{6k+2+i}$ and $E_i^k := J_{k,i}$ of [237] emphasizes the head $E_\infty^\infty : \frac{1}{6}p^3$ of all these.

[4] In Section 2.1.1 we labelled these by their multiplicities $\mu = 9+i$ as in [10], where furthermore $T_{2,k,l} := Y_{k,l}$.

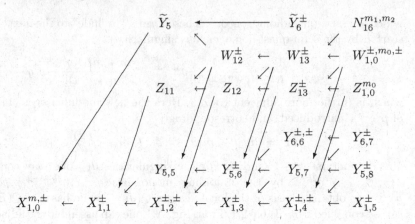

Fig. A.2. Adjacency diagram of planar singularities with vanishing 3-jet. The adjacency $Y_{5,7} \leftarrow W_{1,0}$ only exists if the coefficients of p^4 and q^6 have the same sign.

A.3 Decisive Moduli in Higher Order Terms

For the singularities $J_{2,0}, X_{1,0}, Z_{1,0}, W_{1,0}$ and N_{16} the moduli in the quasi-homogeneous part must not satisfy certain equalities for the singularity to be non-degenerate. The semi-quasi-homogeneous singularities $E_{12}, E_{13}, E_{14}, J_{3,0}$ and those in Table A.1 all have exactly one modulus outside the quasi-homogeneous part. As shown in [293], the versal unfoldings of these singularities are topologically trivial along that modulus. Quasi-homogeneous singularities for which terms of (weighted) higher order become irrelevant when allowing for homeomorphisms establishing right equivalences are called *essentially quasi-homogeneous* in [142]. It is exactly where the quasi-homogeneous part is a simple singularity that this can be achieved diffeomorphically.

Requiring only the singularity itself to be C^0-equivalent to its homogeneous part is not sufficient. Any family $\mathcal{F}_m$ of singularities with constant multiplicity

Table A.1. Exceptional unimodal and further singularities.

[10]	[11, 15, 17]	[237]	standard form
Z_{11}	Z_{12-1}	$Z^1_{(-1)}$	$p^3q + q^5 + m_0 pq^4$
Z_{12}	Z_{12}	$Z^1_{(0)}$	$p^3q + pq^4 + m_0 p^2 q^3$
Z_{13}	Z_{12+1}	$Z^1_{(1)}$	$p^3q + q^6 + m_0 pq^5$
Z_{15}	$Z_{1,0}$	Z^1_0	$p^3q + q^7 + m_0 pq^5 + m_1 pq^6$
W_{12}	W_{12}	$W^1_{(0)}$	$p^4 + q^5 + m_0 p^2 q^3$
W_{13}	W_{12+1}	$W^1_{(1)}$	$p^4 + pq^4 + m_0 q^6$
W_{15}	$W_{1,0}$	W^1_0	$p^4 \pm q^6 + m_0 p^2 q^3 + m_1 p^2 q^4$
N_{16}	N_{16}	N_{16}	$p^5 + q^5 + m_1 p^3 q^2 + m_2 p^2 q^3 + m_3 p^3 q^3$

is homeomorphic to $\mathcal{F}_0 \times \{m\}$, cf. [15, 17]. The standard form

$$\frac{1}{6}p^3 + \frac{1}{10!}q^{10} + \frac{m_0}{7!}pq^7 + \frac{m_1}{8!}pq^8$$

of the singularity E_{18} is an example where this does not imply that the versal unfolding is topologically trivial as well. Indeed, for $m_0 = 0$ the versal unfolding contains a line $\lambda = \lambda(t)$ of parameters for which the singularity E_8 occurs simultaneously at two different points (p, q) near the origin, and this does not happen when $m_0 \neq 0$, see [237]. By the result of [293] the modulus m_1 is topologically irrelevant.

Our starting point was a function $\mathcal{H} : \mathcal{P} \longrightarrow \mathbb{R}$ with a critical point at $z \in \mathcal{P}$. In local co-ordinates $\mathcal{H} : (\mathbb{R}^n, 0) \longrightarrow (\mathbb{R}, 0)$ we may work with a polynomial of degree $\leq (\mu+1)^2$ where μ is the multiplicity of $\mathcal{H}$ in 0 (assuming the singularity to be finitely determined). The unfolding

$$\mathcal{H} + \sum_{\substack{k,l=0 \\ (k,l)\neq 0}}^{\mu+1} \lambda_{kl} p^k q^l$$

is certainly versal, but this huge number of parameters can always be reduced. The universal unfolding is

$$\mathcal{H} + \sum_{j=1}^{\mu-1} \lambda_j h_j$$

where $\mathbb{1} = h_0, h_1, \ldots, h_{\mu-1}$ is a basis of the local algebra $\mathcal{Q}_{\mathcal{H}}$. In case $\mu - 1 > c$ – the co-dimension of $\mathcal{H}$ – we choose the labelling for $\lambda_1, \ldots, \lambda_c$ to be unfolding parameters and $m_i = \lambda_{c+i}$ to be moduli. Then

$$\mathcal{H} + \sum_{i=1}^{\mu-c-1} m_i h_{c+i} \tag{A.3}$$

is a family of $(\mu - c - 1)$-modal singularities. Note that (A.3) might cease to be $(\mu+1)$-determined for certain non-zero values of the moduli, but this does not happen in a sufficiently small neighbourhood of the origin. In Appendix B we explain that the moduli are superfluous with respect to C^0-equivalence, but this leads to additional values of moduli that have to be avoided. Where $m_i = 0$ turns out to be such a value the co-dimension of $\mathcal{H}$ is increased and suitable λ_{c+i} become extra unfolding parameters. The value $m_0 = 0$ in E_{18} is an example for this: the singularity

$$\mathcal{H} = \frac{1}{6}p^3 + \frac{1}{10!}q^{10}$$

has topological co-dimension 16 and its topologically versal unfolding is

$$\mathcal{H} + \sum_{j=1}^{8} \lambda_j q^j + \sum_{j=0}^{7} \lambda_{9+j} pq^j$$

where $\lambda_{16}pq^7$ unfolds the value of the modulus.

B

Stratifications

The polynomial standard forms from singularity theory all have semi-algebraic catastrophe and bifurcation sets. This gives typical examples of stratified sets. Another class of examples is provided by proper group actions which stratify a manifold according to orbit type and isotropy group, cf. [236].

Definition B.1. Let $\mathcal{P}$ be a paracompact Hausdorff space with countable topology and suppose that

$$\mathcal{P} = \bigcup_k S_k$$

is a locally finite partition of $\mathcal{P}$ into strata S_k satisfying the following conditions.

- (i) *If $S_k \cap \bar{S}_l \neq \emptyset$ for two strata then $S_k \subset \bar{S}_l$, and S_k is called a boundary stratum of S_l.*
- (ii) *Every stratum $S_k \subset \mathcal{P}$ is a smooth manifold in the induced topology and if S_k is a boundary stratum of S_l then there is a manifold $M_{k,l}$ containing both as submanifolds.*
- (iii) *Let S_k be a boundary stratum of S_l and suppose that $(x_i)_{i \in \mathbb{N}} \in S_l^{\mathbb{N}}$ converges to some $y \in S_k$. Let furthermore $(y_i)_{i \in \mathbb{N}} \in S_k^{\mathbb{N}}$ converge to y as well and suppose that (in some and thus any chart of $M_{k,l}$) the lines ℓ_i passing through x_i and y_i converge to a limit line ℓ. Suppose finally that the tangent spaces $T_{x_i} S_l$ converge to the linear subspace $\tau \subseteq T_y M_{k,l}$. Then $\ell \subseteq \tau$.*

Then the S_k form a (Whitney)-stratification of $\mathcal{P}$.

In the literature usually only (i) and the first part of (ii) are imposed on stratifications, while the second part of (ii) makes $\mathcal{P}$ a subcartesian space. Condition (iii) on the behaviour of limit tangent spaces is Whitney's condition (B). It implies Whitney's condition (A) where the conclusion is merely $T_y S_k \subseteq \tau$, see [236]. Since all our stratifications are *Whitney*-stratifications we drop this prefix from now on.

B.1 Stratification of Poisson Spaces

Let $\mathcal{P}$ be a Poisson manifold. Letting $S_k \subseteq \mathcal{P}$ denote the submanifold of points in which the Poisson structure has rank $2k$ we obtain a stratification of $\mathcal{P}$. The individual strata are in turn foliated by symplectic leaves. If the S_k already are symplectic leaves (or if their partition into symplectic leaves is locally finite) then the Poisson manifold is a symplectic stratified space.

This latter notion is also defined for Poisson spaces $\mathcal{P}$ that are not necessarily manifolds, but merely stratified spaces. To connect these two structures the Poisson algebra $\mathcal{A} \subseteq C(\mathcal{P})$ consists of functions $f \in \mathcal{A}$ for which all restrictions f_k to strata S_k are smooth, $f_k \in C^\infty(S_k)$, and the embeddings $S_k \hookrightarrow \mathcal{P}$ of the strata are Poisson mappings. Furthermore every stratum is a symplectic manifold.

As discussed in [236] this definition from [265] can be sharpened in such a way that the constructions in [265] still yield symplectic stratified spaces. The smooth structure is then defined by means of a singular atlas instead of singling out an algebra of "smooth" functions and the Poisson structure is defined by means of a section $\Lambda : \mathcal{P} \longrightarrow T\mathcal{P} \otimes T\mathcal{P}$ for which the restriction to each stratum yields a symplectic manifold. Requiring these merely to be Poisson manifolds one obtains the notion of a Poisson stratified space. An example of the latter is the Lie–Poisson space $\mathfrak{sl}_2(\mathbb{R})^*$ considered in Section 2.1.4, where the double cone and its vertex are the 0-dimensional and the two 2-dimensional strata and the complement forms three 3-dimensional strata.

Definition B.2. Let $\mathcal{P}$ be a paracompact Hausdorff space with countable topology and let $\mathcal{A} \subseteq C(\mathcal{P})$ be a subalgebra satisfying the following conditions.

(i) *The subsets $f^{-1}(I) \subseteq \mathcal{P}$ where $f \in \mathcal{A}$ and I runs through the open intervals of $\mathbb{R}$ form a subbasis of the topology of $\mathcal{P}$.*

(ii) *For $h \in C^\infty(\mathbb{R}^n)$, $n \in \mathbb{N}$ and $f_1, \ldots, f_n \in \mathcal{A}$ the composition $g = h \circ (f_1, \ldots, f_n)$ is again an element of $\mathcal{A}$.*

(iii) *A function $f \in C(\mathcal{P})$ for which every $z \in \mathcal{P}$ has an open neighbourhood U on which f coincides with some $g = g_{z,U} \in \mathcal{A}$ is already itself an element of $\mathcal{A}$.*

Then $\mathcal{P}$ is called a differential space and one writes $\mathcal{A} = C^\infty(\mathcal{P})$ for this algebra of smooth functions. In case $\mathcal{P}$ is furthermore locally diffeomorphic to a subset of some $\mathbb{R}^n$ the differential space is subcartesian.

These seem to be the most general spaces on which one can still study dynamical systems, see [87, 97] and references therein. It is therefore natural to require that the Poisson algebra $\mathcal{A} \subseteq C(\mathcal{P})$ of a Poisson space $\mathcal{P}$ makes $\mathcal{P}$ a (subcartesian) differential space.

B.2 Stratification of Singularities

The existence of C^∞-right equivalences between versal unfoldings of simple singularities ultimately relies on the homotopy method used to construct local diffeomorphisms from infinitesimal data, cf. [126, 40, 15, 17]. Where moduli appear these local diffeomorphisms do not match well together. For instance, given two members $p^4 - 3p^2q^2 + q^4$ and $p^4 - 5p^2q^2 + q^4$ of the unimodal family of $X_{1,0}$-singularities, it is straightforward to construct a C^∞-right equivalence in the complement of the origin. But the derivative of an extension to the origin would necessarily preserve the cross ratio between the four critical lines.

A similar problem occurs where one tries to construct a conjugacy between the flows of two dynamical systems around hyperbolic equilibria. Provided that the numbers of (un)stable directions coincide, it is again straightforward to construct a C^∞-conjugacy outside the two equilibria. At an equilibtium the derivative of a conjugacy has to preserve the spectral data of (the linearization of) the equilibrium, turning all eigenvalues into moduli. For this reason one usually is content with a topological conjugacy. Note that the neighbourhood of the equilibrium can be stratified into the equilibrium and its complement, and that taking the limit of a C^∞-conjugacy constructed on the latter provides a homeomorphism that is smooth on "all" strata.

Planar singularities of high co-dimension may have subordinate singularities that already require moduli. This suggests to stratify the product $\mathbb{R}^{\mu+1}$ of $\mathbb{R}^2$ and the parameter space $\mathbb{R}^{\mu-1}$ of the universal unfolding in such a way that one can construct C^∞-equivalences on each stratum that fit together to a C^0-equivalence, cf. [185, 123, 286, 293, 88]. A sufficiently general construction of these stratifications amounts in fact to a stratification of the space of all finitely determined singularities, from which the former are then obtained by restriction to subfamilies.

In [185] such a stratification is constructed for $C^\infty(N)\backslash W(N)$ where N is a compact manifold and $W(N)$ contains the $f \in C^\infty(N)$ which are not finitely determined in at least one of their critical points. A local version of this result provides stratifications of the (finite dimensional) spaces of ℓ-determined ℓ-jets for all $\ell \in \mathbb{N}$, cf. [286]. The universal unfolding of a finitely determined singularity is (multi)-transverse to these stratifications and consequently allows to construct the desired C^0-equivalence between the universal unfolding (with respect to C^∞-equivalence, i.e. containing moduli) and the subunfolding with all moduli kept constant. This shows that the universal unfolding is topologically trivial along the moduli (though with the possible exception of lower-dimensional subsets) and in particular proves Theorem 2.14. For more details see [185, 286], a similar construction is detailed in [123] to prove C^0-stability of smooth mappings (not families) $f : N \longrightarrow P$ between manifolds.

The stratification of all singularities is unique (up to refinement) but not explicitly known. In [293] topological triviality along the first occurring modulus is obtained, here there are no exceptional values in addition to those where the singularity ceases to be finitely determined. As the example of E_{18}

in Appendix A shows, a standard form of a singularity may indeed have values
of the modal parameters where the stratum does change. In [237] the strata
of "low" co-dimension are explicitly computed.

Topological triviality does not behave well under symmetrisation (2.8).
For instance, the value $m = 0$ in the unimodal family

$$X_{1,0}^+ \quad : \qquad \frac{1}{24}p^4 \; + \; \frac{m}{4}p^2q^2 \; + \; \frac{1}{24}q^4$$

is on the same stratum as all other values satisfying $-2 < m < 2$. However,
as shown in [144], the value $m = 0$ has to be excluded for the 2-parameter
universal unfolding within the subspace of $(\mathbb{Z}_2 \times \mathbb{Z}_2)$-symmetric functions. Furthermore, $(\mathbb{Z}_2 \times \mathbb{Z}_2)$-universal unfoldings with modal parameter m satisfying
$-2 < m < 0$ and $0 < m < 2$ turn out not to be C^0-equivalent.

B.3 Cantor Stratifications

An integrable Hamiltonian system X_H gives the phase space $\mathcal{P}$ the structure
of a ramified torus[1] bundle. Correspondingly $\mathcal{P}$ is stratified, with the open
stratum formed by the union of all regular fibres and the 0-dimensional strata
given by the various equilibria. Other strata are given by unions of families of
lower dimensional n-tori, and where bifurcations of these occur we pass to a
boundary stratum.

In this way the stratifications of planar singularities obtained in the previous section can be recovered in integrable Hamiltonian systems where n-tori
undergo a single normal-internal resonance (4.2a) or (4.2b). The local structure of the stratification may also be inherited from that of a quasi-periodic
Hamiltonian Hopf bifurcation at a single normal-internal resonance (4.2c).
Still other local models are given by the ways that n-tori shrink down to tori
of lower dimension. Of a more global type is the question how the strata consisting of (un)stable manifolds of hypo-elliptic n-tori limit to boundary strata
consisting of heteroclinic orbits.

The local structure near strata consisting of n-tori with several normal-
internal resonances (4.2) has yet to be understood. This is part of the following
more general open question. Let $G : \mathcal{P} \longrightarrow \mathbb{R}^l$ be the Poisson mapping consisting of the independent first integrals of X_H. In case $l = d = \frac{1}{2}\dim \mathcal{P}$ and
the G_i commute the Poisson structure on $\mathbb{R}^l$ is trivial, otherwise it is given
by the bracket relations describing the $\{G_i, G_j\}$ as functions of G. The open
question is now to describe the singularities of the mapping G. In the simplest
case $l = 1$ this amount to classifying singularities of functions. For the case
$l = d = 2, \dim \mathcal{P} = 4$ see [27] and references therein.

An important aspect of integrable Hamiltonian systems is always the behaviour under (small) non-inegrable perturbations. As suggested by the results
in Chapter 4 this might be phrased in terms of Cantor stratifications.

---

[1] For definiteness let us assume that the energy level sets are compact.

Definition B.3. Let $\mathcal{P}$ be a stratified space with strata S_k. A Cantor stratification subordinate to the (Whitney)-stratification $(S_k)_k$ is a collection $(C_k)_k$ of Cantor sets $C_k \subseteq S_k$ of Hausdorff dimension $\dim C_k = \dim S_k$ such that if S_k is a boundary stratum of S_ℓ, then C_k consists of Lebesgue density points of C_ℓ.

The condition on the dimension prevents the trivial Cantor sets $C_k = \emptyset$ and the Lebesgue density condition ensures that the various strata still "fit" together. While the latter was omitted in Definition 2.2 of [50], it has anyway been verified for the Cantor stratifications derived in that paper.

The perturbation results in the present notes are formulated and proven locally around families of tori that have a (trivial) product structure $\mathbb{T}^n \times \mathbb{Y}$, with $\mathbb{Y} \subseteq \mathbb{R}^n$ open. A global version of the (classical) KAM theorem (of Lagarangean tori) is obtained in [46]. The global conjugacy is glued together from convex combinations of local conjugacies using a partition of unity. The key ingredient is a unicity result [59] on KAM tori. To this end the set $D_{\tau,\gamma}(\Sigma') := \Sigma'_\gamma$ of Diophantine frequency vectors is shrunken to the subset $D^d_{\tau,\gamma}(\Sigma')$ consisting of its density[2] points which are defined as follows. The frequency vector $\omega \in D_{\tau,\gamma}(\Sigma')$ is a density point if every function $f \in C^\infty(\mathbb{R}^n)$ that vanishes on $D_{\tau,\gamma}(\Sigma')$ has zero derivatives $f^{(n)}(\omega) = 0$ for all $n \in \mathbb{N}_0$ – its ∞-jet in ω vanishes as well. Since the complement $D_{\tau,\gamma}(\Sigma') \backslash D^d_{\tau,\gamma}(\Sigma')$ is of measure zero, this restriction does not alter any measure estimates. The Whitney-smoothness of the KAM-conjugacy Φ on $D^d_{\tau,\gamma}(\Sigma')$ then allows to obtain uniqueness of Φ on $D^d_{\tau,\gamma}(\Sigma')$. See [59] for further details.

As remarked in [46, 59] this result does not immediately carry over to lower dimensional tori. However, such an extension would be very helpful in understanding Hamiltonian systems perturbed from integrability. Indeed, it would allow to match all the local persistence descriptions to obtain a "Cantorised" version of the ramified torus bundle defined by the unperturbed integrable system. This would yield a "global" Cantor stratification of the whole phase space.

The union of persistent invariant tori (with Diophantine frequency vectors that are density points) forms a subcartesian differential space. A possible "global" formulation of KAM theory might thus be that there is a smooth conjugacy between the known dynamics on this differential space and the perturbed dynamics on a subset of $\mathcal{P}$ (which then is necessarily of large measure, carries a (global) Cantor stratification, etc.). The present notes may hopefully serve as a basis towards that goal.

---

[2] These do not coincide with Lebesgue density points, and neither is one a subset of the other.

C

Normal Form Theory

In general terms, a normal form is "one of the prettier representants" of an equivalence class. The equivalence relation in question determines what kind of normal form we are after. Well-known examples are the Jordan normal form for matrices representing an endomorphism and Sylvester's normal form for symmetric matrices representing a quadratic form. Another example is given by singularities of (e.g. planar) functions, where we speak in these notes of "standard forms" to avoid confusion.

An invariant submanifold of a dynamical system poses two types of problems. On the one hand one would like to simplify the flow on the invariant subset, and on the other hand the flow "normal" to the submanifold should be put into normal form. For invariant tori T with conditionally periodic flow the former amounts to finding a co-oridinate system $x : \mathsf{T} \longrightarrow \mathbb{T}^n$ in which the equations of motion read $\dot{x} = \omega$ with frequency vector $\omega \in \mathbb{R}^n$. To treat the latter problem we first consider normal forms around equilibria and around periodic orbits.

C.1 Normal Forms near Equilibria

To study the dynamics locally around an equilibrium, we may restrict to the origin of $\mathbb{R}^{2\ell}$ with standard symplectic structure. The linearization around the origin is defined by the quadratic part

$$\mathcal{H}_0^0(z) \;=\; \frac{1}{2}\langle D^2\mathcal{H}(0) \cdot z, z\rangle$$

of the Hamiltonian $\mathcal{H}$. Note that the linear part $D\mathcal{H}(0) \cdot z \equiv 0$ vanishes completely since the origin is an equilibrium and we may put $\mathcal{H}(0) = 0$ as well. Let us first assume that the linear vector field

$$\Omega \;:=\; X_{\mathcal{H}_0^0} \;=\; DX_{\mathcal{H}}(0)$$

has only simple eigenvalues; in particular 0 is not an eigenvalue.

The normal form we are interested in is that of a Hamiltonian vector field locally at an equilibrium. Thus, the equivalence relation we work with consists of co-ordinate transformations that respect the symplectic structure. To ensure the latter we work with transformations $\varphi_{\mathcal{F}} = \varphi_{t=1}^{\mathcal{F}}$ that are given as time-1-mappings of a Hamiltonian function $\mathcal{F}$. Taylor's formula yields

$$\mathcal{H} \circ \varphi_{\mathcal{F}} = \mathcal{H} + \{\mathcal{H}, \mathcal{F}\} + \int_0^1 (1-t)\{\{\mathcal{H}, \mathcal{F}\}, \mathcal{F}\} \circ \varphi_t^{\mathcal{F}} \, dt \qquad (C.1)$$

for the transformed Hamiltonian $\mathcal{H}$. We expand

$$\mathcal{H} = \sum_{k=0}^{\nu} \frac{1}{k!} \mathcal{H}_k^0 + \mathcal{R}_{\nu} \qquad (C.2)$$

with homogeneous polynomials $\mathcal{H}_k^0 \in \mathcal{G}_{k+2}$ of order $k+2$ and a remainder term satisfying $\mathcal{R}_{\nu}(0) = D\mathcal{R}_{\nu}(0) = \ldots = D^{\nu+2}\mathcal{R}_{\nu}(0) = 0$. Then (C.1) reads

$$\mathcal{H} \circ \varphi_{\mathcal{F}} = \sum_{k=0}^{\nu} \frac{1}{k!} \mathcal{H}_k^0 - \{\mathcal{F}, \mathcal{H}_0^0\} + \sum_{k=1}^{\nu} \frac{1}{k!} \{\mathcal{H}_k^0, \mathcal{F}\} + \ldots$$

where we concentrate on the effects of lowest order. Since

$$\begin{array}{rccc} X_{\mathcal{H}_0^0} : & \mathcal{G}_{k+2} & \longrightarrow & \mathcal{G}_{k+2} \\ & \mathcal{F} & \mapsto & \{\mathcal{F}, \mathcal{H}_0^0\} \end{array} \qquad (C.3)$$

respects the gradation $\bigoplus \mathcal{G}_{k+2}$ while $X_{\mathcal{H}_l^0} : \mathcal{G}_{k+2} \longrightarrow \mathcal{G}_{k+l+2}$ leads to higher order terms, we may inductively choose $\mathcal{F}$ to be in $\mathcal{G}_3, \mathcal{G}_4, \ldots$ to normalize $\mathcal{H}_1^0$ to $\mathcal{H}_0^1$, then $\mathcal{H}_2^0$ to $\mathcal{H}_0^2$ and so on. On each step we have to solve the homological equation

$$\{\mathcal{F}, \mathcal{H}_0^0\} + \mathcal{H}_0^k = \mathcal{H}_k^0 \qquad (C.4)$$

in the unknowns $\mathcal{F}, \mathcal{H}_0^k \in \mathcal{G}_{k+2}$. To this end we need a linear complement of $\operatorname{im} X_{\mathcal{H}_0^0}$ in $\mathcal{G}_{k+2}$. In the presence of an inner product on $\mathcal{G}_{k+2}$ such a complement is given by $(\operatorname{im} X_{\mathcal{H}_0^0})^{\perp}$; this relegates the choice of a complement to the choice of an inner product. However, in the present situation of semi-simple $\mathcal{H}_0^0$ this construction invariably[1] leads to

$$\operatorname{im} X_{\mathcal{H}_0^0} \oplus \ker X_{\mathcal{H}_0^0} = \mathcal{G}_{k+2} . \qquad (C.5)$$

Indeed, since we assume all eigenvalues of the linear mapping

$$\Omega : \mathbb{R}^{2\ell} \longrightarrow \mathbb{R}^{2\ell} \qquad (C.6)$$

---

[1] Different splittings may be obtained by choosing an inner product for which $X_{\mathcal{H}_0^0}$ does not commute with its adjoint, cf. [277, 218].

to be simple, the eigenvalues of the linear operator[2] (C.3) all have algebraic multiplicity equal to geometric multiplicity. Let us illustrate this for the case

$$\mathcal{H}_0^0(q,p) = \sum_{j=1}^{\ell} \alpha_j \frac{p_j^2 + q_j^2}{2} \tag{C.7}$$

of elliptic equilibria. In complex variables $w_j = p_j + iq_j$ the Poisson structure becomes

$$\{w_\iota, \bar{w}_j\} = 2i\delta_{\iota j}$$
$$\{w_\iota, w_j\} = \{\bar{w}_\iota, \bar{w}_j\} = 0$$

and $\mathcal{H}_0^0$ turns into

$$\mathcal{H}_0^0(w) = \sum_{j=1}^{\ell} \alpha_j \frac{w_j \bar{w}_j}{2}$$

whence the differential operator

$$X_{\mathcal{H}_0^0} = \sum_{j=1}^{\ell} i\alpha_j w_j \frac{\partial}{\partial w_j} - i\alpha_j \bar{w}_j \frac{\partial}{\partial \bar{w}_j}$$

takes a diagonal form. The corresponding eigenbasis of (C.3) is given by the monomials

$$\mathcal{M}_m = w_1^{m_1} \cdot \ldots \cdot w_\ell^{m_\ell} \cdot \bar{w}_1^{m_{\ell+1}} \cdot \ldots \cdot \bar{w}_\ell^{m_{2\ell}}$$

with $m \in \mathbb{N}_0^{2\ell}$ satisfying $m_1 + \ldots + m_{2\ell} = k + 2$ and from

$$\{\mathcal{M}_m, \mathcal{H}_0^0\} = \sum_{j=1}^{\ell} i\alpha_j(m_j - m_{\ell+j})\mathcal{M}_m$$

we infer that $\ker X_{\mathcal{H}_0^0}$ is spanned by those monomials for which

$$\sum_{j=1}^{\ell} \alpha_j(m_j - m_{\ell+j}) = 0 . \tag{C.8}$$

Using the splitting (C.5) the solution of the homological equation (C.4) is immediate; for $\mathcal{H}_k^0 \in \mathcal{G}_{k+2}$ we let $\mathcal{H}_0^k$ be the projection to $\ker X_{\mathcal{H}_0^0}$, and since $\mathcal{H}_k^0 - \mathcal{H}_0^k \in \operatorname{im} X_{\mathcal{H}_0^0}$ there must be $\mathcal{F}_k \in \mathcal{G}_{k+2}$ with $\{\mathcal{F}_k, \mathcal{H}_0^0\} = \mathcal{H}_k^0 - \mathcal{H}_0^k$. The composition $\psi = \varphi_{\mathcal{F}_1} \circ \ldots \circ \varphi_{\mathcal{F}_\nu}$ turns (C.2) into

$$\mathcal{H} \circ \psi = \sum_{k=0}^{\nu} \frac{1}{k!}\mathcal{H}_0^k + \mathcal{R}^\nu =: \mathcal{H}^\nu + \mathcal{R}^\nu$$

where the remainder term $\mathcal{R}^\nu$ collects $\mathcal{R}_\nu$ and all higher order terms generated during the normalization procedure.

---

[2] To avoid possible confusion the operator (C.3) is often denoted by $\operatorname{ad}_{\mathcal{H}_0^0}$ in the literature. Here we use the letter Ω for the mapping (C.6) instead.

C.1.1 Elliptic Equilibria

The truncated normal form $\mathcal{H}^\nu$ has aquired a continuous symmetry group. (Moreover, the normalization procedure preserves those symmetries that $\mathcal{H}$ already has, so $\mathcal{H}^\nu$ has these symmetries as well.) In case the (normal) frequencies α_j are non-resonant, i.e. linearly independent over $\mathbb{Z}$ or $\mathbb{Q}$, the aquired symmetry of $\mathcal{H}^\nu$ is invariance under the $\mathbb{T}^\ell$-action generated by the periodic flows of the ℓ Hamiltonian vector fields

$$X_{I_1}, \ldots, X_{I_\ell} \quad \text{with} \quad I_j = \frac{p_j^2 + q_j^2}{2} .$$

Correspondingly, $\mathcal{H}^\nu = \mathcal{H}^\nu(I)$ depends on q and p only in terms of the basic invariants $I_j = \frac{1}{2} w_j \bar{w}_j$. Indeed, every monomial $\mathcal{M}_m$ satisfying (C.8) has $m_j = m_{\ell+j}$ for $j = 1, \ldots, \ell$ and therefore equals the product $(w_1 \bar{w}_1)^{m_1} \cdot \ldots \cdot (w_\ell \bar{w}_\ell)^{m_\ell}$. In particular, we have $\mathcal{H}_0^k = 0$ for k odd.

A resonance $k_1 \alpha_1 + \ldots + k_\ell \alpha_\ell = 0$ between the frequencies yields via (C.8) the additional monomial $\mathcal{M}_{m^k}$ with $m_j^k = k_j$, $m_{\ell+j}^k = 0$ if $k_j \in \mathbb{N}_0$ and with $m_j^k = 0$, $m_{\ell+j}^k = -k_j$ for $k_j < 0$ that is invariant under the torus action; both the real and the imaginary part of $\mathcal{M}_{m^k}$ yield the extra basic invariants

$$J_k = \frac{1}{m^k!} \operatorname{Re} \mathcal{M}_{m^k}$$

$$K_k = \frac{1}{m^k!} \operatorname{Im} \mathcal{M}_{m^k} .$$

While there is no a priori inequality like $I_j \geq 0$ on these new invariants, the syzygy

$$
\begin{aligned}
\frac{J_k^2 + K_k^2}{2} &= \frac{\mathcal{M}_{m^k} \overline{\mathcal{M}}_{m^k}}{2(m^k!)^2} = \frac{1}{2(m^k!)^2} \prod_{j=1}^{\ell} (w_j \bar{w}_j)^{m_j^k + m_{\ell+j}^k} \\
&= \frac{2^{|k_1| + \ldots + |k_\ell| - 1}}{(|k_1|!)^2 \cdot \ldots \cdot (|k_\ell|!)^2} \prod_{j=1}^{\ell} I_j^{m_j^k + m_{\ell+j}^k}
\end{aligned}
\tag{C.9}
$$

shows that the resulting basic invariants are no longer algebraically independent. Note that the order $|k| = |k_1| + \ldots + |k_\ell|$ of the resonance must satisfy $|k| \leq \nu + 2$ for the additional basic invariants to enter the normal form $\mathcal{H}^\nu$.

When there are resonances, the torus action of $\mathbb{T}^\ell$ is no longer effective. In fact, if l denotes the number of independent resonances then this action can be replaced by an effective $\mathbb{T}^{\ell-l}$-action. In particular, for $l = \ell - 1$ we have an S^1-action generated by the periodic(!) flow of $X_{\mathcal{H}_0^0}$. To distinguish normal forms $\mathcal{H}^\nu = \mathcal{H}^\nu(I, J, K)$ involving resonant terms J_k and K_k from $\mathcal{H}^\nu = \mathcal{H}^\nu(I)$ that solely depend on the I_j we speak only of the latter as a Birkhoff normal form and call the former a resonant[3] normal form, cf. [16].

---

[3] In the literature this is also called Gustavson normal form.

C.1.2 Algorithms

Next to the theoretical aspect – "how does the normal form look like ?" – addressed above there is the practical aspect – how does one compute the coefficients of the invariant monomials in $\mathcal{H}^\nu$ for a given Hamiltonian function $\mathcal{H}$? The procedure derived from the splitting (C.5) can be turned into an efficient algorithm, see [188, 53] where furthermore the complexity of the algorithm is measured by the number

$$\sum_{k=1}^{\nu} \frac{\nu(\nu+1)}{2k} - \frac{k-1}{2} \leq \frac{1}{2}\nu(\nu+1)\left(\frac{1}{2} + \ln(\nu+1)\right)$$

of Poisson brackets that have to be computed.

The above algorithm yields the final normalization as a composition of n co-ordinate transformations. A slightly different approach is to look for a single function

$$\mathcal{W} = \sum_{k=1}^{\nu} \frac{1}{k!}\mathcal{W}_k \tag{C.10}$$

generating the normalizing co-ordinate transformation $\varphi_{\mathcal{W}} = \varphi_{t=1}^{\mathcal{W}}$ by means of the time-1-mapping of that one taylored function. The basis of the algorithm is still the splitting (C.5), but the necessary exponentiation is replaced by the Lie–Deprit triangle

$$
\begin{array}{ccccccc}
& & & \mathcal{H}_0^0 & & & \\
& & \swarrow & & & & \\
& & \mathcal{H}_1^0 & \longrightarrow & \mathcal{H}_0^1 & & \\
& \swarrow & & \swarrow & & & \\
& \mathcal{H}_2^0 & \longrightarrow & \mathcal{H}_1^1 & \longrightarrow & \mathcal{H}_0^2 & \\
\swarrow & & \swarrow & & \swarrow & & \\
\mathcal{H}_3^0 & \longrightarrow & \mathcal{H}_2^1 & \longrightarrow & \mathcal{H}_1^2 & \longrightarrow & \mathcal{H}_0^3 \\
\swarrow & & \swarrow & & \swarrow & & \swarrow
\end{array}
$$

with intermediate homogeneous Hamiltonians

$$\mathcal{H}_k^j = \mathcal{H}_{k+1}^{j-1} + \sum_{i=0}^{k} \binom{k}{i} \{\mathcal{H}_i^{j-1}, \mathcal{W}_{k+1-i}\} \ ,$$

see [208, 188, 118, 53] and references therein for more details. The complexity is with $\frac{1}{6}\nu(\nu+1)(\nu+2)$ Poisson bracket computations a bit higher, but can be brought down to $\frac{1}{2}\nu(\nu+1)$ for $\mathcal{H} = \mathcal{H}_0^0 + \mathcal{H}_k^0 + \mathcal{R}_\nu$ with a single homogeneous higher (kth) order terms, see [130]. The number of necessary Poisson bracket computations can also be brought down by performing the computations in the invariants of a symmetry group that the Hamiltonian already has, cf. [145, 118].

In both approaches the homological equation (C.4) is solved by relegating terms $\gamma_m \mathcal{M}_m$ of $\mathcal{H}_k^0$ satisfying (C.8) to $\mathcal{H}_0^k$, while the remaining terms $\gamma_m \mathcal{M}_m$ of $\mathcal{H}_k^0$ are divided by the corresponding non-zero eigenvalue of $X_{\mathcal{H}_0^0}$ to form $\mathcal{F}$. This may lead to small denominators where (C.8) does "nearly" hold, posing numerical problems. The approach in [258, 195] works with a polynomial

$$g(X) \;=\; X^i \,+\, \alpha_1 X^{i-1} \,+\, \dots \,+\, \alpha_{i-1} X \,+\, \alpha_i$$

satisfying $g(X_{\mathcal{H}_0^0}) = 0$ in $\mathrm{End}\,\mathcal{G}_{k+2}$ that may be computed from the characteristic polynomial or from the minimal polynomial of (C.6). For $\alpha_i \neq 0$ the endomorphism (C.3) is bijective and (C.4) is solved with $\mathcal{H}_0^k = 0$ and

$$\mathcal{F}_k \;=\; X_{\mathcal{H}_0^0}^{-1}(\mathcal{H}_k^0) \;=\; \frac{-1}{\alpha_i}\left[X_{\mathcal{H}_0^0}^{i-1}(\mathcal{H}_k^0) + \dots + \alpha_{i-1}\mathcal{H}_k^0 \right].$$

In case $\alpha_i = 0$ we may in the present semi-simple context factor by a suitable power of X to achieve $\alpha_{i-1} \neq 0$ and put

$$\mathcal{F}_k \;=\; \frac{-1}{\alpha_{i-1}}\left[X_{\mathcal{H}_0^0}^{i-2}(\mathcal{H}_k^0) + \dots + \alpha_{i-2}\mathcal{H}_k^0 \right]$$
$$\mathcal{H}_0^k \;=\; \mathcal{H}_k^0 \,-\, X_{\mathcal{H}_0^0}(\mathcal{F}).$$

Since $X_{\mathcal{H}_0^0}^2(\mathcal{H}_k^0) = \{\{\mathcal{H}_k^0, \mathcal{H}_0^0\}, \mathcal{H}_0^0\}$ etc. a lot of Poisson brackets need to be calculated.

A resonant normal form is not unique and can be improved upon, as shows the rotation that achieves $b(\lambda) \equiv 0$ in (2.39). Transformations $\varphi_{\mathcal{F}}$ generated by some $\mathcal{F} \in \ker X_{\mathcal{H}_0^0}$ take normal forms into normal forms, but allow for further improvement as the $\{\mathcal{H}_0^k, \mathcal{F}\}$ do not necessarily vanish. This is called second normalization [80], hypernormal form [218], unique normal form [255] or normalizing by pivoting [234] (see also Section C.1.3 below).

The scaling $(q, p, \mathcal{H}) \mapsto (\varepsilon q, \varepsilon p, \varepsilon^{-2}\mathcal{H})$ "zooms in" to the equilibrium and shows that the influence of the remainder term $\mathcal{R}^\nu$ is of order $\varepsilon^{\nu+1}$, where ν can be chosen as high as the differentiability of $\mathcal{H}$ allows. For C^∞-smooth $\mathcal{H}$ one may use Borel's theorem to e.g. replace the polynomial (C.10) by a C^∞-function $\mathcal{W}$ with Taylor series

$$\mathcal{W} \;=\; \sum_{k=1}^{\infty} \frac{\varepsilon^k}{k!}\mathcal{W}_k$$

whence the transformed Hamiltonian

$$\mathcal{H} \circ \varphi_{\mathcal{W}} \;=\; \sum_{k=1}^{\infty} \frac{\varepsilon^k}{k!}\mathcal{H}_0^k \,+\, \mathcal{R}^\infty$$

has a Taylor series in normal form with an infinitely flat remainder term, i.e. $D^\nu \mathcal{R}^\infty(0) = 0$ for all $\nu \in \mathbb{N}_0$. This procedure from [272] leads to a remainder term that is smaller than any power of ε.

In case $\mathcal{H}$ is analytic one may alternatively perform a large but finite "optimal" number r of normalization steps. For the estimates one needs Diophantine conditions

$$\bigwedge_{h \in \mathbb{Z}^{\ell}} \langle h, \alpha \rangle \neq 0 \quad \Rightarrow \quad |\langle h, \alpha \rangle| \geq \frac{\gamma}{|h|^{\tau}} \tag{C.11}$$

where $\gamma > 0$ and $\tau > \ell - 1$ to arrive at $\mathcal{H}^r + \mathcal{R}^r$ with $\mathcal{H}^r$ in normal form and an exponential bound

$$\|\mathcal{R}^r\| \leq 2e^{\tau+2}\|\mathcal{H}\| \exp\left(-\left(\frac{\varepsilon^*}{\varepsilon}\right)^{1/(\tau+2)}\right)$$

on the remainder terms, see [124, 125]. In case α is too close to a resonance one may replace the condition $\langle h, \alpha \rangle \neq 0$ in (C.11) by $h \notin \mathcal{M}$ where $\mathcal{M}$ is a submodule of $\mathbb{Z}^{\ell}$ containing the module of exact resonances. In particular, the inequalities (C.11) can a posteriori be restricted to $h \in \mathbb{Z}^{\ell}$ satisfying $|h| \leq r + 2$.

C.1.3 Multiple Eigenvalues

The reason we restricted ourselves to the case of simple eigenvalues was that this ensures $X_{\mathcal{H}_0^0}$ to be semi-simple. In fact, this latter condition is all that is needed for the results to remain true. To conclude this section, we now consider $\mathcal{H}_0^0$ to be in non-semi-simple $1{:}{-}1$ resonance, restricting to $\ell = 2$ degrees of freedom. As in Section 2.2.2 we work with the Jordan–Chevalley decomposition

$$X_{\mathcal{H}_0^0} = \alpha X_S + a X_N$$

where $S = q_1 p_2 - q_2 p_1$, $N = \frac{1}{2}(q_1^2 + q_2^2)$ and $a \neq 0$. The standard inner product on $\mathbb{R}^4$ induces the splitting

$$\operatorname{im} X_{\alpha S + a N} \oplus \ker X_{(\alpha S + a N)^*} = \mathcal{G}_{k+2}$$

replacing (C.5), cf. [277]. It is advantageous to fix $a = -1$ for the choice $M = \frac{1}{2}(p_1^2 + p_2^2)$ of the adjoint, keeping in mind that the kernel of a linear operator does not change under multiplication with non-zero scalars. Since X_S and X_M commute we have

$$\ker X_{\alpha S - a M} = \ker X_S \cap \ker X_M$$

which suggests a normalization in two steps. First a resonant normal form $\mathcal{H}^{\nu}$ with respect to the semi-simple part is computed. This introduces an S^1-symmetry generated by the periodic flow of X_S whence $\mathcal{H}^{\nu}$ depends on (q,p) only as a function of S, N, M and $P = p_1 q_1 + p_2 q_2$.

The second part of the normalization transforms $\mathcal{H}^\nu \in \ker X_S$ to $\mathcal{H}_\nu \in \ker X_S \cap \ker X_M$. The splitting

$$(\ker X_S \cap \operatorname{im} X_N) \oplus (\ker X_S \cap \ker X_M) = \ker X_S$$

ensures that the achievements of the first part of the normalization are not undone, i.e. we look for a function $\mathcal{F} = \mathcal{F}(S, N, M, P)$ solving the homological equation

$$\{\mathcal{F}, N\} + \mathcal{H}_\nu = \mathcal{H}^\nu .$$

Here it is helpful that the quotient

$$\mathbb{R}^4/_{S^1} = \left\{ (S, N, M, P) \in \mathbb{R} \times \mathbb{R}^3 \mid 2NM = \tfrac{1}{2}P^2 + \tfrac{1}{2}S^2 , \ N \geq 0, \ M \geq 0 \right\}$$

is embedded in $\mathbb{R} \times \mathfrak{sl}_s(\mathbb{R})$, cf. [198, 218]. From

$$\{S, M\} = 0 , \quad \{N, M\} = P , \quad \{M, M\} = 0 , \quad \{P, M\} = 2M$$

we infer that the normal form $\mathcal{H}_\nu$ is a function in S and M and does not depend on N or P. See [198] for more details.

C.2 Normal Forms near Periodic Orbits

Near a periodic orbit we complete an angle x along that orbit and the conjugate action y to a symplectic co-ordinate system $(x, y, z) \in \mathbb{T} \times \mathbb{R} \times \mathbb{R}^{2\ell}$ (with $y = 0 = z$ at the initial periodic orbit). In these local co-ordinates the Hamiltonian reads

$$H(x, y, z) = \omega y + \frac{1}{2}\langle \Omega(x) \cdot z, z \rangle + R_0(x, y, z)$$

with

$$R_0(x, 0, 0) \equiv 0 , \qquad \frac{\partial R_0}{\partial y}(x, 0, 0) \equiv 0$$

$$D_z R_0(x, 0, 0) \equiv 0 , \qquad D_z^2 R_\theta(x, 0, 0) \equiv 0 .$$

First we transform H into Floquet form, cf. [12, 208], making $\Omega(x) \equiv \Omega$ independent of the angle x. This involves taking the (real) logarithm of the time-1-mapping of $z' = \Omega(x)z$, a real symplectic matrix. If -1 is a Floquet multiplier with geometric multiplicity different from algebraic multiplicity such a logarithm need not exist and we pass to the square of that matrix, the time-2-mapping. This corresponds to passing to a double cover $\mathbb{T} \times \mathbb{R} \times \mathbb{R}^{2\ell}$ with deck transformation

$$\begin{array}{ccc} \mathbb{T} \times \mathbb{R} \times \mathbb{R}^{2\ell} & \longrightarrow & \mathbb{T} \times \mathbb{R} \times \mathbb{R}^{2\ell} \\ (x, y, z) & \mapsto & (x + \tfrac{1}{2}, y, -z) \end{array}$$

where we keep the letters x, y, z for the co-ordinates.

The aim of the normalization procedure is now twofold: make the Hamiltonian x-independent and put the higher order terms in z in the form achieved in the previous section for the given $\mathcal{H}_0^0(z) = \frac{1}{2}\langle \Omega \cdot z, z \rangle$. In the expansion

$$H = \sum_{k=0}^{\nu} \frac{1}{k!} H_k^0 + R_\nu \tag{C.12}$$

we have

$$H_0^0(x, y, z) = \omega y + \mathcal{H}_0^0(z)$$

and $H_k^0 \in \mathcal{G}_{k+2}$ with

$$H_k^0(x, y, z) = \beta_k(x)y^{k-1} + \mathcal{H}_k^0(x, y, z)$$

where $\mathcal{H}_k^0$ is a homogeneous polynomial in z of order $k + 2$ with (x, y)-dependent coefficients. The gradation of $\bigoplus \mathcal{G}_{k+2}$ is respected by

$$X_{H_0^0}(F) = \{F, H_0^0\} = \omega \frac{\partial F}{\partial x} + \{F, \mathcal{H}_0^0\}$$

and we are led to the homological equation

$$\{F, H_0^0\} + H_0^k = H_k^0 . \tag{C.13}$$

For the term $\beta_k(x)y^{k-1}$ we write

$$F_k(x, y, z) = \frac{1}{\omega} B_k(x)y^{k-1} + \mathcal{F}_k(x, y, z)$$

where $B_k'(x) = \beta_k(x) - \beta_k^0$ whence $H_0^k = \beta_k^0 y^{k-1} + \mathcal{H}_0^k$ and y becomes merely a parameter that is easily incorporated in the treatment in Section C.1. Expanding the x-dependent coefficients of z-monimials in Fourier series we see that (C.13) has become an infinite-dimensional problem. In the elliptic case where $\mathcal{H}_0^0$ is given by (C.7) the eigenvalues of

$$X_{H_0^0} : \mathcal{G}_{k+2} \longrightarrow \mathcal{G}_{k+2} \tag{C.14}$$

have no real part, and imaginary part

$$\Delta_{hm} = 2\pi\omega h + \sum_{j=1}^{\ell} \alpha_j(m_j - m_{\ell+j})$$

where $e^{2\pi i h x}$ is the pertinent x-dependent factor of the eigen-monomial. The corresponding splitting

$$\operatorname{im} X_{H_0^0} \oplus \ker X_{H_0^0} = \mathcal{G}_{k+2}$$

exists for other semi-simple $\mathcal{H}_0^0$ as well. Note that the subspace $\ker X_{H_0^0} < \mathcal{G}_{k+2}$ has finite dimension. Indeed, for fixed k there are only finitely many combinations $\sum \alpha_j(m_j - m_{\ell+j})$ with $m \in \mathbb{N}_0^{2\ell}$, $m_1 + \ldots + m_{2\ell} = k + 2$ and therefore only finitely many $h \in \mathbb{Z}$ for which Δ_{hm} vanishes; additional hyperbolic eigenvalues of Ω contribute to the real part of the eigenvalues of (C.14). For the same reason we do not encounter small denominators and the Fourier series

$$\mathcal{F}_k(x, y, z) = \sum_{\Delta_{hm} \neq 0} \frac{\mathcal{H}_{k,m}^{0,h}(y)}{i\Delta_{hm}} e^{2\pi i hx} \mathcal{M}_m(z)$$

converges. In the absence

$$\bigwedge_{h \in \mathbb{Z} \backslash \{0\}} \bigwedge_{\substack{m \in \mathbb{N}_0^{2\ell} \\ |m| \leq n+2}} \Delta_{hm} \neq 0$$

of normal-internal resonances we thus simultaneously achieve our goals to make the truncated normal form

$$H^\nu = \sum_{k=0}^{\nu} \frac{\beta_k^0 y^{k-1} + \mathcal{H}_0^k}{k!}$$

x-independent with $\mathcal{H}_0^k$ recovering the results of Section C.1, cf. [12, 16, 208].

This works *mutatis mutandis* in the non-semi-simple case. Indeed, the first step only involves the semi-simple part of the Jordan–Chevalley decomposition and already yields an x-independent normal form H^ν. Further normalization with respect to the nilpotent part only uses transformations generated by some Hamiltonian function $F = \mathcal{F}(y, z)$. Thus, for

$$H_0^0(y, z) = \omega y + \alpha S(z) + a N(z)$$

we find $\psi : \mathbb{T} \times \mathbb{R} \times \mathbb{R}^4 \longrightarrow \mathbb{T} \times \mathbb{R} \times \mathbb{R}^4$ transforming (C.12) into

$$(H \circ \psi)(x, y, z) = H_0^0(y, z) + \beta^0(y) + \mathcal{H}_\nu(y, S(z), M(z)) + R^\nu(x, y, z)$$

if $2\pi k\omega + l\alpha \neq 0$ for all $(k, l) \in \mathbb{Z} \times \{1, \ldots, \nu + 2\}$. See also [232, 228].

C.3 Normal Forms near Invariant Tori

This time our starting point is a Hamiltonian

$$H(x, y, z) = \sum_{k=0}^{\nu} \frac{1}{k!} H_k^0(x, y, z) + R_\nu(x, y, z) \qquad \text{(C.15)}$$

on $\mathbb{T}^n \times \mathbb{R}^n \times \mathbb{R}^{2\ell}$ for which

$$H_0^0(x, y, z) = \langle \omega, y \rangle + \mathcal{H}_0^0(x, z)$$

already is in Floquet form

$$\mathcal{H}_0^0(x, z) = \frac{1}{2} \langle \Omega \cdot z, z \rangle$$

with the same aim of transforming the higher order terms

$$H_k^0(x, y, z) = \beta_k(x, y) + \mathcal{H}_k^0(x, y, z) \in \mathcal{G}_{k+2}$$

into normal form $H_0^k(y, z) = \beta_k^0(y) + \mathcal{H}_0^k(y, z)$. Already at the first step, to transform

$$\beta_k(x, y) = \sum_{\substack{j \in \mathbb{N}_0^n \\ |j| = k-1}} \sum_{h \in \mathbb{Z}^n} \beta_{k,j}^h e^{2\pi i \langle h, x \rangle} y^j$$

into $\beta_k^0(y) = \sum \beta_{k,j}^0 y^j$ by means of

$$F_k(x, y, z) = B_k(x, y) + \mathcal{F}_k(x, y, z)$$

with

$$B_k(x, y) = \sum_{\substack{j \in \mathbb{N}_0^n \\ |j| = k-1}} \sum_{h \neq 0} \frac{\beta_{k,j}^h}{2\pi i \langle h, \omega \rangle} e^{2\pi i \langle h, x \rangle} y^j \qquad \text{(C.16)}$$

we encounter small denominators. While the right hand side of (C.16) is formally defined for non-resonant ω (all coefficients are well-defined, we do not divide by zero), we need converging Fourier series. We therefore require the (internal) frequency vector ω to satisfy the Diophantine conditions

$$\bigwedge_{h \in \mathbb{Z}^n \setminus \{0\}} |\langle h, \omega \rangle| \geq \frac{\gamma}{|h|^\tau}$$

with $\gamma > 0$ and $\tau > n - 1$. Then B_k defined by (C.16) is a homogeneous polynomial in y with a loss of differentiability of the coefficient functions in x of order $\tau + n + 1 > 2n$.

The assumptions made so far can be summarized by stating that we wish to normalize the analytic Hamiltonian (C.15) in the neighbourhood of a Floquet torus with Diophantine quasi-periodic flow. For semi-simple Ω the linear operator

$$X_{H_0^0} : \quad \mathcal{G}_{k+2} \longrightarrow \mathcal{G}_{k+2}$$
$$F \mapsto \sum \omega_i \frac{\partial F}{\partial x_i} + \{F, \mathcal{H}_0^0\} \qquad \text{(C.17)}$$

again yields the splitting

$$\operatorname{im} X_{H_0^0} \oplus \ker X_{H_0^0} = \mathcal{G}_{k+2} .$$

Restricting once more to elliptic Ω we can avoid normal-internal resonances by extending our Diophantine conditions to

$$\bigwedge_{h \in \mathbb{Z}^n \setminus \{0\}} \quad \bigwedge_{\substack{m \in \mathbb{Z}^\ell \\ |m| \leq \nu+2}} |2\pi \langle h, \omega \rangle + \langle m, \alpha \rangle| \geq \frac{\gamma}{|h|^\tau} .$$

Then the conclusions for the periodic case remain valid, cf. [279, 164, 47]; we obtain a truncated normal form

$$H^\nu = \sum_{k=0}^{\nu} \frac{1}{k!} \left(\sum_{|j|=k-1} \beta_{k,j}^0 y^j + \mathcal{H}_0^k \right)$$

that is simultaneously x-independent and has $\mathcal{H}_0^k$ in the normal form of Section C.1, parametrised by y. For normal forms in the explicit presence of a normal-internal resonance see [49].

D

Proof of the Main KAM Theorem

Our aim is to find a co-ordinate transformation Φ that transforms away – to some extent – the small perturbation P of the integrable normal form N and thus allows to recover the bifurcation scenario. The idea is to construct Φ as a limit of transformations $(\Phi_i)_{i \in \mathbb{N}}$ that approximately solve the homological equation $H \circ \Phi = N$. Such a transformation Φ_i combines parameter shifts in λ and ω with a symplectomorphism that are constructed as a solution of a linear version of the homological equation $H_i \circ \Phi_i = N_{i+1}$ and the resulting error terms are collected in a new perturbation P_{i+1} of the new normal form N_{i+1}. Provided that P_{i+1} is "smaller" than P_i we can repeat the procedure with $H_{i+1} = N_{i+1} + P_{i+1}$ and inductively get the sequence $(\Phi_i)_i$ of transformations. For this KAM iteration scheme we follow the quite universal set-up of [216, 159, 241, 56] in the form given in [50].

At each iteration step we perform two operations. First the terms of lower (weighted) order are made x-independent. To achieve this goal we have to deal with small denominators when solving the homological equation. This imposes restrictions on the domains where Φ_i is defined. Indeed, at each iteration step we can use an ultraviolet cut-off η_i for the order of the resonances so that we only have to deal with finitely many of them whence Φ_i is defined on an open neighbourhood $U_{r_i}(\Sigma'_\gamma)$ of our set Σ'_γ of Diophantine frequencies. In the limit we need $\eta_i \xrightarrow{i \to \infty} \infty$ whence these neighbourhoods shrink down to the Cantor set Σ'_γ. In fact the domain D_i of definition of Φ_i has to shrink in each of its components, and we use a carefully balanced interplay between geometrically and exponentially fast converging sequences (D.6) to obtain (via the Inverse Approximation Lemma of [283]) the desired smoothness of the limit transformation.

The second part of an iteration step has to put the x-independent lower order terms that result from the first step back into normal form. In [34, 139] this is incorporated into solving the homological equation, where it amounts to satisfying certain integrability conditions. The two parts are decoupled in [50, 51, 144] where explicit shear transformations are used to get rid of irrelevant deformations and to recover the universal unfolding of the planar singularity $\mathcal{N}_0$ at hand. Here we decouple the two parts even further and consider a versal unfolding of $\mathcal{N}_0$ that contains more than only the necessary deformations. The simplifying transformations that yield the universal unfolding are applied only once, after the KAM-iteration has resulted in a "final" versal unfolding.

Let us first consider the case that the polynomial $\mathcal{N}_0$ is semi-quasi-homogeneous. Denoting by $\alpha_q \leq \alpha_p$ the two weights we have the weighted order $\|kl\| := \alpha_p k + \alpha_q l$ on indices $k, l \in \mathbb{N}_0$ defining a gradation on $\mathbb{C}[q, p]$. In the expansion

$$\mathcal{N}_0(q, p) \;=\; \sum_{\|kl\|=\alpha_x}^{\alpha_y} a_{kl} p^k q^l \tag{D.1a}$$

the terms with lowest (weighted) order α_x form the non-degenerate quasi-homogeneous part of the semi-quasi-homogeneous singularity $\mathcal{N}_0$, while α_y is the highest occurring order. We use the latter to extend[1] the gradation to $\mathbb{C}[y, q, p]$ by means of the weighted order $\|jkl\| := \alpha_y |j| + \alpha_p k + \alpha_q l$ on indices $j = (j_1, \ldots, j_n) \in \mathbb{N}_0^n$, $k, l \in \mathbb{N}_0$ whence

$$N(y, q, p; 0, \omega) \;=\; \langle \omega, y \rangle + \mathcal{N}_0(q, p) \tag{D.1b}$$

is a polynomial for which the linear term in y joins the terms of highest order. In this way we ensure that a polynomial of (weighted) order α_y has y entering only in linear monomials y_i^1. Our normal form during the KAM-iteration is the unfolding

$$N(x, y, q, p; \lambda, \omega) \;=\; N(y, q, p; 0, \omega) + \sum_{\|kl\|=0}^{\alpha_y} \lambda_{kl} p^k q^l \tag{D.1c}$$

and we distribute weights

$$\alpha_{kl} \;=\; \begin{cases} \alpha_x - \|kl\| \\ 1 \end{cases} \text{if} \quad \begin{matrix} \|kl\| < \alpha_x \\ \|kl\| \geq \alpha_x \end{matrix}$$

to extend the gradation once more. Relabelling the λ_{kl} to $\lambda_0 = \lambda_{00}, \lambda_1, \ldots, \lambda_d$ (where we may assume for $\iota \leq \varsigma \Rightarrow \alpha_\iota \geq \alpha_\varsigma$ to hold true) we obtain the weighted order

$$\|jklm\| \;:=\; \alpha_y |j| + \alpha_p k + \alpha_q l + \sum_{\iota=0}^{d} \alpha_{\lambda_\iota} m_\iota$$

induced by the weight $(\alpha_y, \alpha_q, \alpha_p; \alpha_{\lambda_0}, \ldots, \alpha_{\lambda_d})$ on indices

---

[1] Since x varies in $\mathbb{T}^n$ we do not encounter polynomials in x; with α_x we merely denote the lowest order in $\mathbb{N}_0$ and do not intend to denote a weight.

$$j = (j_1, \ldots, j_n) \in \mathbb{N}_0^n \ , \ k, l \in \mathbb{N}_0 \ , \ m = (m_0, \ldots, m_d) \in \mathbb{N}_0^{d+1} \ .$$

On the ring of all (=formal) power series in y, q, p and λ this defines a gradation

$$\mathcal{A}_\nu := \left\{ F \in \mathbb{C}[y, q, p; \lambda] \ \middle| \ \begin{array}{c} F \text{ quasi-homogeneous of order } \nu \\ \text{with weight } (\alpha_y, \alpha_q, \alpha_p; \alpha_{\lambda_0}, \ldots, \alpha_{\lambda_d}) \end{array} \right\}$$

together with the filtration $\mathcal{F}_n := \prod\limits_{\nu \geq n} \mathcal{A}_\nu$. With this terminology we may write

$$F = G \pmod{\mathcal{F}_{\alpha_y}}$$

if all monomials up to (weighted) order α_y in the power series F and G have equal coefficients.

In case the polynomial singularity $\mathcal{N}_0$ is not semi-quasi-homogeneous, we construct the desired weighted order as follows. Since $\mathcal{N}_0$ is finitely determined it contains a monomial of the form p^β or $p^\beta q$ and a monomial of the form pq^γ or q^γ. This can be used to distribute weights as in Table 2.1, and under all such possibilities we arrange for α_q, α_p to be minimal and to satisfy $\alpha_q \leq \alpha_p$. Recall that the quasi-homogeneous part of lowest order $\alpha_{\hat{x}}$ factors $p^2 q^2$ – otherwise $\mathcal{N}_0$ would be semi-quasi-homogeneous – whence we have again $\alpha_x > \alpha_q + \alpha_p$. The extension to the weighted order $\|jklm\|$ proceeds exactly as above.

At the ith iteration step we construct a diffeomorphism $\Psi_i : D_{i+1} \longrightarrow \Psi_i(D_{i+1}) \subseteq D_i$ with $H_i \circ \Psi_i = H_{i+1}$. Composition then yields

$$\Phi_{i+1} := \Psi_0 \circ \Psi_1 \circ \ldots \circ \Psi_i \ : \ D_{i+1} \longrightarrow D_0 \ .$$

We give the domains D_i the product structure

$$D_i = \mathcal{T}_i \times \mathcal{D}_i \times \mathcal{U}_i \subseteq \mathbb{K}^n \times \mathbb{C}^{n+d+3} \times \mathbb{C}^n$$

where $\mathbb{K} = \mathbb{T} + i\mathbb{R}$ is the complex cylinder to the real circle $\mathbb{T} = \mathbb{R}/\mathbb{Z}$. The reason to consider our Hamiltonians $H_i = N_i + P_i$ on (shrinking) complex domains is that this allows to control derivatives by the supremum norm using Cauchy's inequality. We use an exponentially fast converging sequence $\varepsilon_i \to 0$ to define

$$\mathcal{D}_i = \left\{ (y, q, p; \lambda) \in \mathbb{C}^{n+2+d+1} \ \middle| \ |y| \leq \varepsilon_i^{\alpha_y}, |q| \leq \varepsilon_i^{\alpha_q}, |p| \leq \varepsilon_i^{\alpha_p}, |\lambda_\iota| \leq \varepsilon_i^{\alpha_{\lambda_\iota}} \right\}$$

and also put $\delta_i := \varepsilon_i^{\alpha_y + \xi}$ with $0 < \xi < 1$ for a bound

$$|P_i|_{D_i} \leq \delta_i$$

on the perturbation part P_i of H_i. The aim of the ith iteration step is to achieve this same bound with i replaced by $i+1$. In the angular direction we use a geometrically fast converging sequence $s_i \to 0$ for the open neighbourhoods

$$\mathcal{T}_i = U_{\kappa + s_i}(\mathbb{T}^n)$$

of radius $\kappa + s_i$ in view of the Paley–Wiener estimate on exponential decay of the Fourier coefficients of P_i.

The argumentation becomes more transparent if the sizes r_i of the shrinking neighbourhoods of the set of Diophantine frequency vectors are effectively decoupled from the Diophantine constant $\gamma > 0$. We therefore first restrict to $\gamma = 1$ and afterwards re-install γ by means of a scaling argument, see Section D.3. As the latter requires unbounded frequency domains we have to formulate our result with the supremum norm of the perturbation explicitly taken on the complex domain D_0 (with $\mathcal{U}_0 = U_{r_0}(\Sigma)$ in the third factor).

Theorem D.1. *Let $H = N + P$ be a holomorphic perturbation of the Hamiltonian function (D.1) defined on $D = D_0$. Assume that the ω-dependent coefficient functions $a_{kl} \in \mathcal{O}(\mathcal{U})$ of N_0 satisfy upper bounds*

$$|a_{kl}|_{\mathcal{U}}, \; |Da_{kl}|_{\mathcal{U}} \;\leq\; \Gamma$$

for some constant $\Gamma > 0$ and certain (linear) lower bounds

$$\left| \frac{1}{\sum \nu_{kl} a_{kl}} \right|_{\mathcal{U}} \;<\; \Gamma \tag{D.2}$$

preventing N_0 to leave its stratum. Then there exists $\delta > 0$ such that if

$$|P|_D \;<\; \delta$$

then there exists a Gevrey regular diffeomorphism Φ on $U_\kappa(\mathbb{T}^n) \times \mathbb{C}^n \times \mathbb{C}^2 \times \mathbb{C}^{d+1} \times \Sigma$ such that

1). Φ is holomorphic for fixed ω.
2). Φ is symplectic for fixed (λ, ω).
3). Φ is C^∞-close to the identity.
4). On $U_\kappa(\mathbb{T}^n) \times \mathbb{C}^{n+d+3} \times \Sigma_1' \cap \Phi^{-1}(D)$ one can split $H \circ \Phi = N_\infty + P_\infty$ into an integrable part N_∞ and higher order terms P_∞. Here N_∞ has the same form (D.1) as N. The x-dependence is pushed into the higher order terms, i.e.

$$\frac{\partial^{|j|+k+l+|m|} P_\infty}{\partial y^j \partial p^k \partial q^l \partial \lambda^m}(x, 0, 0, 0; 0, \omega) \;=\; 0$$

for all $(x, \omega) \in \mathbb{T}^n \times \Sigma_1'$ and all j, k, l, m satisfying $\|jklm\| \leq \alpha_y$.

The holomorphicity in x of the limit of the Φ_i follows from

$$\bigcap D_i \;=\; U_\kappa(\mathbb{T}^n) \times \{0\} \times \Sigma_1' \; .$$

We use polynomial truncations in y, q, p, λ to define the limit Φ_∞, with coefficient functions defined on $\mathcal{T}_i \times \mathcal{U}_i$. To obtain Gevrey regularity of the limit of these coefficient functions on $U_\kappa(\mathbb{T}^n) \times \Sigma_1'$ we let $r_i \to 0$ geometrically fast and define

$$\mathcal{U}_i = U_{r_i}(\Sigma_1')$$

as an open neighbourhood of radius r_i.

An additional complication is that a mere polynomial truncation of the Φ_i would cease to preserve the symplectic structure. For this reason we introduce generating functions S_i of the Φ_i, the polynomial truncations $\widetilde{S}_i$ of which generate symplectomorphisms as well. The limit $\widetilde{S}_\infty$ of these then generates the desired Φ_∞. As $\widetilde{S}_\infty$ is polynomial in y, q, p, λ, holomorphic in x and Gevrey regular in ω we conclude that Φ_∞ is Gevrey regular and even holomorphic in x, y, q, p, λ – as stated in Theorem D.1.

D.1 The Iteration Step

During a single step of the KAM iteration we construct a co-ordinate transformation Ψ_i that turns the given Hamiltonian $H_i = N_i + P_i$ into a "new" Hamiltonian $H_{i+1} = N_{i+1} + P_{i+1}$ in such a way that $|P_{i+1}|_{D_{i+1}}$ is much smaller than $|P_i|_{D_i}$. In this process we concentrate on the terms of P_i of lowest (weighted) order while the terms with (weighted) order greater than α_y are directly relegated to P_{i+1}. To define a suitable truncation we write P_i as a Taylor series

$$P_i(x, y, q, p; \lambda, \omega) = \sum_{|j|+k+l+|m|=0}^{\infty} P_{i,jklm}(x, \omega)\, y^j p^k q^l \lambda^m$$

and expand the coefficient functions as Fourier series

$$P_{i,jklm}(x, \omega) = \sum_{h \in \mathbb{Z}^n} P_{i,jklm}^h(\omega)\, \mathrm{e}^{2\pi \mathrm{i}\langle h, x\rangle} \; .$$

For the truncation R_i of P_i we use the (weighted) order a_y and an ultraviolet cut-off η_i that grows geometrically (see (D.6)) to write

$$R_i = \sum_{\|jklm\|=0}^{\alpha_y} \sum_{|h| \leq \eta_i} P_{i,jklm}^h(\omega)\, \mathrm{e}^{2\pi \mathrm{i}\langle h, x\rangle}\, y^j p^k q^l \lambda^m \; . \tag{D.3}$$

Our aim in the KAM iteration step is to turn the truncated perturbed Hamiltonian $N_i + R_i$ into the new integrable part N_{i+1}. To this end the iteration step is split into two parts.

1. We find a function F_i on a suitable domain $D \subseteq D_i$ that solves the homological equation

$$N_i + R_i + \{N_i, F_i\} = \bar{N}_i := N_i + \bar{R}_i \tag{D.4}$$

up to (weighted) order α_y. Here $\bar{R}_i$ denotes the average

$$\bar{R}_i \;=\; \sum_{\|jklm\|=0}^{\alpha_y} P^0_{i,jklm}(\omega)\, y^j p^k q^l \lambda^m$$

of (D.3), whence the "intermediate" $\bar{N}_i$ is already independent of x. The time-1-flow φ^i_1 of the Hamiltonian vector field X_{F_i} then provides a co-ordinate transformation for which

$$
\begin{aligned}
H_i \circ \varphi^i_1 \;&=\; (N_i + R_i) \circ \varphi^i_1 \;+\; (P_i - R_i) \circ \varphi^i_1 \\
&=\; N_i + R_i + \{N_i, F_i\} + \{R_i, F_i\} \\
&\quad + \int_0^1 (1-t)\{\{N_i + R_i, F_i\}, F_i\} \circ \varphi^i_t \, dt \;+\; (P_i - R_i) \circ \varphi^i_1 \\
&=\; \bar{N}_i + \bar{P}_i \;.
\end{aligned}
\tag{D.5}
$$

Here $\bar{P}_i$ contains the terms $\{R_i, F_i\}$, $(P_i - R_i) \circ \varphi^i_1$, the integral and the higher order terms of $\{N_i, F_i\}$ that are relegated to P_{i+1}.

2. Writing the x-independent "intermediate" as

$$\bar{N}_i(x, y, q, p; \lambda, \omega) \;=\; \langle \omega + P^0_{i,1000}(\omega), y \rangle \;+\; \bar{\mathcal{N}}^i_\lambda(q, p; \omega)$$

the second part aims to regain the form (D.1) for the new integrable part N_{i+1}. This can be achieved by means of parameter shifts in λ and ω, thereby turning the family $\bar{\mathcal{N}}^i_\lambda$ of planar singularities into standard form $\mathcal{N}^{i+1}_\lambda$.

We now define the geometrically fast and exponentially fast converging sequences that entered the domains D_i, the bounds of $|P_i|$ and the ultraviolet cut-off. Let $s_0 < \min\{\kappa, \frac{1}{2}\}$ and $\delta_0 < \frac{1}{2}$ be given, then we put

$$
\begin{aligned}
s_i \;&=\; \frac{s_0}{4^i} \\
r_i \;&=\; s_i^{2\tau+2} \\
\eta_i \;&=\; [s_i^{-2}] \\
\delta_i \;&=\; \delta_{i-1}^{\chi/(\alpha_y+\xi)} \delta_{i-1} \;=\; \delta_0^{(1+\chi/(\alpha_y+\xi))^i} \\
\varepsilon_i \;&=\; \delta_i^{1/(\alpha_y+\xi)}
\end{aligned}
\tag{D.6}
$$

with $0 < \chi < \xi < 1$. The constants in the estimates occurring during the iteration step will be absorbed in s_0 and δ_0, leading to inequalities of the form

$$s_0 \leq c\,, \qquad \delta_0 \leq c\,, \qquad \delta_0^\zeta \leq c\, s_0$$

with constants $c > 0$ and exponents $\zeta > 0$. The only exception is an inequality

$$s_0 \;<\; \frac{1}{c - \zeta \ln(\delta_0)}$$

that serves to estimate the higher order Fourier coefficients of P_i. Since $\delta_0^\zeta \ln(\delta_0) \overset{\delta_0 \to 0}{\longrightarrow} 0$ for all $\zeta > 0$ it is possible to find small s_0, δ_0 satisfying all these inequalities – δ_0 is the small positive constant δ sought for in Theorem D.1. Since $(s_i)_i$ decreases geometrically fast and $(\delta_i)_i$ decreases exponentially fast we also include factors $1/s_i$ in the "generic constants", taking care that at each iteration step the total number of such factors remains finite and independent of i.

We are now able to formulate the iteration step in our proof of Theorem D.1. Here and in the following lemmata we refrain from tediously repeating the assumptions we made so far. The following result is the cornerstone of the KAM-procedure, so we give the proof immediately. The proof of the necessary lemmata is deferred to Appendix E.

Proposition D.2. There are $s_0 > 0$ and $\delta_0 > 0$ such that the following holds true. Given $H_i = N_i + P_i$ as above, there is a co-ordinate transformation

$$\Psi_i: \qquad D_{i+1} \qquad \longrightarrow \qquad D_i$$
$$(x, y, q, p; \lambda^+, \omega^+) \quad \mapsto \quad (\psi_{\lambda,\omega}(x, y, q, p); \lambda, \omega)$$

with symplectic $\psi_{\lambda,\omega}$ such that $H_{i+1} := H_i \circ \Psi_i$, defined on D_{i+1}, has the form

$$H_{i+1} = N_{i+1} + P_{i+1}$$

with N_{i+1} in standard form (D.1) and perturbation P_{i+1} satisfying

$$|P_{i+1}|_{D_{i+1}} \leq \delta_{i+1}$$
$$|a_{kl}^{i+1} - a_{kl}^i|_{U_{i+1}} \leq \varepsilon_i .$$

Moreover,

$$\left| \frac{\partial^{|j|+k+l+|m|} P_{i+1}}{\partial y^j \partial p^k \partial q^l \partial \lambda^m} \right|_{D_{i+1}} \leq \varepsilon_{i+1}^{\alpha_y + \xi - \|jklm\|}$$

for all indices satisfying $\|jklm\| \leq \alpha_y$.

The coefficient functions a_{kl}^i are bounded away from special values for which the planar singularity $\mathcal{N}_0^i$ changes its (topological) type. The estimates by ε_i on the differences $|a_{kl}^{i+1} - a_{kl}^i|$ imply that the same is true for the a_{kl}^{i+1} as well, and also for the (existing) limit functions a_{kl}^∞.

D.1.1 The Homological Equation

For the detailed single iteration step proving Proposition D.2 we drop the index i and use the so-called "+"-notation, replacing occurrences of the index $i+1$ by an index $+$. This emphasizes that the constants in our estimates have to be independent of the iteration step i. The generic letter "c" is used where we do not need to remember the value of such a constant, and we also use the shorthand $A \lesssim B$ for $A \leq c \cdot B$.

Most of our estimates rely on Cauchy's inequality and require that we shrink the domain a bit. Therefore we define nested domains

$$D^\iota = U_{\kappa+s-\iota s/\alpha}(\mathbb{T}^n) \times \mathcal{D}^\iota \times U_{r-\iota r/\alpha}(\Sigma_1') \;, \qquad \iota = 1, \ldots, \alpha - 1$$

with

$$\alpha := \left[\frac{\alpha_y}{\alpha_x - \alpha_q - \alpha_p} \right] + 6 \tag{D.7}$$

and

$$\mathcal{D}^\iota = \Big\{ (y, q, p; \lambda) \ \Big| \ |y| \le (2^{\alpha-\iota}\varepsilon_+)^{\alpha_y}, \ |q| \le (2^{\alpha-\iota}\varepsilon_+)^{\alpha_q}, $$
$$|p| \le (2^{\alpha-\iota}\varepsilon_+)^{\alpha_p}, \ |\lambda_j| \le (2^{\alpha-\iota}\varepsilon_+)^{\alpha_{\lambda_j}} \Big\}$$

that satisfy $D_+ \subset D^{\alpha-1} \subset \ldots \subset D^1 \subset D$.

Lemma D.3. (Lemma E.1) *For the truncation (D.3) we have*

$$|R|_{D^1} \ \dot\le \ \delta$$
$$|P - R|_{D^1} \ \dot\le \ \beta^{1-\xi}\delta_+$$

where $\beta = \varepsilon^{\chi/(\alpha_y+\xi)}$.   □

To solve the homological equation (D.4) up to terms of higher (weighted) order we let

$$\overset{\cdot}{F} = \sum_{\nu=0}^{\alpha_y} \sum_{0 < |h| \le \eta} F_\nu^h e^{2\pi \mathrm{i}\langle h, x\rangle}$$

with

$$F_\nu^h = \sum_{\|jklm\|=\nu} F_{jklm}^h(\omega)\, y^j p^k q^l \lambda^m \tag{D.8}$$

be the solution of

$$N + R + \{N, F\} = \bar{N} \qquad (\mathrm{mod}\,\mathcal{F}_{\alpha_y}) \;,$$

i.e. up to weighted order α_y. To understand how the Poisson bracket

$$\{N, F\} = -\frac{\partial N}{\partial y}\frac{\partial F}{\partial x} + \frac{\partial N}{\partial q}\frac{\partial F}{\partial p} - \frac{\partial N}{\partial p}\frac{\partial F}{\partial q}$$
$$= -2\pi \mathrm{i}\langle h, \omega\rangle F + \{\mathcal{N}, F\}$$

can be used to "remove terms from R" we rewrite the y-independent part $\mathcal{N}_\lambda$ of (D.1) as

$$\mathcal{N}(q, p; \lambda, \omega) = \sum_{\mu=\alpha_x}^{\alpha_y+1} \mathcal{N}_\mu(q, p; \lambda, \omega)$$

with

$$\mathcal{N}_{\alpha_x}(q,p;\lambda,\omega) \;=\; \sum_{\|kl\|=\alpha_x} a_{kl}p^k q^l \;+\; \sum_{\|kl\|<\alpha_x} \lambda_{kl}p^k q^l$$

and, for $\mu > \alpha_x$,

$$\mathcal{N}_{\mu}(q,p;\lambda,\omega) \;=\; \sum_{\|kl\|=\mu} a_{kl}p^k q^l \;+\; \sum_{\|kl\|=\mu-1} \lambda_{kl}p^k q^l$$

collecting the terms of weighted order μ. Note that $\mathcal{N}_\mu$ yields a raise of the weighted order ν of F_ν^h by $\mu - \alpha_q - \alpha_p$ to the weighted order $\nu + \mu - \alpha_q - \alpha_p$ of

$$\{\mathcal{N}_\mu', F_\nu^h\} \;=\; \left(\frac{\partial \mathcal{N}_\mu}{\partial q}\frac{\partial}{\partial p} - \frac{\partial \mathcal{N}_\mu}{\partial p}\frac{\partial}{\partial q} \right)(F_\nu^h) \;.$$

Hence, the F_ν^h can be recursively defined by

$$2\pi\mathrm{i}\langle h,\omega\rangle F_\nu^h \;=\; P_\nu^h + \sum_{\mu=\alpha_x}^{\alpha_y+1} \{\mathcal{N}_\mu, F_{\nu-\mu+\alpha_q+\alpha_p}^h\}$$

which leads to

$$F_\nu^h \;=\; \sum_{\iota=0}^{\alpha-6} \left(\frac{1}{2\pi\mathrm{i}\langle h,\omega\rangle} \right)^{\iota+1} \sum_{\mu_1,\dots,\mu_\iota=\alpha_x}^{\alpha_y+1} \{\mathcal{N}_{\mu_1},\dots,\{\mathcal{N}_{\mu_\iota}, P_{\nu(\iota)}^h\}\cdots\} \qquad \text{(D.9)}$$

with the lower index of $P_{\nu(\iota)}^h$ abbreviating

$$\nu(\iota) \;=\; \nu - \sum_{i=1}^{\iota}(\mu_i - \alpha_q - \alpha_p) \qquad\qquad \text{(D.10)}$$

and thus depending on all indices of the second (multi)-summation as well; we put $P_\nu^h = 0$ if ν is negative. It is here that we encounter the small denominators. The important point is that the sum (D.9) is finite. Indeed, we have

$$\nu - \sum_{\iota=1}^{\alpha-5}(\mu_i - \alpha_q - \alpha_p) \;\leq\; \alpha_y - (\alpha-5)(\alpha_x - \alpha_q - \alpha_p) \;\leq\; \alpha_q + \alpha_p - \alpha_x \;<\; 0$$

by the definition (D.7) of α. Since (D.4) is only solved up to (weighted) order α_y, the higher order terms

$$\hat{P} \;=\; \sum_{\mu=\alpha_x}^{\alpha_y+1} \sum_{\nu=\alpha_q+\alpha_p+\alpha_y+1-\mu}^{\alpha_y} \sum_{0<|h|\leq\eta} \{\mathcal{N}_\mu, F_\nu^h e^{2\pi\mathrm{i}\langle h,x\rangle}\} \qquad \text{(D.11)}$$

are relegated to the new perturbation.

Lemma D.4. (Lemma E.2) For the function F we have

$$|F|_{D^{\alpha-5}} \;\leq.\; \delta$$

and

$$\left| \frac{\partial^{|j|+k+l+|m|} F}{\partial y^j \partial p^k \partial q^l \partial \lambda^m} \right|_{D^{\alpha-4}} \;\leq.\; \varepsilon^{-\nu} \delta$$

if $\nu = \|jklm\| \leq \alpha_y$. □

Denote by

$$\|X_F\|_D \;:=\; \max\left\{ |\frac{\partial F}{\partial y}|_D \,,\, \varepsilon^{\alpha_y} |\frac{\partial F}{\partial x}|_D \,,\, \varepsilon^{\alpha_p} |\frac{\partial F}{\partial p}|_D \,,\, \varepsilon^{\alpha_q} |\frac{\partial F}{\partial q}|_D \right\}$$

a weighted norm of the vector field, and for its derivatives

$$\Uparrow D_\nu X_F \Uparrow_D \;:=\; \max_{|j|+k+l\leq\nu}\left\{ |\frac{\partial^{|j|+k+l} G}{\partial y^j \partial p^k \partial q^l}|_D \right\} \quad \text{for} \quad \nu \geq 1$$

where G stands for either of $\partial F/\partial y, \partial F/\partial x, \partial F/\partial p, \partial F/\partial q$.

Lemma D.5. (Lemma E.3) The Hamiltonian vector field X_F satisfies

$$\|X_F\|_{D^{\alpha-4}} \leq. \varepsilon^\xi \,, \qquad \Uparrow D_\nu X_F \Uparrow_{D^{\alpha-4}} \leq. \varepsilon^\xi \quad \forall_{\nu\geq 1} \,.$$ □

Hence, the flow φ_t of X_F satisfies $\|\varphi_t - \mathrm{id}\|_{D^{\alpha-4}} \leq c\,|t|\,\varepsilon^\xi$ as well whence

$$\psi_t : (x,y,q,p;\lambda,\omega) \mapsto (\varphi_t(x,y,q,p);\lambda,\omega)$$

maps $D^{\alpha-2}$ into $D^{\alpha-3}$ for $-1 \leq t \leq 1$. For the time-1-map ψ_1 we define the norm

$$\|\psi_1\|_{C^{jkl}(D)} \;:=\; \max_{0\leq t\leq 1}\left\| \frac{\partial^{|j|+k+l}\psi_t}{\partial y^j \partial p^k \partial q^l} \right\|_D \,.$$

Lemma D.6. (Lemma E.4) The co-ordinate transformation ψ_1 satisfies

$$\|\psi_1 - \mathrm{id}\|_{C^{jkl}(D^{\alpha-2})} \;\leq.\; \varepsilon^\xi \,.$$ □

Remark D.7. In case P is invariant under a discrete symmetry group G, this imposes certain equalities on the coefficients P^h_{jklm} which are in turn inherited by F. Consequently, the resulting time-1-flow φ_1 is G-equivariant.

D.1.2 The Parameter Shift

The first part of the iteration step consisted in solving the small divisor problem (D.4) to construct a symplectomorphism ψ_1 that transforms away the x-dependence of the lower order terms entering in $\bar{N}$. The passage from $\bar{N}$ to N_+ happens essentially in one degree of freedom as

$$\omega_\iota^+ = \omega_\iota + P_{j000}^0 \quad \text{with} \quad j = j(\iota) = (0,\ldots,0,\overset{\iota}{1},0,\ldots,0)$$

turns the y-dependent terms

$$\langle \omega, y \rangle \quad \text{and} \quad \sum_{|j|=1} P_{j000}^0 y^j$$

of $\bar{N} = N + \bar{R}$ into the y-dependent term $\langle \omega^+, y \rangle$ of N_+. Similarly, we use the passage from λ to λ^+ to account for the y-independent terms in $\bar{R}$. For instance, the constant terms are assembled in

$$\lambda_0^+ = \lambda_0 + \sum_{\|000m\|=0}^{\alpha_y} P_{000m}^0 \lambda^m .$$

Furthermore, the coefficients $a_{kl}(\omega^+ - P_{1000}^0)$ in $\mathcal{N}$ are relabelled $a_{kl}^+(\omega^+)$. This yields the new standard form $\mathcal{N}_+$ and the parameter transformation $\lambda^+ \mapsto \lambda$ (with a non-singular Jacobian).

Remark D.8. Also the second part of the iteration step respects possible symmetries, cf. Remark D.7. Indeed, while equivariance with respect to a discrete symmetry group excludes certain terms in $\mathcal{N}$, it is the same symmetry that forces those coefficients P_{0klm}^0 to vanish that no longer can be accounted for.

D.1.3 Estimates of the Iteration Step

We now compose our map $\Psi : D^{\alpha-1} \longrightarrow D^1 \subseteq D$ (which we later restrict to D_+) from $\psi_1 : D^{\alpha-2} \longrightarrow D^1$ and the mapping

$$\phi : \quad \begin{array}{ccc} D^{\alpha-1} & \longrightarrow & D^{\alpha-2} \\ (x,y,q,p;\lambda^+,\omega^+) & \mapsto & (x,y,q,p;\lambda,\omega) \end{array} .$$

The inequalities

$$|\lambda_\iota - \lambda_\iota^+| \leq \frac{\delta}{\varepsilon^{\alpha_y - \alpha_{\lambda_\iota}}} < (1 - 2\beta)^{\alpha_{\lambda_\iota}} \varepsilon^{\alpha_{\lambda_\iota}}$$

$$|P_{1000}^0| \leq \frac{\delta}{\varepsilon^{\alpha_y}} < \frac{r}{12}$$

imply that indeed the λ- and ω-components are mapped from $D^{\alpha-1}$ to $D^{\alpha-2}$. Similar to Lemma D.6 we can estimate Ψ:

Lemma D.9. (Lemma E.5) The co-ordinate transformation Ψ satisfies

$$\|\Psi - \mathrm{id}\|_{C^{jkl}(D^{\alpha-1})} \leq \varepsilon^\xi . \qquad \square$$

The new perturbation P_+ is given by $(\bar{P} + \hat{P}) \circ \phi$ with $\bar{P}$ defined in (D.5) and $\hat{P}$ defined in (D.11).

Lemma D.10. (Lemma E.6) The perturbation P_+ can be estimated by

$$|P_+|_{\mathcal{D}_+} \;\leq\; |P_+|_{\mathcal{D}^{\alpha-1}} \;\leq\; \beta^{(1-\xi)(\xi-\chi)}\, \varepsilon^{\chi}\, \delta \;<\; \delta_+. \qquad\qquad \square$$

As the domain $\mathcal{D}_+$ is again smaller than $\mathcal{D}^{\alpha-1}$, we have for $\nu = \|jklm\| \leq \alpha_y$

$$\left| \frac{\partial^{|j|+k+l+|m|} P_+}{\partial y^j \partial p^k \partial q^l \partial \lambda^m} \right|_{\mathcal{D}_+} \;\leq\; \varepsilon_+^{-\nu} \delta_+$$

by Cauchy's inequality. This concludes the proof of Proposition D.2.

D.2 Iteration and Convergence

Composing $\Phi_{i+1} := \Psi_0 \circ \Psi_1 \circ \ldots \circ \Psi_i$ the co-ordinate transformations given by Proposition D.2, we obtain a change of variables that transforms H_0 into $H_{i+1} = N_{i+1} + P_{i+1}$. Our aim is to find a "limit" Φ_∞ and this necessitates a bound on the C^ν-norm

$$\|\Phi_i\|_{C^\nu(D_i)} \;=\; \max_{|j|+k+l \leq \nu} \left\| \frac{\partial^{|j|+k+l} \Phi_i}{\partial y^j \partial p^k \partial q^l} \right\|_{D_i}.$$

Lemma D.11. (Lemma E.7) The co-ordinate transformation Φ_i satisfies

$$\bigwedge_{i \in \mathbb{N}} \|\Phi_i\|_{C^\nu(D_i^{\alpha-1})} \;\leq\; c$$

with a suitable constant $c = c(\nu)$. $\qquad\qquad \square$

To construct a limit of the Φ_i we have to pass to a polynomial truncation of fixed degree to obtain Φ_∞ by means of coefficient functions defined on

$$\bigcap_i \mathcal{T}_i \times \mathcal{U}_i \;=\; \overline{U_\kappa(\mathbb{T}^n)} \times \Sigma_1'.$$

Since we want to preserve the symplectic structure we follow [139, 50] and pass to a generating function S_i of Φ_i, which we truncate to $\widetilde{S}_i$ and the limit $\widetilde{S}_\infty$ then generates the desired limit transformation Φ_∞.

Lemma D.12. (Lemma E.8) The function $\Phi_i : (x, y, q, p; \lambda, \omega) \mapsto (X, Y, Q, P)$ admits a generating function $S_i = S_i(x, Y, q, P; \lambda, \omega)$. $\qquad\qquad \square$

We define the truncation $\widetilde{S}_i$ of S_i to be of order $\alpha_y + 1$ in $(Y, q, P; \lambda)$. This ensures that the estimates implied by Proposition D.2 remain valid after the transformed Hamiltonian $H_0 \circ \Phi_i$ is replaced by the Hamiltonian transformed by the symplectomorphism generated by $\widetilde{S}_i$. Furthermore we drop all terms that involve more than one derivative with respect to parameters λ_ι. On the other hand we do not truncate in x or ω. To be precise, we write

$$\Phi_i(x,y,q,p;\lambda,\omega) \;=\; ((x,y,q,p)+W_i(x,y,q,p;\lambda,\omega)\,;\lambda+\tilde{\Lambda}_i(\lambda,\omega),\omega+\tilde{\Omega}_i(\lambda,\omega))$$

and let $\mathcal{F}_i : \mathcal{D}_i \longrightarrow \mathcal{D}_0$ denote the transformation of $(x,y,q,p;\lambda,\omega)$ into

$$(x\,,\,y+W_i^2(x,y,q,p;\lambda,\omega)\,,q\,,\,p+W_i^3(x,y,q,p;\lambda,\omega)\,;\,\lambda,\,\omega) \;\overset{!}{=}\; (x,Y,q,P;\lambda,\omega)$$

and $\mathcal{G}_i := \mathcal{F}_i^{-1}$. The truncations $\widetilde{S}_i$ are polynomials in Y, q, P and λ, the coefficients of which are holomorphic functions in x and ω. To truncate we write S_i as a Taylor series at $\mathcal{F}_i(x,0,0,0;0,\omega) =: (x,Y_i,0,P_i;0,\omega)$. Therefore,

$$S_i^{jklm}(x,\omega) \;=\; \frac{\partial^{|j|+k+l+|m|}S_i}{\partial Y^j\partial P^k\partial q^l\partial\lambda^m}(x,Y_i,0,P_i;0,\omega)\,,$$

and we define

$$\widetilde{S}_i(x,y,q,p;\lambda,\omega) \;:=\; \sum_{|j|+k+l=0}^{\alpha_y+1}\sum_{|m|=0}^{\min(|j|+k+l,1)} S_i^{lijh}(x,\omega)\cdot(Y-Y_i)^j(P-P_i)^k q^l \lambda^m\,.$$

Lemma D.13. (Lemma E.9) *The finitely many sequences $(\widetilde{S}_i^{jklm})_{i\in\mathbb{N}}$ of coefficient functions are uniformly convergent on $\overline{U_\kappa(\mathbb{T}^n)}\times\Sigma_1'$ with differences $\widetilde{S}_{i+1}-\widetilde{S}_i$ converging exponentially fast to zero.*   □

We finally have computed all the necessary estimates. We conclude the proof of Theorem D.1 using the Inverse Approximation Lemma. The improvement in [283] of this classical result allows to achieve Gevrey regularity instead of C^∞-smoothness, both in the sense of Whitney.

Lemma D.14. (Inverse Approximation Lemma) *On the geometrically shrinking neighbourhoods $\mathcal{U}_i$ of Σ_1' let $(F_i)_{i\in\mathbb{N}_0}$ be a sequence of analytic functions with $F_0 = 0$. Assume there are constants $M > 0$ and $\sigma > 1$ such that*

$$\bigwedge_{i\in\mathbb{N}} \;|F_i-F_{i-1}|_{\mathcal{U}_i} \;\leq\; M\,\delta_0^{\sigma^i}\,.$$

Then there exists a unique function $F : \Sigma_1' \longrightarrow \mathbb{R}$, which is Gevrey regular on Σ_1' in the sense of Whitney, and a constant C such that

$$\bigwedge_{\beta\in\mathbb{N}_0^n} \;\left|\frac{1}{\beta!}D^\beta F\right|_{\Sigma_1'} \;\leq\; C^{|\beta|+1}(\beta!)^{(\ln 4)/\sigma}$$

and

$$\left|D^\beta F - D^\beta F_i\right|_{\Sigma_1'} \;\overset{i\to\infty}{\longrightarrow}\; 0\,.\qquad\qquad\qquad □$$

For a proof see [283]. Since $\Sigma_1' \subseteq \mathbb{R}^n$ is a closed subset, Whitney's Extension Theorem yields a C^∞-smooth function on $\mathbb{R}^n$ that coincides on Σ_1' with F. As shown in [29] we may choose this extension to be Gevrey regular as well.

Applying the above to the sequences $(\lambda_\iota^i)_i$ and $(\omega_\iota^i)_i$ yields the limit functions $\tilde{\Lambda}_\infty$ and $\tilde{\Omega}_\infty$ within the co-ordinate transformation Φ_∞. For the symplectic part $\mathrm{id} + W_i$ we use the Gevrey regular coefficients $\tilde{S}_\infty^{jklm}$ of the generating function

$$\tilde{S}_\infty \; : \; U_\kappa(\mathbb{T}^n) \times \mathbb{C}^n \times \mathbb{C}^2 \times \mathbb{C}^{d+1} \times \Sigma_1' \; \longrightarrow \; \mathbb{C}$$

of Φ_∞ which is polynomial in $Y, q, P; \lambda$ and holomorphic in x and can be extended Gevrey-regularly to all of Σ. The definition $P_\infty := H_0 \circ \Phi_\infty - N_\infty$ concludes the proof of Theorem D.1.

D.3 Scaling Properties

To prove of the main KAM Theorem we have to re-install the Diophantine constant γ by means of a suitable scaling. The latter is defined by means of the co-ordinate transformation

$$(x, y, q, p; \lambda, \omega) \; \mapsto \; (X, Y, Q, P; \Lambda, \Omega)$$

with

$$
\begin{aligned}
X &= x \\
\Omega &= \gamma^{-1}\omega \\
Y &= \gamma^{-\alpha_y+1}y \\
Q &= \gamma^{-\alpha_q}q \\
P &= \gamma^{-\alpha_p}p \\
\Lambda_\iota &= \gamma^{-\alpha_{\lambda_\iota}}\lambda_\iota \; .
\end{aligned}
$$

In the new co-ordinates the standard form (D.1) becomes

$$\gamma^{\alpha_y}\hat{N}(X, Y, Q, P; \Lambda, \Omega) \; = \; \langle \Omega, Y \rangle + \gamma^{\alpha_y}\hat{N}_\Lambda(Q, P; \Omega)$$

where the coefficient functions $\hat{a}_{kl}$ of $\hat{N}$ are related to those of N through

$$\hat{a}_{kl}(\Omega) \; = \; a_{kl}(\gamma\Omega) \; .$$

D.4 Parameter Reduction

The final step in our proof of the main KAM Theorem consists in reducing the $d+1$ parameters in the versal unfolding (D.1c). First of all, we may simply omit the constant term $\lambda_0 = \lambda_{00}$. The monomials $p^k q^l$ that do not belong to the basis of the local algebra $\mathcal{Q}_{\mathcal{N}_0}$ could be removed by means of a (planar) right equivalence. However, such a diffeomorphism does not necessarily preserve the symplectic structure.

For the lower order monomials we can work with explicit polynomial co-ordinate changes ϕ that do preserve the symplectic structure. In the semi-quasi-homogeneous case these are the monomials $p^k q^l$ with $\|kl\| < \alpha_x$ and a more general characterisation is that these do not figure in the central singularity $\mathcal{N}_0$ defined by (D.1a).

It is more transparent to write ϕ as a composition $\phi = \phi_1 \circ \phi_2$ of transformations that change only one co-ordinate. Here ϕ_1 is a shear transformation

$$\phi_1 : \begin{cases} q_1 = q \\ p_1 = p + Q(q, \lambda) \end{cases}$$

where Q is a polynomial in q of degree $[\alpha_p/\alpha_q]$. For $\alpha_q < \alpha_p$ this is combined with a translation

$$\phi_2 : \begin{cases} q_2 = q_1 + T(\lambda) \\ p_2 = p_1 \end{cases}$$

which allows in total for $[\alpha_p/\alpha_q] + 1$ parameters to be transformed away. Note that in the semi-quasi-homogeneous case this is indeed the number of monomials $p^k q^l$ of weighted order $\leq \alpha_x$ that are neither included in the basis of $\mathcal{Q}_{\mathcal{N}_0}$ nor in the Jacobi ideal of $\mathcal{N}_0$.

In case $\alpha_q = \alpha_p$ both ϕ_1 and ϕ_2 are affine transvections, i.e. Q is of degree one in q and $T(\lambda)$ is replaced by a first order polynomial $T(p, \lambda)$ in p as well. This allows for the "extra" 4 parameters to be transformed away. Here the extra conditions (D.2) are used as they make sure that the implicit conditions on the linear terms in Q and T can be made explicit. This also applies to a leading term of Q of degree α_p/α_q when $\alpha_q < \alpha_p$.

For simple singularities we have $\alpha_x = \alpha_y$ and there are only two parameters $\lambda_{kl} = \lambda_j, \lambda_d$ left. This yields the two coefficients $a = a_j + \lambda_j$ and $b = a_d + \lambda_d$ in A_μ, D_μ, E_μ. Further reduction in the presence of moduli is detailed in Chapter 4.

D.5 Possible Generalizations

We chose to present our results in the real analytic category, though these are easily generalized to the context of infinite and finite differentiability. The corresponding proofs rely on a kind of diagonal procedure, cf. [298, 299, 240, 73] and references therein. The key idea is to use the result in the real analytic category, approximating the system on every step of the KAM-iteration by a real analytic system.

We also restricted to "maximal" lower dimensional tori. It is straightforward to replace the integrable Hamiltonian (D.1) by

$$\langle \omega, y \rangle + \langle Aw, w \rangle + \mathcal{N}_\lambda(q, p; \omega)$$

with an invertible infinitesimally symplectic $(m \times m)$-matrix A. The hyperbolic part of A does not pose additional problems and can e.g. be dealt with by means of a centre manifold. Therefore we concentrate on the normally elliptic case

$$\langle Aw, w \rangle = \sum_{j=1}^{m} \alpha_j (u_j^2 + v_j^2)$$

where $w = (u_1, \ldots, u_m, v_1, \ldots, v_m)$. The normal frequencies α_j are further parameters the coefficient functions a_{kl} may depend upon. The α_j also become involved in further Diophantine conditions

$$\bigwedge_{k \in \mathbb{Z}^n \setminus \{0\}} \bigwedge_{\substack{\ell \in \mathbb{Z}^m \\ |\ell| \leq 2}} |2\pi\langle k, \omega \rangle + \langle \ell, \alpha \rangle| \geq \frac{\gamma}{|k|^\tau} \qquad (D.12)$$

needed in the analogue of Theorems 4.10 and 4.21. This complicates the proof of the analogues of Theorem 4.11 and Corollary 4.22, and in fact one can only get a slightly weaker result. As long as the n actions y only had to play the rôle of the $n + d \leq 2n - 1$ parameters ω and λ, it was sufficient to pose conditions on the curvature of the frequency mapping. Since the number m of normal frequencies α_j may easily exceed the number n of internal frequencies ω_i one now has to work with higher derivatives as well. This enforces the restriction $\tau > nL - 1$ on the Diophantine constant τ, where L is the order of the highest derivative needed.

A crucial observation during the proof was that the sum (D.9) is finite. Indeed, when dealing with normally elliptic tori the corresponding sum is infinite and necessitates the additional Diophantine conditions (D.12). At that point, the weight $\alpha_{\lambda_3} \neq 0$ for the unfolding parameter λ_3 which is multiplied with $\frac{1}{2}q^2$ is important: already in the parabolic case this results in

$$\left\{ \mathcal{N}_0(q, p), \frac{\lambda_3}{2} q^2 \right\} = \left\{ \frac{a}{2} p^2, \frac{\lambda_3}{2} q^2 \right\} = a\lambda_3 pq$$

having increased order $\alpha_{\lambda_3} + \alpha_q + \alpha_p = 2\alpha_y - \alpha_q - \alpha_p > \alpha_y$. This inequality becomes sharp for the quasi-periodic centre-saddle bifurcation, where the weights are $(\alpha_y, \alpha_q, \alpha_p; \alpha_{\lambda_0}, \alpha_{\lambda_1}, \alpha_{\lambda_2}, \alpha_{\lambda_3}, \alpha_{\lambda_4}, \alpha_{\lambda_5}) = (6, 2, 3; 6, 4, 3, 2, 1, 1)$.

The progressive decoupling of the KAM-part and the singularity-theoretic part from [34, 139] via [50, 51, 144] to the present proof lead to simplifications that allowed for improved results. A further such decoupling is achieved in [284] where a similar bifurcation problem of invariant tori in the dissipative context is solved by means of the Translated Torus Theorem of Rüssmann [250] and Herman. While this is formulated for mappings, the corresponding result for vector fields is developed in [285]. It would be interesting to know how this approach can be used to simplify and generalize the proof given here.

E

Proofs of the Necessary Lemmata

We now provide the omitted proofs of the nine lemmata needed in the course of the proof of Theorem D.1. We continue to use the notations introduced in Appendix D and repeat only the most crucial definitions.

Lemma E.1. (Lemma D.3) For the truncation

$$R = \sum_{\|jklm\|=0}^{\alpha_y} \sum_{|h|\leq\eta} P_{jklm}^h(\omega) \, e^{2\pi i \langle h,x \rangle} \, y^j p^k q^l \lambda^m$$

of P we have

$$|R|_{D^1} \leq \delta$$
$$|P-R|_{D^1} \leq \beta^{1-\xi}\varepsilon^\chi \delta$$

where $\delta = |P|_D$ and $\beta = \varepsilon^{\chi/(\alpha_y+\xi)}$.

Proof. Since $|R|_{D^1} \leq |R-P|_{D^1} + |P|_{D^1}$ we only have to show the second inequality. Note that

$$P - R = \sum_{\substack{\nu\leq\alpha_y \\ |h|>\eta}} P_\nu^h e^{2\pi i \langle h,x \rangle} + \sum_{\substack{\nu>\alpha_y \\ h\in\mathbb{Z}^n}} P_\nu^h e^{2\pi i \langle h,x \rangle} \ .$$

For the first term we use the Paley–Wiener estimate $|P_{jklm}^h(\omega)y^j p^k q^l \lambda^m| \leq |P|_D \, e^{-2\pi|h|(\kappa+s)}$ and an upper bound of the number of vectors $h \in \mathbb{Z}^n$ with $|h| = \nu$ to derive

$$\left| \sum_{\substack{\|jklm\|\leq\alpha_y \\ |h|>\eta}} P_{\|jklm\|}^h e^{2\pi i \langle h,x \rangle} \right| \leq \sum_{|h|>\eta} |P|_D \, e^{-2\pi|h|(\kappa+s)} e^{2\pi|\mathrm{Im}\,x|\,|h|}$$

$$\leq \delta \sum_{\nu=\eta+1}^\infty \nu^n e^{-(\pi/6)\nu s} \leq \delta \int_\eta^\infty \zeta^n e^{-(\pi/6)\zeta s} \, d\zeta$$

$$\leq \eta^n \, e^{-\pi\eta s/6} \delta \leq s^{-2n} \, e^{-\pi/(6s)} \delta \leq \varepsilon\delta$$

where the last inequality is given by Lemma 5.10 of [33]. For the second term we use $\varepsilon_+ = \beta\varepsilon$ and obtain

$$
\left|\sum_{\substack{\|jklm\|>\alpha_y \\ h\in\mathbb{Z}^n}} P^h_{\|jklm\|} e^{2\pi i\langle h,x\rangle}\right| \;\leq
$$

$$
\left|\sum_{\substack{\|jklm\|>\alpha_y \\ h\in\mathbb{Z}^n}} \int \frac{\partial^{|j|+k+l+|m|}}{\partial y^j\,\partial p^k\,\partial q^l\,\partial\lambda^m} P^h_{\|jklm\|} e^{2\pi i\langle h,x\rangle}\;dy^j dp^k dq^l d\lambda^m\right|
$$

$$
\leq \frac{(\beta\varepsilon)^{\alpha_y+1}}{\varepsilon^{\alpha_y+1}} \left|\sum_{\nu=0}^{\infty}\sum_{|h|\in\mathbb{Z}^n} P^h_\nu e^{2\pi i\langle h,x\rangle}\right| \;\leq\; \beta^{\alpha_y+1}\delta
$$

where $\int$ stands for

$$
\underbrace{\int_0^{y_1}\cdots\int_0^{y_1}}_{j_1}\cdots\underbrace{\int_0^{y_n}\cdots\int_0^{y_n}}_{j_n}\underbrace{\int_0^{p}\cdots\int_0^{p}}_{k}\underbrace{\int_0^{q}\cdots\int_0^{q}}_{l}\underbrace{\int_0^{\lambda_0}\cdots\int_0^{\lambda_0}}_{m_0}\cdots\underbrace{\int_0^{\lambda_d}\cdots\int_0^{\lambda_d}}_{m_d}
$$

with $\|jklm\| = \alpha_y + 1$. $\square$

Lemma E.2. (Lemma D.4) *For the function*

$$
F \;=\; \sum_{\nu=0}^{\alpha_y}\sum_{0<|h|\leq\eta} F^h_\nu e^{2\pi i\langle h,x\rangle}
$$

with coefficient functions $F^h_\nu = F^h_\nu(y,q,p;\lambda,\omega)$ defined by (D.8) we have

$$
|F|_{D^{\alpha-5}} \;\leq\; |P|_D
$$

and

$$
\left|\frac{\partial^{|j|+k+l+|m|} F}{\partial y^j\,\partial p^k\,\partial q^l\,\partial\lambda^m}\right|_{D^{\alpha-4}} \;\leq\; \varepsilon^{-\nu}\,|P|_D
$$

if $\nu = \|jklm\| \leq \alpha_y$.

Proof. For Poisson brackets with $\mathcal{N}_\mu$ we have the inequality

$$
|\{\mathcal{N}_\mu, G\}|_{D^\iota} \;\leq\; \varepsilon^{\mu-\alpha_q-\alpha_p}\,|G|_{D^{\iota-1}}\;. \tag{E.1}
$$

From this and the Paley–Wiener estimate we obtain

$$|F|_{D^7} \leq \sum_{\nu=0}^{\alpha_y} \sum_{0<|h|\leq\eta} \sum_{\iota=0}^{\alpha-6} \left|\frac{1}{2\pi i\langle h,\omega\rangle}\right|^{\iota+1}$$

$$\times \left|\{\mathcal{N}_{\mu_1},\dots,\{\mathcal{N}_{\mu_\iota}, P^h_{\nu-2\iota}\}\cdots\}e^{2\pi i\langle h,x\rangle}\right|_{D^7}$$

$$\leq \sum_{\nu=0}^{\alpha_y} \sum_{0<|h|\leq\eta} \sum_{\iota=0}^{\alpha-6} \sum_{\mu_1,\dots,\mu_\iota=\alpha_x}^{\alpha_y+1} |2\pi\langle h,\omega\rangle|^{-\iota-1}\, \varepsilon^{\nu(\iota)}\, |P|_{D^1}\, e^{-(\pi/4)|h|s}$$

with exponent $\nu(\iota)$ defined in (D.10). To estimate the small denominators we take $w \in \Sigma'_1$ with $|\omega - w| \leq r/2$ and use $\eta = [r^{-1/(\tau+1)}]$ to derive

$$|\langle h,\omega\rangle| \geq |\langle h,w\rangle| - |\omega - w|\cdot|h| \geq |h|^{-\tau} - (r/2)\eta \geq \frac{1}{2}|h|^{-\tau}$$

whence the inequality

$$\sum_{h\in\mathbb{Z}} |h|^{(\alpha-5)\tau}\, e^{-(\pi/4)|h|s} \leq. \; s^{-(\alpha-4)\tau}$$

yields the first inequality. The second inequality then follows from Cauchy's estimate. □

Lemma E.3. (Lemma D.5) *Under the conditions of Lemma E.2 the Hamiltonian vector field X_F satisfies*

$$\|X_F\|_{D^{\alpha-4}} \leq. \varepsilon^\xi , \qquad \Uparrow D_\nu X_F \Uparrow_{D^{\alpha-4}} \leq. \varepsilon^\xi \; \forall_{\nu\geq 1} .$$

Proof. This is an immediate consequence of Lemma E.2 and $\delta = \varepsilon^{\alpha_y+\xi}$. Indeed, the expressions $(\partial^{|j|+k+l}/\partial y^j \partial p^k \partial q^l)G$ with $G \in \{\partial F/\partial y, \partial F/\partial x, \partial F/\partial p, \partial F/\partial q\}$ are either partial derivatives of F of weighted order $\leq \alpha_y$, or vanish identically. □

Lemma E.4. (Lemma D.6) *Under the conditions of Lemma E.2 the time-1-flow φ_1 of X_F satisfies*

$$\|\varphi_1 - \mathrm{id}\|_{C^{jkl}(D^{\alpha-2})} \leq. \varepsilon^\xi .$$

Proof. Note that φ_t^F satisfies the integral equation

$$\varphi_t^F = \mathrm{id} + \int_0^t X_F \circ \varphi_{\tilde{t}}^F\, d\tilde{t}$$

from which we derive for the (total) derivatives of order $\mu \geq 2$

$$D^\mu\varphi_F = 0 + \int_0^1 \sum c_{\ell\iota}^\mu((D^\ell X_F) \circ \varphi_t^F) \bullet (D^{\iota_1}\varphi_t^F,\dots,D^{\iota_\ell}\varphi_t^F)\, dt \qquad (E.2)$$

where the sum is taken over $\ell = 1, \ldots, \mu$ and $\iota_1 \geq \ldots \geq \iota_\ell \geq 1$ with $\iota_1 + \ldots + \iota_\ell = \mu$. Partial derivatives on the left hand side that do not differentiate with respect to x involve on the right hand side only partial derivatives with at most one differentiation with respect to x. For the first derivative we have

$$\Uparrow D\varphi_F \Uparrow_{D^{\alpha-2}} \leq 1 + \Uparrow DX_F \Uparrow_{D^{\alpha-3}} \cdot \Uparrow D\varphi_F \Uparrow_{D^{\alpha-2}} .$$

When $|j| + k + l = 1$ Lemma E.3 yields

$$\left\| \frac{\partial^{|j|+k+l}\varphi_F}{\partial y^j \partial p^k \partial q^l} \right\|_{D^{\alpha-2}} \leq \Uparrow D\varphi_F \Uparrow_{D^{\alpha-2}} \leq \frac{1}{1 - c\varepsilon^\xi} \leq 2 ,$$

if ε_0 is sufficiently small. This immediately implies the desired result for $|j| + k + l = 1$. Inductively, assuming the proper bound on $(\partial^{|j|+k+l}/\partial y^j \partial p^k \partial q^l)\varphi_F$ holds for $|j| + k + l \leq \mu - 1$, from (E.2) we have

$$\left\| \frac{\partial^\mu(\varphi_F - \mathrm{id})}{\partial y^j \partial p^k \partial q^l} \right\|_{D^{\alpha-2}} \leq. \Uparrow D_\mu X_F \Uparrow_{D^{\alpha-3}} \Uparrow D\varphi_F \Uparrow^\mu_{D^{\alpha-2}}$$
$$+ \Uparrow D_{\mu-1} X_F \Uparrow_{D^{\alpha-3}} \varepsilon^\xi .$$
$$+ \Uparrow DX_F \Uparrow_{D^{\alpha-3}} \|\partial^\mu \varphi_F\|_{D^{\alpha-2}} .$$

Here $\partial^\mu \varphi_F$ denote partial derivatives of order μ that do not differentiate with respect to x; those that do have been included in the second term. It follows that

$$\left\| \frac{\partial^\mu(\varphi_F - \mathrm{id})}{\partial y^j \partial p^k \partial q^l} \right\|_{D^{\alpha-2}} \leq. \frac{2^\mu + \tilde{c}}{1 - c\varepsilon^\xi} \varepsilon^\xi$$

for sufficiently small ε_0. □

Lemma E.5. (Lemma D.9) *The co-ordinate transformation Ψ defined in Section D.1.2 satisfies*

$$\|\Psi - \mathrm{id}\|_{C^{jkl}(D^{\alpha-1})} \leq. \varepsilon^\xi .$$

Proof. Since ϕ is a linear map, the estimate follows from Lemma E.4 and the chain rule. □

Lemma E.6. (Lemma D.10) *The perturbation $P_+ = (\bar{P} + \hat{P}) \circ \phi$ with $\bar{P}$ defined in (D.5) and $\hat{P}$ defined in (D.11) can be estimated by*

$$|P_+|_{\mathcal{D}_+} \leq |P_+|_{D^{\alpha-1}} \leq. \beta^{(1-\xi)(\xi-\chi)} \varepsilon^\chi \delta < \delta_+ .$$

Proof. We have to estimate all terms by $\delta_+ = \varepsilon^\chi \delta$. Using the Cauchy inequality we immediately have

$$|\{R, F\} \circ \phi|_{D^{\alpha-1}} \leq |\{R, F\}|_{D^{\alpha-2}}$$
$$\leq. \varepsilon^{-\alpha_y} |R|_D |F|_{D^{\alpha-5}}$$
$$\leq. \varepsilon^{\alpha_y + 2\xi}$$

and

$$\Big| \int_0^1 (1-t)\{\{N+R,F\},F\} \circ \varphi_t^F \, dt\Big|_{D^{\alpha-2}}$$

$$\leq \quad |\{\{N+R,F\},F\}|_{D^{\alpha-3}} \cdot \int_0^1 (1-t)dt$$

$$\leq. \quad \varepsilon^{-\alpha_y} |\{N+R,F\}|_{D^{\alpha-4}} |F|_{D^{\alpha-5}}$$

$$\leq. \quad \varepsilon^{-\alpha_y} |F|^2_{D^{\alpha-5}}$$

$$\leq. \quad \varepsilon^{\alpha_y+2\xi} \ .$$

Furthermore Lemma E.1 yields

$$|(P-R) \circ \varphi_F \circ \phi|_{D^{\alpha-1}} \quad \leq. \quad \beta^{1-\xi} \varepsilon^{\chi} \delta$$

and finally we obtain

$$\Big|\hat{P}\Big|_{D^{\alpha-2}} \quad \leq. \quad \sum_{\mu=\alpha_x}^{\alpha_y+1} \sum_{\nu=\alpha_q+\alpha_p+\alpha_y+1-\mu}^{\alpha_y} \sum_{0<|h|\leq\eta} \varepsilon^{\mu-\alpha_q-\alpha_p} \Big| F_\nu^h e^{2\pi i\langle h,x\rangle} \Big|_{D^{\alpha-3}}$$

$$\leq. \quad \varepsilon\delta \ . \qquad \qquad \Box$$

Lemma E.7. (Lemma D.11) *The co-ordinate transformation Φ_i defined in Section D.2 satisfies for every $i \in \mathbb{N}$ a bound*

$$\|\Phi_i\|_{C^\nu(D_i^{\alpha-1})} \quad \leq \quad c_\nu \ .$$

Proof. The estimate in this lemma holds for $\Psi_i = \psi_1^i \circ \phi^i$ by Lemma E.5. From $\Phi_{i+1} = \Phi_i \circ \Psi_i$ it follows that $(\partial/\partial z)\Phi_{i+1} = ((\partial/\partial z)\Phi_i \circ \Psi_i) \bullet D\Psi_i$ and we have to estimate products of the form

$$\Big(\frac{\partial^{|j|+k+l}\Phi_i}{\partial y^j \partial p^k \partial q^l} \circ \Psi_i \Big) \bullet (D^{\iota_1}\Psi_i, \dots, D^{\iota_\ell}\Psi_i)$$

with $\ell = |j| + k + l$. The estimates for Φ_i can be proven inductively. $\qquad \Box$

Lemma E.8. (Lemma D.12) *The function $\Phi_i : (x,y,p,q;\lambda,\omega) \mapsto (X,Y,P,Q)$ admits a generating function $S_i = S_i(x,Y,q,P;\lambda,\omega)$.*

Proof. The symplectomorphism Φ_i is composed from finitely many time one maps (and parameter shifts) and therefore homotopic to the identity. Thus, the closed one-form

$$\sum_{\iota=1}^n (y_\iota - Y_\iota)\, dx_\iota \ + \ (X_\iota - x_\iota)\, dY_\iota \ + \ (Q-q)\, dP \ + \ (p-P)\, dq$$

is exact and can therefore be written as dS_i. $\qquad \Box$

Lemma E.9. (Lemma D.13) *For $\|jklm\| \leq \alpha_y$ and $|m| \leq 1$ the sequence $(\widetilde{S}_i^{jklm})_{i \in \mathbb{N}}$ of functions on $\overline{U_\kappa(\mathbb{T}^n)} \times \Sigma_1'$ is uniformly convergent with exponential estimates on the differences $\widetilde{S}_{i+1} - \widetilde{S}_i$.*

Proof. For $k \geq 1$ we can use $\partial S_i / \partial P = W_i^4 \circ \mathcal{G}_i$. We immediately get

$$
\begin{aligned}
| S_{i+1}^{0100}(x,\omega) - S_i^{0100}(x,\omega) | &= | W_{i+1}^4(x,0,0,0,0,\omega) - W_i^4(x,0,0,0,0,\omega) | \\
&\leq \; \|\Phi_i\|_{C^2(D_i)} \, |(\Psi_i - \mathrm{id})(x,0,0,0,0,\omega)| \; \leq. \; \varepsilon_i^\xi
\end{aligned}
$$

and this exponential decay yields a limit $\widetilde{S}_\infty^{0100}$. For higher derivatives we use the chain rule to write $S_i^{jklm}(x,\omega)$ as a sum of terms

$$
\Big(\frac{\partial^\mu W_i^4}{\partial y^{j'} \partial p^{k'} \partial q^{l'} \partial \lambda^{m'}} \circ \mathcal{G}_i \Big)(D^{\iota_1}\mathcal{G}_i, \ldots, D^{\iota_\mu}\mathcal{G}_i)
$$

with $(j',k',l',m') \leq (j,k,l,m)$ in all components and $\iota_1 + \ldots + \iota_\mu = |j| + k + l - 1$. Using $D\mathcal{G}_{i+1} - D\mathcal{G}_i = D\mathcal{G}_i(D\mathcal{F}_i - D\mathcal{F}_{i+1})D\mathcal{G}_{i+1}$ we obtain again an exponential decay for the difference $S_{i+1}^{jklm}(x,\omega) - S_i^{jklm}(x,\omega)$ and thus a limit coefficient function $\widetilde{S}_\infty^{jklm}(x,\omega)$.

By means of $\partial S_i / \partial Y = W_i^1 \circ \mathcal{G}_i$ and $\partial S_i / \partial q = -W_i^3 \circ \mathcal{G}_i$ we obtain this same result for $|j| \geq 1$ and $l \geq 1$ as well.

We are left with the coefficient functions $S_i^{0000}(x,\omega)$. Their averages vanish and they are determined by $\partial S_i / \partial x = -W_i^2 \circ \mathcal{G}_i$. We conclude

$$
\begin{aligned}
| S_{i+1}^{0000}(x,\omega) - S_i^{0000}(x,\omega) | &\leq \max_{i=1,\ldots,n} | \int_0^{x_i} W_i^{2,i} - W_i^{2,i} \circ \mathcal{G}_i \, d\tilde{x}_i | \\
&\leq |W_i^2 - W_i^2 \circ \mathcal{G}_i| \; \leq. \; \varepsilon_i^\xi \qquad\qquad \square
\end{aligned}
$$

Glossary

action angle variables: In an *integrable* Hamiltonian system with d degrees of free-dom the level sets of regular values of the d integrals are *Lagrangean subman-ifolds*. In case these level sets are compact they are (unions of) d-tori. In the neighbourhood of such an invariant torus one can find *Darboux co-ordinates* (φ, I) such that the integrals (and in particular the Hamiltonian) depend only on the actions I. The values of (φ, I) form a product set $\mathbb{T}^d \times \mathbb{Y}$, with $\mathbb{Y} \subseteq \mathbb{R}^d$ open. Every invariant d-torus $I = const$ in the domain of this chart is para-metrised by the angles φ.

adjacent: A *singularity* $\mathcal{K}$ is adjacent to $\mathcal{H}$ if $\mathcal{H}$ is contained in the *universal un-folding* of $\mathcal{K}$.

Arnol'd diffusion: The existence of orbits in nearly integrable systems that connect distant parts of the phase space. Because of Nekhoroshev's theorem such motion must be very slow, and because of KAM theory the majority of initial condi-tions does not lead to such motions (but to quasi-periodic motions instead). In the literature this expression is sometimes also used for a special mecha-nism proposed by Arnol'd in [7] where such orbits are constructed on the basis of normally hyperbolic tori that are generated when the perturbation destroys resonant tori.

Birkhoff normal form: In case there are no *normal resonances* between the *nor-mal frequencies* α_j of a *centre* there is a formal change of variables that turns the Hamiltonian into a power series in the invariants $I_j = \frac{1}{2}(x_j^2 + y_j^2)$. Correspondingly, there is a "real" co-ordinate change into Birkhoff normal form $H = \sum \alpha^{(k)} I^k + R$ up to order n with a remainder term satisfying $D^k R(0) = 0 \ \forall_{k \leq n}$, and this latter result only requires absence of normal reso-nances $k_j \alpha_j = 0$ up to order $|k| \leq n$.

Cantor set, Cantor dust, Cantor family, Cantor fibration, Cantor stratification: Cantor dust is a separable locally compact space that is perfect, i.e. every point is in the closure of its complement, and totally disconnected. It can be shown that this determines Cantor dust up to homeomorphy. We use the term "Cantor set" for subsets of $\mathbb{R}^n$ that are locally homeomorphic to a product of Cantor dust and a closed set.

On the real line $\mathbb{R}$ one can define Cantor dust of positive measure by excluding around each rational number p/q an interval of size $2\gamma/q^\tau$, $\gamma > 0, \tau > 2$. Similar

Diophantine conditions define Cantor sets in $\mathbb{R}^n$. Since these Cantor sets have positive measure their (Hausdorff)-dimension is n. We also use expressions like "Cantor family of tori" or "(ramified) Cantor fibration" for a (Whitney)-smooth collection of tori, parametrised over a Cantor set. Where the unperturbed system is stratified according to the co-dimension of occurring (bifurcating) tori, this leads to a Cantor stratification. See also Definition B.3.

Casimir element: A function $f : \mathcal{P} \longrightarrow \mathbb{R}$ is a Casimir element of the *Poisson algebra* $C^\infty(\mathcal{P})$ if $\{f, g\} = 0$ for all $g \in C^\infty(\mathcal{P})$. This induces a Poisson structure on the level sets $f^{-1}(a)$, $a \in \mathbb{R}$. In case every point has a small neighbourhood on which the only Casimir elements are the constant functions, the Poisson structure on $\mathcal{P}$ is non-degenerate, i.e. $\mathcal{P}$ is a *symplectic manifold*.

centre: An equilibrium of a vector field is called a centre if all eigenvalues of the linearization are nonzero and on the imaginary axis.

centre-saddle bifurcation: Under variation of a parameter λ a *centre* and a *saddle* meet and vanish.

Chern class: Let $\rho : \mathcal{P} \longrightarrow B$ be a torus bundle with fibre $\mathbb{T}^n$ and denote by $\mathcal{G}$ the locally constant sheaf of first homotopy groups $\pi_1(\mathbb{T}^n)$ of the fibres. The Chern class of the torus bundle is an element of $H^2(B, \mathcal{G})$ that measures the obstruction to the existence of a global section of ρ – such a section exists if and only if the Chern class vanishes. An example with a non-vanishing Chern class is the Hopf fibration $S^3 \longrightarrow S^2$.

co-isotropic: For a subspace $U < V$ of a symplectic vector space (V, ω) are equivalent: (*i*) U contains its ω-orthogonal complement, (*ii*) $U^{\perp \omega}$ is *isotropic*. If U satisfies one and thus both of these conditions it is called a co-isotropic subspace. A submanifold $\mathcal{B} \subseteq \mathcal{P}$ of a symplectic manifold $(\mathcal{P}, \omega)$ is called co-isotropic if all tangent spaces $T_y \mathcal{B} < T_y \mathcal{P}$ are co-isotropic subspaces.

conditionally periodic: A motion $t \mapsto \alpha(t) \in \mathcal{P}$ is conditionally periodic if there are frequencies $\omega_1, \dots, \omega_k \in \mathbb{R}$ and a smooth embedding $F : \mathbb{T}^k \longrightarrow \mathcal{P}$ such that $\alpha(t) = F(e^{2\pi i \omega_1 t}, \dots, e^{2\pi i \omega_k t})$. We can think of the motion as a superposition of the periodic motions $t \mapsto F(1, \dots, 1, e^{2\pi i \omega_j t}, 1, \dots, 1)$. If the frequencies are rationally independent, the motion $t \mapsto \alpha(t) \in \mathcal{P}$ lies dense on $\mathrm{im} F$ and this embedded torus is an invariant torus. In case there are *resonances* among the frequencies the motion is restricted to a subtorus.

A flow on a torus is parallel or conditionally periodic if there exist co-ordinates in which the vector field becomes constant.

conjugacy: Two flows φ_t and ψ_t of vector fields X and Y are (topologically) conjugate if there is a homeomorphism η with $\eta \circ \varphi_t = \psi_t \circ \eta$. In case the conjugacy η is a diffeomorphism one may differentiate this equation and obtains $Y = T\eta \circ X \circ \eta^{-1}$.

conjugate: (*a*) For flows: see *conjugacy*.   (*b*) Two co-ordinates Q and P of a *Darboux* chart are (canonically) conjugate if $\omega(X_Q, X_P) = \pm 1$, *i.e.* X_Q and X_P span for every point x in the domain U of the chart a hyperbolic plane in $T_x U$.

connection bifurcation: Under variation of a parameter λ the stable and the unstable manifolds of two saddles approach and pass, coinciding for an isolated parameter value. This takes place in one degree of freedom where the Hamiltonian $\frac{1}{2}p^2 + \frac{1}{2}q^2 - q^4 - \lambda q$ defines a standard example. In n degrees of freedom the saddles are replaced by normally hyperbolic $(n-1)$-tori and the system has to be *integrable*. While the connection bifurcation is robust in one degree of freedom, in $n \geq 2$ degrees of freedom a small *generic* perturbation may lead to transverse *heteroclinic orbits* and to *tangency bifurcations.*

Darboux co-ordinates, Darboux basis: In a Darboux basis $\{e_1, \ldots, e_n, f_1, \ldots, f_n\}$ of a symplectic vector space (V, ω) the symplectic product takes the simple form $\omega(e_i, e_j) = 0 = \omega(f_i, f_j)$ and $\omega(e_i, f_j) = \delta_{ij}$. In Darboux co-ordinates $(q_1, \ldots, q_n, p_1, \ldots, p_n)$ of a symplectic manifold $(\mathcal{P}, \omega)$ the symplectic form becomes $\omega = \sum \mathrm{d}q_i \wedge \mathrm{d}p_i$.

degree of freedom: In so-called simple mechanical systems the *phase space* is the cotangent bundle of the configuration space and the dimension of the latter encodes "in how many directions the system can move". For *symplectic manifolds* this notion is immediately generalized to one half of the dimension of the phase space. *Poisson spaces* are foliated by their symplectic leaves and the number of degrees of freedom is defined to be one half of the rank of the Poisson structure.

differential space: A topological space $\mathcal{P}$ for which one singles out a subalgebra $\mathcal{A} \subseteq C(\mathcal{P})$ of "smooth functions" that is locally defined, sufficiently rich to reproduce the topology and that is consistent with respect to compositions with smooth functions on Euclidian spaces. One then writes $\mathcal{A} = C^\infty(\mathcal{P})$. See also Definition B.2.

Diophantine condition, Diophantine frequency vector: A frequency vector $\omega \in \mathbb{R}^n$ is called Diophantine if there are constants $\gamma > 0$ and $\tau > n - 1$ with

$$\bigwedge_{k \in \mathbb{Z}^n \setminus \{0\}} |\langle k, \omega \rangle| \geq \frac{\gamma}{|k|^\tau} \, .$$

The Diophantine frequency vectors satisfying this condition for fixed γ and τ form a Cantor set of half lines. As the "Diophantine parameter" γ tends to zero (while τ remains fixed), these half lines extend to the origin. The complement in any compact set of frequency vectors satisfying a Diophantine condition with fixed τ has a measure of order $O(\gamma)$ as $\gamma \to 0$.

distinguished parameter: Let H depend on the parameters I and α. If we allow only re-parametrisations $(I, \alpha) \mapsto (J, \beta)$ of the form $J = J(I, \alpha)$ and $\beta = \beta(\alpha)$, we make I an internal or distinguished parameter, cf. [288]. This is appropriate if, for example, H is a reduced Hamiltonian, I the (fixed) value of the *momentum mapping* of the original system and α an external parameter the original system depends upon. See also Section 1.1.2.

dual bifurcation: Where the bifurcation scenario depends on a sign $\pm$ the "second choice" is called the dual case. An alternative division is that into *supercritical* and *subcritical* case.

elliptic: A periodic orbit/invariant torus is elliptic if all *Floquet multipliers/exponents* are on the unit circle/imaginary axis. An elliptic equilibrium is called a *centre.*

energy shell: The set of points that can be attained given a certain energy. If the phase space is a symplectic manifold, this set is given by the pre-image of that energy value under the Hamiltonian. For more general Poisson spaces this pre-image has to be intersected with a symplectic leaf.

energy-momentum mapping: Let $\Gamma : G \times \mathcal{P} \longrightarrow \mathcal{P}$ be a symplectic action of the Lie group G on the *symplectic manifold* $\mathcal{P}$ with *momentum mapping* $J : \mathcal{P} \longrightarrow \mathfrak{g}^*$. For a Hamiltonian function H that is invariant under the action Γ we call $(H, J) : \mathcal{P} \longrightarrow \mathbb{R} \times \mathfrak{g}^*$ the energy-momentum mapping.

equivalence: The meaning of "f is equivalent to g" depends on the context. For $f, g \in C^\infty(\mathcal{P})$ it means that there is a diffeomorphism η of $\mathcal{P}$ with $g \circ \eta = f$. Sometimes this is called "right equivalence" and the term "left-right equivalence" is used if transformations are allowed in the range $\mathbb{R}$ as well. See also Definitions 2.3 and 2.12. For vector fields one usually works with *topological equivalence.*

ergodic: A Hamiltonian system is ergodic if the relative measure within the *energy shell* of any invariant set is either zero or one.

essentially quasi-homogeneous: A *semi-quasi-homogeneous* planar *singularity* for which the C^∞-*moduli* of higher (weighted) order are topologically irrelevant.

filtration: A filtration of an algebra $\mathcal{A}$ is a series $(\mathcal{F}_n)_{n \in \mathbb{N}_0}$ of subalgebras with $\mathcal{F}_n \geq \mathcal{F}_m$ for $n \leq m$ and $\mathcal{F}_0 = \mathcal{A}$. For a *graded* algebra $\mathcal{A} = \prod_{k=0} \mathcal{A}_k$ a filtration is given by $\mathcal{F}_n = \prod_{k \geq n} \mathcal{A}_k$.

finitely determined: A singularity that is *equivalent* to its $(\mu+1)$-*jet*, where $\mu < \infty$ is the *multiplicity* (or Milnor number).

Floquet exponent: The eigenvalues of the normal matrix of an invariant n-torus in *Floquet form.* For $n = 1$ exponentiation of these yields *Floquet multipliers.*

Floquet form: An invariant torus is in Floquet form if the *normal linear* behaviour does not depend on the toral angles.

Floquet multipliers: Let γ be a periodic orbit of the vector field X with flow φ_t and denote by τ the period of γ. For $x \in \gamma$ the eigenvalues of the linearization $D\varphi_\tau$ are called Floquet multipliers of γ (they do not depend on the particular choice $x \in \gamma$). The tangent vector spanning $T_x\gamma$ is an eigenvector of $D\varphi_\tau$ to the eigenvalue 1. The other eigenvalues (counted with multiplicity) are also the eigenvalues of a *Poincaré mapping* defined on a *Poincaré section* through x. In case the vector field is Hamiltonian the Floquet multipliers occur in pairs (λ, λ^{-1}) of the same multiplicity and the multipliers ± 1 have even multiplicity. The "second" generalized eigenvector to the eigenvalue 1 is transverse to the energy shell of γ.

frequency-halving bifurcation: In the *supercritical* case the invariant torus loses its stability, giving rise to a stable torus with one of its frequencies halved. In the *dual* or *subcritical* case stability is lost through collision with an unstable invariant torus with one of its frequencies halved.

generic: A property is generic if the set it defines contains a countable intersection of open dense sets. The genericity conditions that we impose on our normal forms are usually only finitely many inequalities and thus define open dense sets themselves. Where the perturbation has to be generic this often does involve infinitely many conditions (and it is therefore much more involved to check that all these are indeed satisfied).

genuine first and second order resonance: A *centre* in $k_1 : k_2 : \ldots : k_\ell$ resonance in ℓ degrees of freedom for which there are $\ell - 1$ independent *resonances* of order ≤ 3 (resp. ≤ 4).

germ: The germ of H in z consists of all K that coincide with H on some neighbourhood of z. To prove results on germs one usually keeps shrinking the neighbourhood.

gradation: A gradation of a finite dimensional algebra $\mathcal{A}$ is a direct sum representation

$$\mathcal{A} = \bigoplus_{k=0}^{n} \mathcal{A}_k$$

with $\mathcal{A}_k \cdot \mathcal{A}_m \subseteq \mathcal{A}_{k+m}$. For infinite dimensional algebras (like the algebra $\mathbb{R}[x]$ of polynomials) one might need infinite direct sums ($\mathbb{R}[x]$ is graded as a direct sum of polynomials with degree k), or even infinite (direct) products (as for the algebra $\mathbb{R}[[x]]$ of formal power series in one real variable).

group *action:* A mapping $\Gamma : G \times \mathcal{P} \longrightarrow \mathcal{P}$ is a group action of the Lie group G on the phase space $\mathcal{P}$ if $\Gamma_e = \mathrm{id}$ and $\Gamma_{g \cdot h} = \Gamma_g \circ \Gamma_h$ for all Lie group elements g and h. For $G = \mathbb{R}$ one can recover the generating vector field by means of $X = \frac{\mathrm{d}}{\mathrm{d}t}\Gamma_t$.

Hamiltonian flip bifurcation: On the cone $\mathcal{B} \subseteq \mathbb{R}^3$ defined by $2uv = \frac{1}{2}w^2$, $u \geq 0, v \geq 0$ the singular equilibrium $(u, v, w) = 0$ loses its stability giving rise to a regular *centre,* cf. Fig. 2.4. In the *dual* case $(u, v, w) = 0$ gains stability and a *saddle* splits off.

Hamiltonian Hopf bifurcation: In a two-degree-of-freedom system a *centre* loses its stability: the two pairs of purely imaginary eigenvalues meet in a *Krein collision.* In a three-degree-of-freedom system the same can happen to two pairs $e^{\pm\lambda}, e^{\pm\mu}$ of *Floquet multipliers* in a one-parameter family of periodic orbits – parametrised by the energy. For a quasi-periodic Hamiltonian Hopf bifurcation one needs at least four degrees of freedom.

Hamiltonian pitchfork bifurcation: In the *supercritical* case the equilibrium/periodic orbit/invariant torus loses its stability, giving rise to an additional pair of stable equilibria/periodic orbits/invariant tori. In the *dual* or *subcritical* case stability is lost as a pair of unstable equilibria/periodic orbits/invariant tori shrinks down.

Hamiltonian system: Newton's second law states $F = m\ddot{q}$. Suppose that F is a conservative force, with potential V. We write the equations of motion as a system of first order differential equations

$$\dot{q} = \frac{1}{m}p$$
$$\dot{p} = -\frac{\partial V}{\partial q}$$

that has the total energy $H(q, p) = \langle p, p \rangle / (2m) + V(q)$ as a first integral. This can be generalized. Given a Hamiltonian function $H(q, p)$ we have the Hamiltonian vector field

$$\dot{q} = \frac{\partial H}{\partial p}$$
$$\dot{p} = -\frac{\partial H}{\partial q}$$

with first integral H. Moreover, we can replace $\mathbb{R}^{2n}$ by a *symplectic manifold* or by a *Poisson space.*

heteroclinic orbit: A point z is heteroclinic to two equilibria a and b if z lies in the intersection of the *stable manifold* of a and the *unstable manifold* of b. This implies that the whole orbit of z consists of heteroclinic points. This notion immediately generalizes to orbits heteroclinic to two invariant sets, e.g. periodic orbits or invariant tori.

Hilbert basis: The (smooth) invariants of an action of a compact group are all given as functions of finitely many "basic" invariants.

homoclinic orbit: A point z is homoclinic to an equilibrium a if z lies in the intersection of the *stable* and *unstable manifolds* of a. This implies that the whole orbit of z consists of homoclinic points. This notion immediately generalizes to orbits homoclinic to invariant sets, e.g. periodic orbits or invariant tori.

hypo-elliptic: We call an equilibrium hypo-elliptic if its linearization has both elliptic and hyperbolic eigenvalues, all nonzero. Similarly for periodic orbits and invariant tori.

hyperbolic: A periodic orbit/invariant torus is hyperbolic if no eigenvalue of the symplectic *normal linearization* is on the imaginary axis. A hyperbolic equilibrium is called a *saddle.*

integrable system: A *Hamiltonian system* with d degrees of freedom is (Liouville)-integrable if it has d functionally independent commuting integrals of motion. Locally this implies the existence of a (local) torus action.

iso-energetic Poincaré mapping: For a Hamiltonian system a *Poincaré mapping* leaves the *energy shells* invariant. Restricting to the intersection $\Sigma \cap \{H = h\}$ of the *Poincaré section* with an energy shell we obtain an iso-energetic Poincaré mapping.

isotropic: For a subspace $U < V$ of a symplectic vector space (V, ω) are equivalent: *(i)* U is contained in its ω-orthogonal complement, *(ii)* every basis of U can be extended to a *Darboux basis* of V. If U satisfies one and thus both of these conditions it is called an isotropic subspace. A submanifold $\mathcal{B} \subseteq \mathcal{P}$ of a symplectic manifold $(\mathcal{P}, \omega)$ is called isotropic if all tangent spaces $T_x\mathcal{B} < T_x\mathcal{P}$ are isotropic subspaces.

ℓ-jet: The Taylor polynomial of order ℓ – up to co-ordinate changes.

Krein collision: Two pairs of purely imaginary *Floquet exponents* meet in a double pair on the imaginary axis and split off to form a complex quartet $\pm\Re\pm i\Im$. This is a consequence of a transversality condition on the linear terms at 1:−1 *resonance*; an additional non-degeneracy condition on the non-linear part ensures that a ((quasi-)periodic) *Hamiltonian Hopf bifurcation* takes place.

Lagrangean submanifold: For a subspace $U < V$ of a symplectic vector space (V, ω) are equivalent: *(i)* U is *isotropic* and *co-isotropic,* *(ii)* U is a maximal isotropic subspace, *(iii)* U is a minimal co-isotropic subspace, *(iv)* V has a *Darboux basis* $\{e_1, \ldots, e_n, f_1, \ldots, f_n\}$ with $\text{span}\{e_1, \ldots, e_n\} = U$. If U satisfies one and thus all of these conditions it is called a Lagrangean subspace. A submanifold $\mathcal{B} \subseteq \mathcal{P}$ of a symplectic manifold $(\mathcal{P}, \omega)$ is called Lagrangean if all tangent spaces $T_x\mathcal{B} < T_x\mathcal{P}$ are Lagrangean subspaces.

Lie–Poisson structure: Let $\mathfrak{g}$ be a Lie algebra with structure constants Γ_{ij}^k. Then

$$\{\mu_i, \mu_j\} = \pm \sum \Gamma_{ij}^k \mu_k$$

defines two *Poisson structures* on the dual space $\mathfrak{g}^*$.

linear centraliser unfolding: The *universal unfolding* Ω_λ of a matrix $\Omega_0 \in M_{n\times n}(\mathbb{R})$ within $M_{n\times n}(\mathbb{R})$ obtained as $\Omega_\lambda = \Omega_0 + \sum \lambda_i \Gamma_i$ where the Γ_i span $\ker(\text{Ad}_{\Omega_0^T}) = \{ \Gamma \in M_{n\times n}(\mathbb{R}) \mid \Gamma \circ \Omega_0^T = \Omega_0^T \circ \Gamma \}$.

local algebra: The quotient of all germs in z by the Jacobi ideal, generated by the partial derivatives of a given *singularity.* As explained in [15] this is the "algebra of functions on the infinitesimal pre-image of the point z".

local bifurcation: Bifurcations of equilibria can be studied within a small neighbourhood in the product of phase space and parameter space. The same is true for fixed points of a discrete dynamical system, but when suspended to a flow the corresponding bifurcating periodic orbits obtain a *semi-local* character.

local group action: The defining properties $\Gamma_e = \mathrm{id}$ and $\Gamma_{g \cdot h} = \Gamma_g \circ \Gamma_h$ of a *group action* of the Lie group G on the phase space $\mathcal{P}$ also characterise a local group action, but the mapping Γ may only be defined on a neighbourhood of $\{e\} \times \mathcal{P} \subseteq G \times \mathcal{P}$.

modulus: Two planar singularities are called right *equivalent* if there is a (smooth) co-ordinate change of the plane that transforms these singularities into each other. Invariants obstructing such an equivalence that vary continuously with the singularity are called moduli.

momentum mapping: Let $\Gamma : G \times \mathcal{P} \longrightarrow \mathcal{P}$ be a symplectic action of the Lie group G on the *symplectic manifold* $\mathcal{P}$. A mapping $J : \mathcal{P} \longrightarrow \mathfrak{g}^*$ into the dual space of the Lie algebra of G is a momentum mapping for the action if $X_{\hat{J}(\xi)} = \xi_{\mathcal{P}}$

for all $\xi \in \mathfrak{g}$. Here $\hat{J}(\xi) : \mathcal{P} \longrightarrow \mathbb{R}$ is defined by $\hat{J}(\xi)(z) = J(z) \cdot \xi$ and $\xi_{\mathcal{P}}$ is the infinitesimal generator of the action corresponding to ξ. The momentum mapping J is called Ad^*-equivariant provided that $J \circ \Gamma_g = \mathrm{Ad}^*_{g^{-1}} \circ J$.

monodromy: Let $\rho : \mathcal{P} \longrightarrow B$ be a torus bundle with fibre $\mathbb{T}^k$ and choose a connection on this bundle. Then a closed curve in B with base boint y gives rise to an automorphism of the torus $\rho^{-1}(y)$. The induced homomorphism $\pi_1(B, y) \longrightarrow GL(H^1(\rho^{-1}(y), \mathbb{Z}))$ does not depend on the particular choice of the connection and is called the monodromy of (the connected component containing $\rho^{-1}(y)$ of) $\mathcal{P}$. For a product of k connected circle bundles the monodromy becomes a mapping $\pi_1(B) \longrightarrow GL_k(\mathbb{Z})$.

Morse function: A function $H \in C^2(\mathcal{P})$ for which all *singularities* are quadratic and have different values.

multiplicity: The (maximal) number of critical points "contained" in a *singularity*. See also Definition 2.10.

Newton diagram: For a planar *singularity* the integer points $(j, k) \in \mathbb{N}_0^2$ represent monomials $p^k q^j$ and in the *non-degenerate* case finitely many of these can be chosen as representants of a basis of the *local algebra*, cf. Fig. 2.2.

non-degenerate integrable Hamiltonian system, function: In *action angle variables* (φ, I) the Hamiltonian H only depends on the action variables I and the equations of motion become $\dot{\varphi} = \omega(I)$, $\dot{I} = 0$, with frequencies $\omega(I) = \partial H / \partial I$. The system is non-degenerate at the invariant torus $\{I = I_0\}$ if $D^2 H(I_0)$ is invertible. In this case ω defines near $\{I = I_0\}$ an isomorphism between the actions and the angular velocities. Other conditions ensuring that most tori have *Diophantine* frequency vectors are iso-energetic non-degeneracy or Rüssmann-like conditions on higher derivatives.

non-degenerate singularity: A *singularity* for which every unfolding has only finitely many critical points. This means that the *multiplicity* is finite and is equivalent to *finite determinacy* and implies that there exists a *universal unfolding*.

normal frequency: Given an elliptic invariant torus of a Hamiltonian system, one can define the normal linearization on the symplectic normal bundle, see [159, 56]. The eigenvalues of the normal linearization being $\pm i\alpha_1, \ldots, \pm i\alpha_m$, we call the α_j the normal frequencies. Under the exponential mapping the eigenvalues of the normal linearization of a periodic orbit are mapped to *Floquet multipliers*.

normal hyperbolicity, normally hyperbolic manifold: An invariant submanifold of a dynamical system is called normally hyperbolic if the normal bundle admits a splitting into attracting and repelling directions with uniform bounds. For Hamiltonian systems one also speaks of hyperbolic invariant n-tori if the symplectic normal bundle admits a splitting into attracting and repelling directions. The normal bundle itself is then split into the symplectic normal bundle and the n directions conjugate to the *isotropic* tangent space. In particular, the n-parameter family of tori parametrised by the actions conjugate to the toral angles does form a normally hyperbolic manifold.

normal linearization: The linearization within the normal bundle of an invariant submanifold. In the Hamiltonian context these are often *isotropic* and we further restrict to the symplectic normal linear behaviour, e.g. to identify *elliptic* and *hyperbolic* invariant tori.

normal resonance: A *centre* in ℓ degrees of freedom is in $k_1:k_2:\ldots:k_\ell$ resonance if the linearization is *conjugate* to the superposition of ℓ oscillators with frequencies $k_1\omega, k_2\omega, \ldots, k_\ell\omega$ for some $\omega \in \mathbb{R}$.

orbit cylinder: An embedding $\Gamma : S^1 \times \,]a, b[\longrightarrow \mathcal{P}$ such that for all $e \in \,]a, b[$ the set $\gamma_e = \Gamma(S^1 \times \{e\})$ is a periodic orbit of the Hamiltonian system X_H on $\mathcal{P}$. In case Γ is transverse to every *energy shell* the parameter e can be chosen to be the energy of γ_e.

parabolic: We call an equilibrium of a one-degree-of-freedom system parabolic if its linearization is nilpotent but nonzero. An invariant torus is parabolic if its symplectic *normal linearization* has a parabolic equilibrium. In particular the four *Floquet multipliers* of a parabolic periodic orbit in two degrees of freedom are all equal to 1.

phase space: By Newton's second law the equations of motion are second order differential equations. The trajectory is completely determined by the initial positions and the initial velocities, or, equivalently, the initial momenta. The phase space is the set of all possible combinations of initial positions and initial momenta.

pinched torus: The compact *(un)stable manifold* of a saddle in two degrees of freedom with a quartet $\pm\Re \pm i\Im$ of hyperbolic eigenvalues resembles a torus $\mathbb{T}^2 = S^1 \times S^1$ with one of the fibres $\{x\} \times S^1$ reduced to a point.

Poincaré mapping: To a point $z \in \Sigma$ on a *Poincaré section* with time of first return $T(z)$ the Poincaré mapping assigns the point $\varphi_{T(z)}(z) \in \Sigma$. This yields a diffeomorphism between domain and range.

Poincaré section: A hypersurface Σ that is transverse to the orbits of the flow φ_t. For points $z \in \Sigma$ one calls $\min\{T > 0 \mid \varphi_T(z) \in \Sigma\}$ the time of first return.

Poisson space, Poisson structure: A Poisson algebra $\mathcal{A}$ is a real Lie algebra that is also a commutative ring with unit. These two structures are related by Leibniz' rule $\{f \cdot g, h\} = f \cdot \{g, h\} + g \cdot \{f, h\}$.

A Poisson manifold $\mathcal{P}$ has a Poisson bracket on $C^\infty(\mathcal{P})$ that makes $C^\infty(\mathcal{P})$ a Poisson algebra. If there are locally no *Casimir elements* other than constant functions this leads to a *symplectic* structure on $\mathcal{P}$, see [16, 81].

Poisson spaces naturally arise in *singular reduction*, this motivates us to allow varieties $\mathcal{P}$ where the Poisson bracket is defined on a suitable subalgebra $\mathcal{A}$ of $C(\mathcal{P})$.

Given a Hamiltonian function $H \in \mathcal{A}$ one obtains for $f \in \mathcal{A}$ the equations of motion $\frac{\mathrm{d}}{\mathrm{d}t} f = \{f, H\}$. For canonically *conjugate* co-ordinates (q, p) on $\mathcal{P}$, i.e. with $\{q_i, q_j\} = 0 = \{p_i, p_j\}$ and $\{q_i, p_j\} = \delta_{ij}$, this amounts to

$$\dot{q} \;=\; \frac{\partial H}{\partial p}$$
$$\dot{p} \;=\; -\frac{\partial H}{\partial q}\,.$$

Poisson symmetry: A symmetry that preserves the *Poisson structure.*

proper degeneracy: For the application of the KAM-theorem to a perturbation of an *integrable system* it is necessary that the integrable system is *non-degenerate,* so that the frequencies have maximal rank as function of the actions. If there are global conditions relating the frequencies, so that the *conditionally periodic* motion can be described by a smaller number of these, the system is properly degenerate. *Superintegrable systems* are a particular example.

quasi-homogeneous: A homogeneous polynomial H satisfies $H(e^{\tau} z) = e^{d\tau} H(z)$ where d is the order of H. For quasi-homogeneous polynomials one allows for different weights α_{z_j} for the co-ordinates z_j. See also Definition 2.11.

quasi-periodic: A *conditionally periodic* motion that is not periodic. The closure of the trajectory is an invariant k-torus T with $k \geq 2$.

A parallel or conditionally periodic flow on a k-torus is called quasi-periodic if the frequencies $\omega_1, \ldots, \omega_k$ are rationally independent.

ramified torus bundle: Let a differentiable mapping $f : \mathcal{P} \longrightarrow \mathbb{R}^m$ be given. According to Sard's Lemma almost all values $a \in \mathbb{R}^m$ are regular. The connected components of the sets $f^{-1}(a)$, $a \in \mathrm{im} f$ regular, define a foliation of an open subset of the $(n + m)$-dimensional manifold $\mathcal{P}$.

In our settings the components of f are the first integrals of an *integrable* Hamiltonian system with compact level sets. Then the regular fibres are n-tori, and their union is a torus bundle. The topology of this bundle is determined by the topology of the base space B, the *monodromy* and the *Chern class* of the bundle. In many examples that we encounter, the connected components of B are contractible, whence monodromy and Chern class are trivial.

For the geometry of the bundle one also wants to know how the singular fibres are distributed: where n-tori shrink to normally elliptic $(n-1)$-tori and where they are separated by stable and unstable manifolds of normally hyperbolic $(n-1)$-tori. The singular fibres are not necessarily manifolds, but may be stratified into X_H-invariant strata which are possibly non-compact.

One can continue and look for the singularities of these 'regular singular leaves'; the tori of dimension $\leq n-2$ and the normally *parabolic* $(n-1)$-tori in which normally elliptic and normally hyperbolic $(n-1)$-tori meet in a quasi-periodic *centre-saddle bifurcation* or a *frequency-halving bifurcation.* The next "layer" is given by quasi-periodic *Hamiltonian Hopf bifurcations* and bifurcations of higher co-dimension.

reduced phase space: Let $\Gamma : G \times \mathcal{P} \longrightarrow \mathcal{P}$ be a symplectic action of the Lie group G on the *symplectic manifold* $\mathcal{P}$ with Ad^*-equivariant *momentum mapping* $J : \mathcal{P} \longrightarrow \mathfrak{g}^*$. For a regular value $\mu \in \mathfrak{g}^*$ let the action of the isotropy group $G_{\mu} = \{ g \in G \mid \mathrm{Ad}^*_{g^{-1}}(\mu) = \mu \}$ on $J^{-1}(\mu)$ be free and proper. Then the quotient $J^{-1}(\mu)/G_{\mu}$ is again a symplectic manifold, the reduced phase space. A Γ-invariant Hamiltonian function H on $\mathcal{P}$ leads to a reduced Hamiltonian function H_{μ} on the reduced phase space.

reducible: Periodic orbits are always reducible to *Floquet form.* For invariant tori reducibility is a consequence of *integrability.*

relative equilibrium: Let H be a Hamiltonian function that is invariant under the *symplectic* action $G \times \mathcal{P} \longrightarrow \mathcal{P}$ and let $\mu \in \mathfrak{g}^*$ be a regular value of the Ad*-equivariant *momentum mapping* $J : \mathcal{P} \longrightarrow \mathfrak{g}^*$. Also assume that the isotropy group G_μ under the Ad* action on $\mathfrak{g}^*$ acts freely and properly on $J^{-1}(\mu)$. Then X_H induces a Hamiltonian flow on the *reduced phase space* $\mathcal{P}_\mu = J^{-1}(\mu)/G_\mu$. The phase curves of the given Hamiltonian system on $\mathcal{P}$ with momentum constant $J = \mu$ that are taken by the projection $J^{-1}(\mu) \longrightarrow \mathcal{P}_\mu$ into equilibrium positions of the reduced Hamiltonian system are called relative equilibria or stationary motions (of the original system).

remove the degeneracy: A perturbation of a *superintegrable system* removes the degeneracy if it is sufficiently mild to define an "intermediate system" that is still integrable and sufficiently wild to make that intermediate system *non-degenerate*. See also Definition 5.2.

resonance: If the frequencies of an invariant torus with *conditionally periodic* flow are rationally dependent this torus divides into invariant subtori. Such resonances $\langle h, \omega \rangle = 0$, $h \in \mathbb{Z}^k$, define hyperplanes in ω-space and, by means of the frequency mapping, also in phase space. The smallest number $|h| = |h_1| + \ldots + |h_k|$ is the order of the resonance. *Diophantine conditions* describe a measure-theoretically large complement of a neighbourhood of the (dense!) set of all resonances.

reversible: A vector field X is reversible with respect to an involution γ if $D\gamma \circ X \circ \gamma = -X$. For a Hamiltonian vector field X_H this is guaranteed if γ preserves H and multiplies the *Poisson structure* by -1 or if γ preserves the Poisson bracket and satisfies $H \circ \gamma = -H$.

saddle: An equilibrium of a vector field is called a saddle if the linearization has no eigenvalues on the imaginary axis. On a small neighbourhood of a saddle the flow is topologically *conjugate* to its linearization.

semi-local bifurcation: Bifurcations of n-tori can be studied in a tubular neighbourhood. For $n = 1$ a *Poincaré section* turns the periodic orbit into a fixed point of the *Poincaré mapping* and the bifurcation obtains a *local* character.

semi-quasi-homogeneous: A singularity for which the "lowest order terms" form a *quasi-homogeneous* polynomial that is *non-degenerate*. Because of the last condition not every quasi-homogenous polynomial is semi-quasi-homogeneous. See also Definition 2.11.

simple: A *singularity* is simple if it has no *moduli*. See also Definition 2.7.

singular reduction: If $\Gamma : G \times \mathcal{P} \longrightarrow \mathcal{P}$ is a *Poisson symmetry*, then the group action on $\mathcal{P}$ makes $\mathcal{B} = \mathcal{P}/G$ a *Poisson space* as well. Fixing the values of the resulting *Casimirs* yields the *reduced phase space*, which turns out to have singular points where the action Γ is not free.

singularity: A critical point of the (Hamiltonian) function, in one degree of freedom this yields an equilibrium. See also the first paragraph of Appendix A.

solenoid: Given a sequence $f_j : S^1 \longrightarrow S^1$ of coverings $f_j(\zeta) = \zeta^{\alpha_j}$ of the circle S^1 the solenoid $\Sigma_a \subseteq (S^1)^{\mathbb{N}_0}$, $a = (\alpha_j)_{j \in \mathbb{N}_0}$ consists of all $z = (\zeta_j)_{j \in \mathbb{N}_0}$ with $\zeta_j = f_j(\zeta_{j+1}) \; \forall_{j \in \mathbb{N}_0}$.

stable manifold: For an equilibrium $a \in \mathcal{P}$ of the vector field X with flow φ_t the stable manifold is the set $W_s(a) = \{ z \in \mathcal{P} \mid \lim_{t \to \infty} \varphi_t(z) = a \}$ of points that are asymptotic to this equilibrium. Given an invariant subset $A \subseteq \mathcal{P}$ of the flow this notion generalizes to the stable manifold $W_s(A)$ of points that are asymptotic to A. The unstable manifold $W_u(A) = \{ z \in \mathcal{P} \mid \lim_{t \to -\infty} d(\varphi_t(z), A) = 0 \}$ is defined by time reversal.

stratification: The decomposition of a topological space into smaller pieces satisfying certain boundary conditions. With the exception of *Cantor stratifications* all stratifications in these notes are Whitney-stratifications, see Definition B.1.

structurally stable: A system is structurally stable if it is *topologically equivalent* to all nearby systems. A family is structurally stable if for every nearby family there is a re-parametrisation such that all corresponding systems are topologically equivalent. See also Definition 2.2.

subcartesian: Essentially the stratified subsets of $\mathbb{R}^n$, see Definition B.2.

subcritical: Where the bifurcation scenario depends on a sign $\pm$ the subcritical case is the one where additional critical elements are created as the equilibrium/periodic orbit/invariant torus gains its stability. Alternatively one speaks of the *dual* case.

supercritical: Where the bifurcation scenario depends on a sign $\pm$ the supercritical case is the one where additional critical elements are created as the equilibrium/periodic orbit/invariant torus loses its stability, making this a "soft" loss of stability.

superintegrable system: A Hamiltonian system with d degrees of freedom is superintegrable if it has $d + 1$ functionally independent integrals of motion such that each of the first $d - 1$ of them commutes with all $d + 1$. Such a *properly degenerate* system admits generalized action angle co-ordinates $(\varphi_1, \ldots, \varphi_{d-1}, I_1, \ldots, I_{d-1}, q, p)$, see [219]. In case the non-degeneracy condition $\det D^2 H(I) \neq 0$ is satisfied almost everywhere the system is "minimally superintegrable". In the other extreme of a "maximally superintegrable" system all motions are periodic.

symplectic manifold: A $2n$-dimensional manifold $\mathcal{P}$ with a non-degenerate closed two-form ω, i.e. $d\omega = 0$ and $\omega(u, v) = 0 \ \forall_{v \in T\mathcal{P}} \Rightarrow u = 0$. A diffeomorphism ψ of symplectic manifolds that respects the two-form(s) is called a symplectomorphism. Given a Hamiltonian function $H \in C^\infty(\mathcal{P})$ one obtains through $\omega(X_H, ..) = dH$ the Hamiltonian vector field X_H. For every $x \in \mathcal{P}$ there are co-ordinates (q, p) around x with $\omega = dq \wedge dp$. In these *Darboux co-ordinates* X_H reads

$$\dot{q} = \frac{\partial H}{\partial p}$$

$$\dot{p} = -\frac{\partial H}{\partial q} \ .$$

syzygy: A constraining equation that is identically fulfilled by the elements of a *Hilbert basis*.

tangency bifurcations: Generically stable and *unstable manifolds* of fixed points of a *symplectic* mapping intersect transversely. Within generic one-parameter families they may also touch – intersect with matching tangents – giving rise to a tangency bifurcation.

(topologically) equivalent: Two flows φ_t and ψ_t are equivalent if there is a homeomorphism η that maps trajectories of φ_t into trajectories of ψ_t – preserving the time direction. In other words, there is a monotonous re-parametrisation $\tau(t, y)$ of the flow ψ_τ such that the two flows become topologically *conjugate*: $\eta(\varphi_t(z)) = \psi_{\tau(t, \eta(z))}(\eta(z)) \ \forall_z$. See also Definition 2.1. In case the phase space is not a smooth manifold we require η to map singular points to singular points.

ultraviolet cut-off: An upper bound of the order of the Fourier coefficients.

unfolding: Every family $(f_\lambda)_{\lambda \in \Lambda}$ containing g for à particular parameter value λ_0, i.e. with $f_{\lambda_0} = g$, is an unfolding of g; even the constant family $f_\lambda = g \ \forall_{\lambda \in \Lambda}$ is an unfolding of g. However, we are mostly interested in *structurally stable* or versal unfoldings.

unimodal: A *singularity* with exactly one *modulus*.

universal unfolding: An *unfolding* g_μ of h is universal if to every other unfolding f_λ of h there is a parameter change $\mu(\lambda)$ such that $g_{\mu(\lambda)}$ and f_λ are *equivalent*.

(un)stable manifold: In Hamiltonian systems with one degree of freedom the *stable manifold* and the unstable manifold of an equilibrium often coincide and thus consist of *homoclinic orbits*. In such a case we call it an (un)stable manifold. This carries over to the stable and the unstable manifold of a periodic orbit or an invariant torus in higher degrees of freedom if the system is *integrable*.

References

1. E. van der Aa and F. Verhulst: Asymptotic integrability and periodic solutions of a Hamiltonian system in 1:2:2-resonance; *SIAM J. Math. Anal.* **15**(5), p. 890–911 (1984)
2. E. van der Aa and M. de Winkel: Systems in 1:2:ω-resonance ($\omega = 5$ or 6); *Int. J. Non-Lin. Mech.* **29**(2), p. 261–270 (1994)
3. R. Abraham and J.E. Marsden: *Foundations of Mechanics*, 2nd ed.; Benjamin (1978)
4. J.M. Arms, R.H. Cushman and M.J. Gotay: A Universal Reduction Procedure for Hamiltonian Group Actions; p. 33–51 in *The Geometry of Hamiltonian Systems, Berkeley 1989* (ed. T.S. Raţiu), Springer (1991)
5. V.I. Arnol'd: Proof of a theorem of A. N. Kolmogorov on the invariance of quasi-periodic motions under small perturbations of the Hamiltonian; *Russ. Math. Surv.* **18**(5), p. 9–36 (1963)
6. V.I. Arnol'd: Small denominators and problems of stability of motion in classical and celestial mechanics; *Russ. Math. Surv.* **18**(6), p. 85–191 (1963)
7. V.I. Arnol'd: Instability of dynamical systems with several degrees of freedom; *Sov. Math. Dokl.* **5**, p. 581–585 (1964)
8. V.I. Arnol'd: On matrices depending on parameters; *Russ. Math. Surv.* **26**(2), p. 29–43 (1971)
9. V.I. Arnol'd: Normal forms of functions near degenerate critical points, the Weyl groups A_k, D_k, E_k, and Lagrangian singularities; *Funct. Anal. Appl.* **6**, p. 254–272 (1972)
10. V.I. Arnol'd: Normal forms of functions in neighbourhoods of degenerate critical points; *Russ. Math. Surv.* **29**(2), p. 11–49 (1974)
11. V.I. Arnol'd: Critical points of smooth functions and their normal forms; *Russ. Math. Surv.* **30**(5), p. 1–75 (1975)
12. V.I. Arnol'd: *Geometrical Methods in the Theory of Ordinary Differential Equations*; Springer (1983)
13. V.I. Arnol'd: *Mathematical Methods of Classical Mechanics*, 2nd ed.; Springer (1989)
14. V.I. Arnol'd, V.S. Afrajmovich, Yu.S. Il'yashenko and L.P. Shil'nikov: Bifurcation Theory; Part I in *Dynamical Systems V* (ed. V.I. Arnol'd) Springer (1993)

15. *V.I. Arnol'd, S.M. Gusein-Zade and A.N. Varchenko:* Singularities of Differentiable Maps, Vol. 1; Birkhäuser (1985)

16. *V.I. Arnol'd, V.V. Kozlov and A.I. Neishtadt:* Mathematical Aspects of Classical and Celestial Mechanics; in *Dynamical Systems III* (ed. V.I. Arnol'd) Springer (1988)

17. *V.I. Arnol'd, V.A. Vasil'ev, V.V.Goryunov and O.V. Lyashko:* Singularity theory I: Singularities, local and global theory; in *Dynamical Systems VI* (ed. V.I. Arnol'd) Springer (1993)

18. *I. Baldomá:* Contribution to the study of invariant manifolds and the splitting of separatrices of parabolic points; Ph.D. thesis, Universitat de Barcelona (2001)

19. *I. Baldomá and E. Fontich:* Exponentially Small Splitting of Invariant Manifolds of Parabolic Points; *Mem. AMS* **167** #792, p. 1–83 (2004)

20. *L. Bates and M. Zou:* Degeneration of Hamiltonian monodromy cycles; *Nonlinearity* **6**(2), p. 313–335 (1993)

21. *M. Beer:* Endliche Bestimmtheit und universelle Entfaltungen von Keimen mit Gruppenoperation; Diplomarbeit, Universität Regensburg (1976)

22. *M.L. Bertotti:* Localization of Closed Orbits of Nonlinear Hamiltonian Systems with $1, -2$ Resonance Near an Equilibrium; *Boll. U.M.I.*(7) **1**-B, p. 965–978 (1987)

23. *Yu.N. Bibikov:* Local theory of nonlinear analytic ordinary differential equations; LNM **702**, Springer (1979)

24. *Yu.N. Bibikov:* Construction of Invariant Tori of Systems of Differential Equations with a Small Parameter; *Transl. AMS, Ser.2* **155**, p. 19–46, (1993)

25. *B.D. Birkhoff:* Nouvelles recherches sur les systemes dynamiques; *Mem. Pont. Acad. Sci. Novi Lyncaei, Ser.3* **1**, p. 85–216 (1935)

26. *B.D. Birkhoff:* What is the ergodic theorem; *Amer. Math. Monthly* **49**, p. 222–226 (1942)

27. *A.V. Bolsinov and A.T. Fomenko:* Integrable Hamiltonian Systems – Geometry, Topology, Classification; Chapman & Hall/CRC (2004)

28. *S.V. Bolotin and D.V. Treschëv:* Remarks on the definition of hyperbolic tori of Hamiltonian systems; *Reg. & Chaot. Dyn.* **5**(4), p. 401–412 (2000)

29. *J. Bonet, R.W. Braun, R. Meise and B.A. Taylor:* Whitney's extension theorem for nonquasianalytic classes of ultradifferentiable functions; *Studia Math.* **99**(2), p.155–184 (1991)

30. *J. Bourgain:* Construction of quasi-periodic solutions for Hamiltonian perturbations of linear equations and applications to nonlinear PDE; *Int. Math. Res. Notices* **1994**(11), p. 475–497 (1994)

31. *J. Bourgain:* On Melnikov's persistency problem; *Math. Res. Lett.* **4**, p. 445–458 (1997)

32. *B.J.L. Braaksma and H.W. Broer:* Quasiperiodic flow near a codimension one singularity of a divergence free vector field in dimension four; p. 74–142 in *bifurcation, théorie ergodique et applications, Dijon 1981* (ed. Anonymous), *Astérisque* **98-99** (1982)

33. *B.J.L. Braaksma and H.W. Broer:* On a quasi-periodic Hopf bifurcation; *Ann. Inst. H. Poincaré, Analyse non linéaire* **4**(2), p. 115–168 (1987)

34. *B.J.L. Braaksma, H.W. Broer and G.B. Huitema:* Toward a quasi-periodic bifurcation theory; *Mem. AMS* **83** #421, p. 83–167 (1990)

35. *T.J. Bridges:* Bifurcation of periodic solutions near a collision of eigenvalues of opposite signature; *Math. Proc. Camb. Phil. Soc.* **108**, p. 575–601 (1990)

36. *T.J. Bridges:* Stability of periodic solutions near a collision of eigenvalues of opposite signature; *Math. Proc. Camb. Phil. Soc.* **109**, p. 375–403 (1991)

37. *T.J. Bridges, R.H. Cushman and R.S. MacKay:* Dynamics near an Irrational Collision of Eigenvalues for Symplectic Mappings; p. 61–79 in *Normal Forms and Homoclinic Chaos, Waterloo 1992* (eds. W.F. Langford and W. Nagata) Fields Institute Communications **4** AMS (1995)

38. *T.J. Bridges and J.E. Furter:* Singularity Theory and Equivariant Symplectic Maps; LNM **1558**, Springer (1993)

39. *E. Brieskorn:* Die Hierarchie der 1-modularen Singularitäten; *Manuscr. Math.* **27**, p. 183–219 (1979)

40. *Th. Bröcker and L. Lander:* Differentiable Germs and Catastrophes; Cambridge Univ. Press (1975)

41. *H.W. Broer:* Unfolding of Singularities in Volume Preserving Vector Fields; Ph.D. Thesis, Rijksuniversiteit Groningen (1979)

42. *H.W. Broer:* Quasiperiodic flow near a codimension one singularity of a divergence free vector field in dimension three; p. 75–89 in *Dynamical Systems and turbulence, Warwick 1979/1980* (eds. D. Rand and L.-S. Young) LNM **898**, Springer (1981)

43. *H.W. Broer, S.-N. Chow, Y. Kim and G. Vegter:* A normally elliptic Hamiltonian bifurcation; *Z. angew. Math. Phys.* **44**, p. 389–432 (1993)

44. *H.W. Broer, M.C. Ciocci and H. Hanßmann:* The quasi-periodic reversible Hopf bifurcation; preprint #1348, Mathematisch Instituut, Universiteit Utrecht (2006)

45. *H.W. Broer, M.C. Ciocci and A. Vanderbauwhede:* Normal 1:1 resonance of invariant tori in reversible systems: a study in quasiperiodic bifurcation theory – in preparation

46. *H.W. Broer, R. Cushman, F. Fassò and F. Takens:* Geometry of KAM tori for nearly integrable Hamiltonian systems; preprint, Rijksuniversiteit Groningen (2004)

47. *H.W. Broer, H. Hanßmann and J. Hoo:* The quasi-periodic Hamiltonian Hopf bifurcation; preprint, Rijksuniversiteit Groningen (2004)

48. *H.W. Broer, H. Hanßmann, J. Hoo and V. Naudot:* Nearly-integrable perturbations of the Lagrange top: applications of KAM-theory; p. 286–303 in: *Dynamics & Stochastics: Festschrift in Honor of M.S. Keane* (eds. D. Denteneer, F. den Hollander and E. Verbitskiy) Lecture Notes **48**, Inst. of Math. Statistics (2006)

49. *H.W. Broer, H. Hanßmann, À. Jorba, J. Villanueva and F.O.O. Wagener:* Normal-internal resonances in quasi-periodically forced oscillators: a conservative approach; *Nonlinearity* **16**, p. 1751–1791 (2003)

50. *H.W. Broer, H. Hanßmann and J. You:* Bifurcations of normally parabolic tori in Hamiltonian systems; *Nonlinearity* **18**, p. 1735–1769 (2005)

51. *H.W. Broer, H. Hanßmann and J. You:* Umbilical Torus Bifurcations in Hamiltonian Systems; *J. Diff. Eq.* **222**, p. 233–262 (2006)

52. *H.W. Broer, J. Hoo and V. Naudot:* Normal Linear Stability of Quasi-Periodic Tori; *J. Diff. Eq.* (to appear)

53. *H.W. Broer, I. Hoveijn, G. Lunter and G. Vegter:* Bifurcations in Hamiltonian systems: Computing Singularities by Gröbner Bases; LNM **1806**, Springer (2003)

54. *H.W. Broer and G.B. Huitema:* A proof of the isoenergetic KAM-theorem from the "ordinary" one; *J. Diff. Eq.* **90**(1), p. 52–60 (1991)

55. H.W. Broer, G.B. Huitema and M.B. Sevryuk: Quasi-Periodic Motions in Families of Dynamical Systems: Order amidst Chaos; LNM **1645**, Springer (1996)

56. H.W. Broer, G.B. Huitema and F. Takens: Unfoldings of quasi-periodic tori; Mem. AMS **83** #421, p. 1–82 (1990)

57. H.W. Broer, G. Lunter and G. Vegter: Equivariant singularity theory with distinguished parameters: Two case studies of resonant Hamiltonian systems; p. 64–80 in Time-reversal symmetry in dynamical systems, Warwick 1996 (ed. J.S.W. Lamb), Physica D **112** (1998)

58. H.W. Broer, R. Roussarie and C. Simó: Invariant circles in the Bogdanov–Takens bifurcation for diffeomorphisms; Erg. Th. Dyn. Syst. **16**, p. 1147–1172 (1996)

59. H.W. Broer and F. Takens: Unicity of KAM tori; preprint, Rijksuniversiteit Groningen (2005)

60. H.W. Broer, F. Takens and F.O.O. Wagener: Integrable and non-integrable deformations of the skew-Hopf bifurcation; Reg. & Chaot. Dyn. **4**(2), p. 17–43 (1999)

61. H.W. Broer and G. Vegter: Bifurcational Aspects of Parametric Resonance; Dynamics Rep., new ser. **1**, p. 1–53 (1992)

62. H.W. Broer and F.O.O. Wagener: Quasi-periodic stability of subfamilies of an unfolded skew-Hopf bifurcation; Arch. Rat. Mech. An. **152**, p. 283–326 (2000)

63. C.A. Buzzi and M.A. Teixeira: Time-reversible Hamiltonian vector fields with symplectic symmetries; J. Dynamics Diff. Eq. **16**, p. 559–574 (2004)

64. H.E. Cabral and K.E. Meyer: Stability of equilibria and fixed points of conservative systems; Nonlinearity **12**, p. 1351–1362 (1999)

65. A. Chenciner: Bifurcations des points elliptiques-I. Courbes invariantes; Publ. Math., Inst. Hautes Etud. Sci. **61**, p. 67–127 (1985)

66. A. Chenciner: Bifurcations des points elliptiques-II. Orbites périodiques et ensembles de Cantor invariants; Inv. Math. **80**, p. 81–106 (1985)

67. A. Chenciner: Bifurcations des points elliptiques-III. Orbites périodiques de "petites" périodes et élimination résonnante des couples de courbes invariantes; Publ. Math., Inst. Hautes Etud. Sci. **66**, p. 5–91 (1987)

68. A. Chenciner and G. Iooss: Bifurcations de tores invariants; Arch. Rat. Mech. An. **69**(2), p. 109–198 (1979)

69. C.-Q. Cheng: Birkhoff–Kolmogorov–Arnold–Moser tori in convex Hamiltonian systems; Comm. Math. Phys. **177**, p. 529–559 (1996)

70. C.-Q. Cheng: Lower Dimensional Invariant Tori in the Regions of Instability for Nearly Integrable Hamiltonian Systems; Comm. Math. Phys. **203**, p. 385–419 (1999)

71. L. Chierchia and C. Falcolini: A direct proof of a Theorem by Kolmogorov in Hamiltonian Systems; Ann. Sc. Norm. Sup. Pisa, Cl. Sci., Ser.IV **21**(4), p. 541–593 (1994)

72. L. Chierchia and G. Gallavotti: Smooth prime integrals for quasi-integrable Hamiltonian systems; Nuovo Cimento B **67**, p. 277–295 (1982)

73. L. Chierchia and D. Qian: Moser's theorem for lower dimensional tori; J. Diff. Eq. **206**(1), p. 55–93 (2004)

74. R.C. Churchill, M. Kummer and D.L. Rod: On Averaging, Reduction, and Symmetry in Hamiltonian Systems; J. Diff. Eq. **49**, p. 359–414 (1983)

75. M.C. Ciocci: *Bifurcations of periodic points in families of reversible mappings and of quasi-periodic solutions in reversible systems*; Ph.D. thesis, Universiteit Gent (2003)

76. M.C. Ciocci: Persistence of quasi periodic orbits in families of reversible systems with a 1:1 resonance; p. 720–725 in: *Equadiff 2003, Hasselt 2003* (eds. F. Dumortier et al.) World Scientific (2005)

77. M.C. Ciocci, A. Litvak-Hinenzon and H.W. Broer: Survey on dissipative KAM theory including quasi-periodic bifurcation theory; p. 303–355 in *Geometric Mechanics and Symmetry: the Peyresq Lectures* (eds. J. Montaldi and T.S. Raţiu) Cambridge Univ. Press (2005)

78. C. Cotter: *The 1:1 semisimple resonance*; Ph.D. thesis, University of California at Santa Cruz (1986)

79. D. Cox, J. Little and D. O'Shea: *Ideals, Varieties and Algorithms: An Introduction to Computational Algebraic Geometry and Commutative Algebra*; Springer (1992)

80. R. Cushman: A Survey of Normalization Techniques Applied to Perturbed Keplerian Systems; *Dynamics Rep., new ser.* **1**, p. 54–112 (1992)

81. R.H. Cushman and L.M. Bates: *Global Aspects of Classical Integrable Systems*; Birkhäuser (1997)

82. R. Cushman, H.R. Dullin, A. Giacobbe, D.D. Holm, M. Joyeux, P. Lynch, D.A. Sadovskií and B.I. Zhilinskií: Quantum Realization of the 1:1:2 Resonant Swing-Spring with Monodromy; *Phys. Rev. Lett.* **93**, p. 024302-1–024302-4 (2004)

83. R. Cushman, S. Ferrer and H. Hanßmann: Singular reduction of axially symmetric perturbations of the isotropic harmonic oscillator; *Nonlinearity* **12**, p. 389–410 (1999)

84. R. Cushman and J.C. van der Meer: The Hamiltonian Hopf bifurcation in the Lagrange top; p. 26–38 in *Géométrie Symplectique et Mécanique, La Grande Motte 1988* (ed. C. Albert) LNM **1416**, Springer (1990)

85. R. Cushman and D.A. Sadovskií: Monodromy in the hydrogen atom in crossed fields; *Physica D* **142** p. 166–196 (1998)

86. R. Cushman and J. Sanders: Invariant Theory and Normal Form of Hamiltonian Vectorfields with Nilpotent Linear Part; p. 353–371 in *Oscillation, Bifurcation and Chaos, Toronto 1986* (eds. F.V. Atkinson, W.F. Langford and A.B. Mingarelli), CMS Conference Proceedings **8**, AMS (1987)

87. R. Cushman and J. Śniatycki: Lectures on Reduction Theory for Hamiltonian systems; in: *Utrecht spring summer school on Lie groups* (eds. E. van der Ban and J.A.C. Kolk) Springer (to appear)

88. J.M. Damon: *Topological triviality and versality for subgroups of $\mathcal{A}$ and $\mathcal{K}$*; *Mem. AMS* **75** #389 (1988)

89. G.F. Dell'Antonio and B.M. D'Onofrio: On the number of periodic solutions of an Hamiltonian system near an equilibrium point; *Boll. U.M.I.*(6) **3**-B, p. 809–835 (1984)

90. A. Delshams, R. de la Llave and T. Martínez-Seara: A geometric mechanism for diffusion in Hamiltonian systems overcoming the large gap problem: announcement of results; *Electron. Res. Announc. AMS* **9**, p. 125–134 (2003)

91. A. Delshams, R. de la Llave and T. Martínez-Seara: A Geometric Mechanism for Diffusion in Hamiltonian Systems Overcoming the Large Gap Problem: Heuristics and Rigorous Verification on a Model; *Mem. AMS* **179** #844, p. 1–141 (2006)

92. A. Deprit and A. Deprit-Bartholomé: Stability of the Lagrange points; *Astron. J.* **72**, p. 173–179 (1967)

93. J.J. Duistermaat: Oscillatory integrals, Lagrange immersions, and unfoldings of singularities; *Comm. Pure Appl. Math.* **27**, p. 207–281 (1974)

94. J.J. Duistermaat: Bifurcations of periodic solutions near equilibrium points of Hamiltonian systems; p. 57–105 in *Bifurcation Theory and Applications, Montecatini 1983* (ed. L. Salvadori) LNM **1057**, Springer (1984)

95. J.J. Duistermaat: Non-integrability of the 1:1:2-resonance; *Ergod. Th. & Dynam. Sys.* **4**, p. 553–568 (1984)

96. J.J. Duistermaat: The monodromy of the Hamiltonian Hopf bifurcation; *Z. angew. Math. Phys.* **49**, p. 156–161 (1998)

97. J.J. Duistermaat: Dynamical Systems with Symmetry; in: *Utrecht spring summer school on Lie groups* (eds. E. van der Ban and J.A.C. Kolk) Springer (to appear)

98. H.R. Dullin, A. Giacobbe and R. Cushman: Monodromy in the resonant swing spring; *Physica D* **190**, p. 15–37 (2004)

99. H.R. Dullin and A.V. Ivanov: Vanishing twist in the Hamiltonian Hopf bifurcation; *Physica D* **201**, p. 27–44 (2005)

100. H.R. Dullin and A.V. Ivanov: Another look at the saddle-centre bifurcation: Vanishing twist; *Physica D* **211**, p. 47–56 (2005)

101. H.R. Dullin, A.V. Ivanov and J.D. Meiss: Normal Forms for Symplectic Maps with Twist Singularities; preprint #2134/564, Loughborough University (2005)

102. H.R. Dullin, J.D. Meiss and D. Sterling: Generic twistless bifurcations; *Nonlinearity* **13**, p. 203–224 (2000)

103. H.R. Dullin and Vũ Ngọc San: Vanishing twist near focus-focus points; *Nonlinearity* **17**, p. 1777–1785 (2004)

104. K. Efstathiou: *Metamorphoses of Hamiltonian systems with symmetries*; LNM **1864**, Springer (2005)

105. K. Efstathiou, R.H. Cushman and D.A. Sadovskií: Hamiltonian Hopf bifurcation of the hydrogen atom in crossed fields; *Physica D* **194**, p. 250–274 (2004)

106. K. Efstathiou and D.A. Sadovskií: Perturbations of the 1:1:1 resonance with tetrahedral symmetry: on a three degree of freedom generalization of the two degree of freedom Hénon–Heiles Hamiltonian; *Nonlinearity* **17**, p. 415–446 (2004)

107. K. Efstathiou, D.A. Sadovskií and R.H. Cushman: Linear Hamiltonian Hopf bifurcation for point-group-invariant perturbations of the 1:1:1 resonance; *Proc. Royal Soc. London* **459** #2040, p. 2997–3019 (2003)

108. L.H. Eliasson: Absolutely Convergent Series Expansions for Quasi Periodic Motions; *Math. Phys. El. J.* **2** #4, p. 1–33 (1996)

109. L.H. Eliasson, S.B. Kuksin, S. Marmi and J.-C. Yoccoz: *Dynamical Systems and Small Divisors, Cetraro 1998* (eds. S. Marmi and J.-C. Yoccoz) LNM **1784**, Springer (2002)

110. A. Elipe, V. Lanchares, T. López-Moratella and A. Riaguas: Nonlinear Stability in Resonant Cases: A Geometrical Approach; *J. Nonlinear Sci.* **11**, p. 211–222 (2001)

111. F. Fassò: Superintegrable Hamiltonian systems: Geometry and Perturbation; p. 93–121 in *Symmetry and Perturbation Theory, Cala Gonone 2004* (ed. G. Gaeta) *Acta Appl. Math.* **87**, p. 93 - 121 (2005)

112. *F. Fassò and A. Giacobbe:* Geometric structure of "broadly integrable" Hamiltonian systems; *J. Geom. Phys.* **44**, p. 156–170 (2002)

113. *J. Féjoz:* Dynamique séculaire globale du problème plan des trois corps et application à l'existence de mouvements quasipériodiques; Ph.D. thesis, Université Paris XIII (1999)

114. *J. Féjoz:* Quasiperiodic Motions in the Planar Three-Body Problem; *J. Diff. Eq.* **183**, p. 303–341 (2002)

115. *J. Féjoz:* Global secular dynamics in the planar three-body problem; *Cel. Mech. & Dyn. Astr.* **84**, p. 159–195 (2002)

116. *J. Féjoz:* Démonstration du "théorème d'Arnold" sur la stabilité du système planétaire (d'apres M. Herman); *Ergod. Th. & Dynam. Sys.* **24**, p. 1521–1582 (2004)

117. *E. Fermi:* Beweis, daß ein mechanisches Normalsystem im Allgemeinen quasi-ergodisch ist; *Phys. Z.* **24**, p. 261–265 (1923)

118. *S. Ferrer, H. Hanßmann, J. Palacián and P. Yanguas:* On perturbed oscillators in 1-1-1 resonance: the case of axially symmetric cubic potentials; *J. Geom. Phys.* **40**, p. 320–369 (2002)

119. *D. Flockerzi:* Generalized bifurcation of higher-dimensional tori; *J. Diff. Eq.* **55**(3), p. 346–367 (1984)

120. *D.M. Galin:* Versal Deformations of Linear Hamiltonian Systems; *AMS Transl.* **118**, p. 1–12 (1982)

121. *G. Gallavotti:* Twistless KAM tori, quasi flat homoclinic intersections, and other cancellations in the perturbation series of certain completely integrable Hamiltonian systems. A review; *Rev. Math. Phys.* **6**(3), p. 343–411 (1994)

122. *V. Gelfreich and L. Lerman:* Almost invariant elliptic manifold in a singularly perturbed Hamiltonian system; *Nonlinearity* **15**, p. 447–457 (2002)

123. *C.G. Gibson, K. Wirthmüller, A.A. du Plessis and E.J.N. Looijenga:* Topological stability of smooth mappings; LNM **552**, Springer (1976)

124. *A. Giorgilli and L. Galvani:* Rigorous estimates for the series expansions of Hamiltonian perturbation theory; *Celest. Mech.* **37**, p. 95–112 (1985)

125. *A. Giorgilli, A. Delshams, E. Fontich, L. Galvani and C. Simó:* Effective Stability for a Hamiltonian System near an Elliptic Equilibrium Point, with an Application to the Restricted Three Body Problem; *J. Diff. Eq.* **77**, p. 167–198 (1989)

126. *M. Golubitsky and V. Guillemin:* Stable Mappings and Their Singularities; Springer (1973)

127. *M. Golubitsky, J.E. Marsden, I. Stewart and M. Dellnitz:* The Constrained Liapunov–Schmidt Procedure and Periodic Orbits; p. 81–127 in *Normal Forms and Homoclinic Chaos, Waterloo 1992* (eds. W.F. Langford and W. Nagata) Fields Institute Communications 4 AMS (1995)

128. *M. Golubitsky and I. Stewart:* Generic bifurcation of Hamiltonian systems with symmetry; *Physica D* **24**, p. 391–405 (1987)

129. *J. Guckenheimer and P. Holmes:* Nonlinear Oscillations, Dynamical Systems, and Bifurcations of Vector Fields, 2nd ed.; Springer (1986)

130. *S. Gutiérrez-Romero, J.F. Palacián and P. Yanguas:* A universal procedure for normalizing n-degree-of-freedom polynomial Hamiltonian systems; *SIAM J. Math. Anal.* **65**(4), p. 1130–1152 (2005)

131. *G. Haller and S. Wiggins:* Whiskered tori and chaos in resonant Hamiltonian normal forms; p. 129–149 in *Normal Forms and Homoclinic Chaos, Waterloo*

1992 (eds. W.F. Langford and W. Nagata) Fields Institute Communications **4** AMS (1995)

132. G. Haller and S. Wiggins: Geometry and chaos near resonant equilibria of 3-DOF Hamiltonian systems; *Physica D* **90**, p. 319–365 (1996)

133. G. Haller: Chaos near Resonance; Applied Mathematical Sciences **138**, Springer (1999)

134. H. Hanßmann: Normal Forms for Perturbations of the Euler Top; p. 151–173 in *Normal Forms and Homoclinic Chaos, Waterloo 1992* (eds. W.F. Langford and W. Nagata) Fields Institute Communications **4** AMS (1995)

135. H. Hanßmann: *Quasi-periodic Motions of a Rigid Body — A case study on perturbations of superintegrable systems*; Ph.D. thesis, Rijksuniversiteit Groningen (1995)

136. H. Hanßmann: Equivariant Perturbations of the Euler Top; p. 227–253 in *Nonlinear Dynamical Systems and Chaos, Groningen 1995* (eds. H.W. Broer et al.) Birkhäuser (1996)

137. H. Hanßmann: Quasi-periodic Motions of a Rigid Body I – Quadratic Hamiltonians on the Sphere with a Distinguished Parameter; *Reg. & Chaot. Dyn.* **2**(2), p. 41–57 (1997)

138. H. Hanßmann: The reversible umbilic bifurcation; p. 81–94 in *Time-reversal symmetry in dynamical systems, Warwick 1996* (ed. J.S.W. Lamb), *Physica D* **112** (1998)

139. H. Hanßmann: The Quasi-Periodic Centre-Saddle Bifurcation; *J. Diff. Eq.* **142**(2), p. 305–370 (1998)

140. H. Hanßmann: A Survey on Bifurcations of Invariant Tori; p. 109–121 in *New Advances in Celestial Mechanics and Hamiltonian systems, Guanajuato 2001* (eds. J. Delgado et al.) Kluwer/Plenum (2004)

141. H. Hanßmann: Hamiltonian Torus Bifurcations Related to Simple Singularities; p. 679–685 in *Dynamic Systems and Applications, Atlanta 2003* (eds. G.S. Ladde, N.G. Medhin and M. Sambandham) Dynamic Publishers (2004)

142. H. Hanßmann: *Local and Semi–Local Bifurcations in Hamiltonian Dynamical Systems — Results and Examples*; Habilitationsschrift, RWTH Aachen (2004)

143. H. Hanßmann: Perturbations of integrable and superintegrable Hamiltonian systems; p. 1527–1536 in *Fifth Euromech Nonlinear Dynamics Conference, Eindhoven 2005* (eds. D.H. van Campen, M.D. Lazurko and W.P.J.M. van den Oever) Technische Universiteit Eindhoven (2005)

144. H. Hanßmann: On Hamiltonian bifurcations of invariant tori with a Floquet multiplier −1; *Dynamical Systems* **21**(2), p. 115–145 (2006)

145. H. Hanßmann and P. Holmes: On the global dynamics of Kirchhoff's equations: Rigid body models for underwater vehicles; p. 353–371 in *Global Analysis of Dynamical Systems, Leiden 2001* (eds. H.W. Broer, B. Krauskopf and G. Vegter) Inst. of Phys. (2001)

146. H. Hanßmann and J.C. van der Meer: On the Hamiltonian Hopf bifurcations in the 3*D* Hénon–Heiles family; *J. Dynamics Diff. Eq.* **14**, p. 675–695 (2002)

147. H. Hanßmann and J.C. van der Meer: On non-degenerate Hamiltonian Hopf bifurcations in 3DOF systems; p. 476–481 in: *Equadiff 2003, Hasselt 2003* (eds. F. Dumortier et al.) World Scientific (2005)

148. H. Hanßmann and J.C. van der Meer: Algebraic methods for determining Hamiltonian Hopf bifurcations in three-degree-of-freedom systems; *J. Dynamics Diff. Eq.* **17**(3), p. 453–474 (2005)

149. *H. Hanßmann and B. Sommer:* A degenerate bifurcation in the Hénon–Heiles family; *Cel. Mech. & Dyn. Astr.* **81**(3), p. 249–261 (2001)

150. *D.C. Heggie:* Bifurcation at complex instability; *Cel. Mech.* **35**, p. 357–382 (1985)

151. *M.W. Hirsch, C.C. Pugh and N. Shub: Invariant manifolds;* LNM **583**, Springer (1977)

152. *D.D. Holm and P. Lynch:* Stepwise precession of the resonant swinging spring; *SIAM J. Appl. Dynam. Syst.* **1**, p. 44–64 (2002)

153. *J. Hoo: Quasi-periodic bifurcations in a strong resonance: combination tones in gyroscopic stabilization;* Ph.D. thesis, Rijksuniversiteit Groningen (2005)

154. *E. Hopf:* A Mathematical Example Displaying Features of Turbulence; *Comm. (Pure) Appl. Math.* **1**, p. 303–322 (1948)

155. *E. Hopf:* Repeated branching through loss of stability, an example; p. 49–56 in *Proceedings of the conference on differential equations, Maryland 1955* (ed. J.B. Diaz) Univ. Maryland book store (1956)

156. *I. Hoveijn: Aspects of resonance in dynamical systems;* Ph.D. thesis, Universiteit Utrecht (1992)

157. *I. Hoveijn:* Versal Deformations and Normal Forms for Reversible and Hamiltonian Linear Systems; *J. Diff. Eq.* **126**(2), p. 408–442 (1996)

158. *I. Hoveijn and F. Verhulst:* Chaos in the 1:2:3 Hamiltonian normal form; *Physica D* **44**(3), p. 397–406 (1990)

159. *G.B. Huitema: Unfoldings of Quasi-Periodic Tori;* Ph.D. thesis, Rijksuniversiteit Groningen (1988)

160. *D. Huang and Z. Liu:* On the persistence of lower-dimensional invariant hyperbolic tori for smooth Hamiltonian systems; *Nonlinearity* **13**(1), p. 189–202 (2000)

161. *H.H. de Jong: Quasiperiodic breathers in systems of weakly coupled pendulums;* Ph.D. Thesis, Rijksuniversiteit Groningen (1999)

162. *À. Jorba, R. de la Llave and M. Zou:* Lindstedt series for lower dimensional tori; p. 151–167 in *Hamiltonian Systems with Three or More Degrees of Freedom, S'Agaró 1995* (ed. C. Simó) NATO Adv. Sci. Inst. C **533**, Kluwer (1999)

163. *À. Jorba and J. Villanueva:* On the persistence of lower dimensional invariant tori under quasi-periodic perturbations, *J. Nonlinear Sci.* **7**, p. 427–473 (1997)

164. *À. Jorba and J. Villanueva:* On the normal behaviour of partially elliptic lower-dimensional tori of Hamiltonian systems; *Nonlinearity* **10**, p. 783–822 (1997)

165. *H. Koçak:* Normal Forms and Versal Deformations of Linear Hamiltonian Systems; *J. Diff. Eq.* **51**, p. 353–407 (1984)

166. *A.N. Kolmogorov:* О сохранении условно-периодических движений при малом изменении функции Гамильтона; Докл. акад. наук СССР **98**, p. 527–530 (1954) – translated as: Preservation of conditionally periodic movements with small change in the Hamilton function; p. 51–56 in *Stochastic Behavior in Classical and Quantum Hamiltonian Systems, Como 1977* (eds. G. Casati and J. Ford) Lect. Notes Phys. **93**, Springer (1979) – reprinted p. 81–86 in *Chaos* (ed. Bai Lin Hao) World Scientific (1984)

167. *R. Krikorian:* Réductibilité presque partout des flots fibrés quasi-périodiques à valeurs dans des groupes compactes; *Ann. Sci. École Norm. Sup. 4* **32**(2), p. 187–240 (1999)

168. *S.B. Kuksin: Nearly integrable infinite-dimensional Hamiltonian systems;* LNM **1556**, Springer (1993)

228   References

169. *M. Kummer:*   On Resonant Nonlinearly Coupled Oscillators with Two Equal Frequencies; *Comm. Math. Phys.* **48**, p. 53–79 (1976)

170. *M. Kummer:*   On Resonant Classical Hamiltonians with Two Equal Frequencies; *Comm. Math. Phys.* **58**, p. 85–112 (1978)

171. *M. Kummer:*   On resonant Hamiltonian systems with finitely many degrees of freedom; p. 19–31 in *Local and global methods of nonlinear dynamics, Silver Spring 1984* (eds. A.V. Sáenz, W.W. Zachary and R. Cawley) Lect. Notes Phys. **252**, Springer (1986)

172. *M. Kummer:*   On Resonant Classical Hamiltonians with n Frequencies; *J. Diff. Eq.* **83**, p. 220–243 (1990)

173. *Yu. Kuznetsov: Elements of applied bifurcation theory;* Applied Mathematical Sciences **112**, Springer (1995)

174. *V. Lanchares, A.I. Pascual, J. Palacián, P. Yanguas and J.P. Salas:* Perturbed ion traps: A generalization of the three-dimensional Hénon–Heiles problem; *Chaos* **12**(1), p. 87–99 (2002)

175. *Y. Li and Y. Yi:* Persistence of invariant tori in generalized Hamiltonian systems; *Ergod. Th. & Dynam. Syst.* **22**(4), p. 1233–1261 (2002)

176. *B.B. Lieberman:* Existence of Quasi-Periodic Solutions to the Three-Body Problem; *Celest. Mech.* **3**, p. 408–426 (1971)

177. *B.B. Lieberman:* Quasi-Periodic Solutions of Hamiltonian Systems; *J. Diff. Eq.* **11**, p. 109–137 (1972)

178. *A. Litvak-Hinenzon: Parabolic Resonances in Hamiltonian Systems;* Ph.D. thesis, The Weizmann Institute of Science, Rehovot (2001)

179. *A. Litvak-Hinenzon:* The mechanism of parabolic resonance orbits; p. 738–743 in: *Equadiff 2003, Hasselt 2003* (eds. F. Dumortier et al.) World Scientific (2005)

180. *A. Litvak-Hinenzon and V. Rom-Kedar:* Parabolic resonances in near integrable Hamiltonian systems; p. 358–368 in *Stochaos: Stochastic and Chaotic Dynamics in the Lakes* (eds. D.S. Broomhead et al.), Amer. Inst. Phys. (2000)

181. *A. Litvak-Hinenzon and V. Rom-Kedar:* Parabolic resonances in 3 degree of freedom near-integrable Hamiltonian systems; *Physica D* **164**, p. 213–250 (2002)

182. *A. Litvak-Hinenzon and V. Rom-Kedar:* Resonant tori and instabilities in Hamiltonian systems; *Nonlinearity* **15**, p. 1149–1177 (2002)

183. *A. Litvak-Hinenzon and V. Rom-Kedar:* On Energy Surfaces and the Resonance Web; *SIAM J. Appl. Dynam. Syst.* **3**(4), p. 525–573 (2004)

184. *Z.-J. Liu and P. Xu:* On Quadratic Poisson Structures; *Lett. Math. Phys.* **26**, p. 33–42 (1992)

185. *E.J.N. Looijenga: Structural stability of families of C^∞-functions;* Ph.D. thesis, Universiteit van Amsterdam (1974)

186. *J. Los:* Dédoublement de courbes invariantes sur le cylindre: petit diviseurs; *Ann. Inst. H. Poincaré, Analyse non linéaire* **5**(1), p. 37–95 (1988)

187. *J. Los:* Non-normally hyperbolic invariant curves for maps in $\mathbb{R}^3$ and doubling bifurcation; *Nonlinearity* **2**, p. 149–174 (1989)

188. *G. Lunter: Bifurcations of Hamiltonian systems: computing singularities by Gröbner bases;* Ph.D. thesis, Rijksuniversiteit Groningen (1999)

189. *R.S. MacKay: Renormalisation in Area-preserving Maps;* World Scientific (1993)

190. *J.-P. Marco and D. Sauzin:* Stability and instability for Gevrey quasi-convex near-integrable Hamiltonian systems; *Publ. Math., Inst. Hautes Etud. Sci.* **96**, p. 199–275 (2003)

191. *L. Markus and K.R. Meyer:* Generic Hamiltonian dynamical systems are neither integrable nor ergodic; *Mem. AMS* #144, p. 1–52 (1974)

192. *L. Markus and K.R. Meyer:* Periodic orbits and solenoids in generic Hamiltonian dynamical systems; *Am. J. Math.* **102**(1), p. 25–92 (1980)

193. *J.E. Marsden and T.S. Raţiu:* Introduction to Mechanics and Symmetry; Springer (1994)

194. *J.E. Marsden and A. Weinstein:* Reduction of symplectic manifolds with symmetry; *Rep. Math. Phys.* **5**(1), p. 121–130 (1974)

195. *S. Mayer, J. Scheurle and S. Walcher:* Practical normal form computations for vector fields; *Z. angew. Math. Mech.* **84**(7), p. 472–482 (2004)

196. *M. Mazzocco:* KAM theorem for generic analytic perturbations of the Euler system; *Z. Angew. Math. Phys.* **48**, p. 193–219 (1997)

197. *J.C. van der Meer:* Nonsemisimple 1:1 resonance at an equilibrium; *Celest. Mech.* **27**, p. 131–149 (1982)

198. *J.C. van der Meer:* The Hamiltonian Hopf bifurcation; LNM **1160**, Springer (1985)

199. *J.C. van der Meer:* Hamiltonian Hopf bifurcation with symmetry; *Nonlinearity* **3**, p. 1041–1056 (1990)

200. *J.C. van der Meer:* Degenerate Hamiltonian Hopf bifurcations; p. 159–176 in *Conservative systems and quantum chaos, Waterloo 1992*; (eds. L.M. Bates and D.L. Rod) *Fields Institute Communications* **8** AMS (1996)

201. *J.C. van der Meer and R. Cushman:* Constrained normalization of Hamiltonian systems and perturbed Keplerian motion; *Z. Angew. Math. Phys.* **37**(3), p. 402–424 (1986)

202. *J.C. van der Meer and R. Cushman:* Orbiting dust under radiation pressure; p. 403–414 in *Differential Geometric Methods in Theoretical Physics, Clausthal, 1986* (ed. H. Doebner) World Scientific (1987)

203. *J.D. Meiss:* Class renormalization: Islands around islands; *Phys. Rev. A* **34**(3), p. 2375–2383 (1986)

204. *V.K. Mel'nikov:* On some cases of conservation of conditionally periodic motions under a small change of the Hamiltonian function; *Sov. Math. Dokl.* **6**(6), p. 1592–1596 (1965)

205. *K.R. Meyer:* Generic bifurcation of periodic points; *Trans. AMS* **149**, p. 95–107 (1970)

206. *K.R. Meyer:* Symmetries and integrals in mathematics; p. 259–272 in *Dynamical Systems* (ed. M.M. Peixoto), Academic Press (1973)

207. *K.R. Meyer:* Generic Bifurcation in Hamiltonian Systems; p. 62–70 in *Dynamical Systems — Warwick 1974* (ed. A. Manning) LNM **468**, Springer (1975)

208. *K.R. Meyer and G.R. Hall:* Introduction to Hamiltonian Dynamical Systems and the N-Body Problem; Applied Mathematical Sciences **90**, Springer (1992)

209. *K.R. Meyer and D.S. Schmidt:* The stability of the Lagrange triangular point and a theorem of Arnold; *J. Diff. Eq.* **62**(2), p. 222–236 (1986)

210. *K.R. Meyer and G.R. Sell:* Melnikov Transforms, Bernoulli Bundles, and Almost Periodic Perturbations; *Trans. AMS* **314**(1), p. 63–105 (1989)

211. *A. Mielke:* Hamiltonian and Lagrangian Flows on Center Manifolds – with Applications to Elliptic Variational Problems; LNM **1489**, Springer (1991)

212. *A.S. Mishchenko and A.T. Fomenko:* Generalized Liouville method of integration of Hamiltonian systems; *Funct. Anal. Appl.* **12**(2), p. 113–121 (1978)

213. *A. Morbidelli and A. Giorgilli:* Superexponential Stability of KAM Tori; *J. Stat. Phys.* **78**(5/6), p. 1607–1617 (1995)

214. *A. Morbidelli and A. Giorgilli:* On a connection between KAM and Nekhoroshev's theorems; *Physica D* **86**(3), p. 514–516 (1995)

215. *J. Moser:* On Invariant Curves of Area-Preserving Mappings of an Annulus; *Nachr. Akad. Wiss. Göttingen, II. Math.-Phys. Kl.* **1962**(1), p. 1–20 (1962)

216. *J. Moser:* Convergent Series Expansion for Quasi-Periodic Motions; *Math. Ann.* **169**(1), p. 136–176 (1967)

217. *J. Moser:* Periodic Orbits near an Equilibrium and a Theorem by Alan Weinstein; *Comm. Pure Appl. Math.* **29**, p. 727–747 (1976)

218. *J. Murdock:* Normal forms and unfoldings for local dynamical systems; Springer (2003)

219. *N.N. Nekhoroshev:* Action-angle variables and their generalizations; *Trans. Mosc. Math. Soc.* **26**, p. 180–198 (1972)

220. *N.N. Nekhoroshev:* An exponential estimate of the time of stability of nearly-integrable Hamiltonian systems; *Russ. Math. Surv.* **32**(6), p. 1–65 (1977)

221. *N.N. Nekhoroshev:* An exponential estimate of the time of stability of nearly integrable Hamiltonian systems. II; p. 1–58 in *Topics in Modern Mathematics, Petrovskii Seminar No.5* (ed. O.A. Oleinik), Consultants Bureau (1985)

222. *A.I. Neishtadt:* Estimates in the Kolmogorov theorem on conservation of conditionally periodic motions; *J. Appl. Math. Mech.* **45**, p. 766–772 (1982)

223. *S. Newhouse, J. Palis and F. Takens:* Bifurcations and stability of families of diffeomorphisms; *Publ. Math., Inst. Hautes Etud. Sci.* **57**, p. 5–71 (1983)

224. *Nguyen Tien Zung:* Kolmogorov condition for integrable systems with focus-focus singularities; *Phys. Lett. A* **215**, p. 40–44 (1996)

225. *Nguyen Tien Zung:* A note on focus-focus singularities; *Diff. Geom. Appl.* **7**(2), p. 123–130 (1997)

226. *L. Niederman:* Prevalence of exponential stability among nearly-integrable Hamiltonian systems; preprint #2004-39, Département Mathématique, Université de Paris-Sud (2004)

227. *M. Ollé, J.R. Pacha and J. Villanueva:* Dynamics and bifurcation near the transition from stability to complex instability; p. 185–197 in *New Advances in Celestial Mechanics and Hamiltonian systems, Guanajuato 2001* (eds. J. Delgado et al.) Kluwer/Plenum (2004)

228. *M. Ollé, J.R. Pacha and J. Villanueva:* Dynamics close to a non semi-simple 1:−1 resonance periodic orbit; *Discr. Cont. Dyn. Syst. B* **5**, p. 799–816 (2005)

229. *M. Ollé, J.R. Pacha and J. Villanueva:* Quantitative estimates on the normal form around a non-semi-simple 1:−1 resonant periodic orbit; *Nonlinearity* **18**, p. 1141–1172 (2005)

230. *J.-P. Ortega and T.S. Raţiu:* Momentum Maps and Hamiltonian Reduction; Progress in Mathematics **222**, Birkhäuser (2004)

231. *J.C. Oxtoby:* Measure and category. A survey of the analogies between topological and measure spaces; Graduate Texts in Mathematics **2**, Springer (1971)

232. *J.R. Pacha:* On the quasi-periodic Hamiltonian Andronov–Hopf bifurcation; Ph.D. thesis, Universitat Politècnica de Catalunya (2002)

233. *J. Palacián and P. Yanguas:* Reduction of polynomial Hamiltonians by the construction of formal integrals; *Nonlinearity* **13**, p. 1021–1054 (2000)

234. J. Palacián and P. Yanguas: Invariant Manifolds of Spatial Restricted Three-Body Problems: the Lunar Case; p. 199–224 in *New Advances in Celestial Mechanics and Hamiltonian systems, Guanajuato 2001* (eds. J. Delgado et al.) Kluwer/Plenum (2004)

235. G.W. Patrick, M. Roberts and C. Wulff: Stability of Poisson Equilibria and Hamiltonian Relative Equilibria by Energy Methods; *Arch. Rat. Mech. An.* **173**(3), p. 301–344 (2004)

236. M.J. Pflaum: Analytic and Geometric Study of Stratified Spaces; LNM **1768**, Springer (2001)

237. A. du Plessis and T. Wall: *The Geometry of Topological Stability*; LMS Monographs, new ser. **9**, Clarendon (1995)

238. V. Poénaru: *Singularités C^∞ en Présence de Symétrie*; LNM **510**, Springer (1976)

239. H. Poincaré: *Les Méthodes Nouvelles de la Mécanique Céleste*; Gauthier-Villars (1892,1893,1899) – reprint: Dover (1957)

240. J. Pöschel: Integrability of Hamiltonian Systems on Cantor Sets; *Comm. Pure Appl. Math.* **35**, p. 653–696 (1982)

241. J. Pöschel: On Elliptic Lower Dimensional Tori in Hamiltonian Systems; *Math. Z.* **202**, p. 559–608 (1989)

242. T. Poston and I. Stewart: *Catastrophe Theory and Its Applications*; Pitman (1978)

243. R.J. Rimmer: Symmetry and bifurcation of fixed points of area preserving maps; *J. Diff. Eq.* **29**, p. 329–344 (1978)

244. R.J. Rimmer: Generic bifurcations for involutary area preserving maps; *Mem. AMS* **41** #272, p. 1–165 (1983)

245. B. Rink: *Geometry and dynamics in Hamiltonian lattices*; Ph.D. thesis, Universiteit Utrecht (2003)

246. B. Rink: A Cantor set of tori with monodromy near a focus-focus singularity; *Nonlinearity* **17**, p. 347–356 (2004)

247. R.C. Robinson: Generic properties of conservative systems; *Am. J. Math.* **92**, p. 562–603 (1970)

248. R.C. Robinson: Generic properties of conservative systems II; *Am. J. Math.* **92**, p. 897–906 (1970)

249. M. Rudnev: Hamilton–Jacobi method for a simple resonance; preprint, University of Bristol (2003)

250. H. Rüssmann: Über invariante Kurven differenzierbarer Abbildungen eines Kreisringes; *Nachr. Akad. Wiss. Göttingen, II. Math.-Phys. Kl.* **1970**(5), p. 67–105 (1970)

251. H. Rüssmann: Invariant tori in non-degenerate nearly integrable Hamiltonian systems; *Reg. & Chaot. Dyn.* **6**(2), p. 119–204 (2001)

252. R. Sacker: *On invariant surfaces and bifurcations of periodic ordinary differential equations*; Ph.D. thesis, New York Univ. (1964)

253. D.A. Sadovskií and J.B. Delos: Bifurcation of periodic orbits of Hamiltonian systems: An analysis using normal form theory; *Phys. Rev. E* **54**(2), p. 2033–2070 (1996)

254. J. Sanders: Are higher order resonances really interesting? *Celest. Mech.* **16**, p. 421–440 (1978)

255. J. Sanders: Normal form theory and spectral sequences; *J. Diff. Eq.* **192**, p. 536–552 (2003)

256. *J.A. Sanders and F. Verhulst:* Averaging Methods in Nonlinear Dynamical Systems; Springer (1985)

257. *D. Schaeffer:* A Regularity Theorem for Conservation Laws; Adv. Math. **11** p. 368–386 (1973)

258. *J. Scheurle and S. Walcher:* On Normal Form Computations; p. 309–325 in: Geometry, Mechanics and Dynamics: Volume in honor of the 60th birthday of J.E. Marsden (eds. P. Newton, P. Holmes and A. Weinstein) Springer (2002)

259. *R. Schrauwen:* Series of Singularities and Their Topology; Ph.D. thesis, Rijksuniversiteit te Utrecht (1990)

260. *G.R. Sell:* Bifurcation of higher-dimensional tori; Arch. Rat. Mech. An. **69**(3), p. 199–230 (1979)

261. *M.B. Sevryuk:* Invariant Tori of Intermediate Dimensions in Hamiltonian Systems; Reg. & Chaot. Dyn. **2**(3/4), p. 30–40 (1997)

262. *M.B. Sevryuk:* The classical KAM theory at the dawn of the twenty-first century; Moscow Math. J. **3**(3), p. 1113–1144 (2003)

263. *D. Siersma:* Classification and deformation of singularities; Ph.D. thesis, Universiteit van Amsterdam (1974)

264. *D. Siersma:* Singularities of functions on boundaries, corners, etc.; Quart. J. Math. Oxford (2) **32**, p. 119–127 (1981)

265. *R. Sjamaar and E. Lerman:* Stratified symplectic spaces and reduction; Ann. Math. **134**, p. 375–422 (1991)

266. *A.G. Sokol'skii:* On the stability of an autonomous Hamiltonian system with two degrees of freedom in the case of equal frequencies; J. appl. Math. Mech. **38**(5), p. 741–749 (1974-5)

267. *A.G. Sokol'skii:* Proof of the stability of Lagrangian solutions at a critical relation of masses; Sov. Astron. Lett. **4**(2), p. 79–81 (1978)

268. *B.S. Sommer:* A KAM Theorem for the Spatial Lunar Problem; Ph.D. thesis, RWTH Aachen (2003)

269. *I. Stewart:* Quantizing the classical cat; Nature **430**, p. 731–732 (2004)

270. *F. Takens:* A C^1-counterexample of Mosers's twist theorem; Indag. math. **33**, p. 378–386 (1971)

271. *F. Takens:* Introduction to Global Analysis; Communications **2** of the Mathematical Institute, Rijksuniversiteit Utrecht (1973)

272. *F. Takens:* Forced oscillations and bifurcations; p. 1–59 in Applications of Global Analysis; Communications **3** of the Mathematical Institute, Rijksuniversiteit Utrecht (1974) – reprinted p. 2–61 in Global Analysis of Dynamical Systems, Leiden 2001 (eds. H.W. Broer, B. Krauskopf and G. Vegter) Inst. of Phys. (2001)

273. *F. Takens and F.O.O. Wagener:* Resonances in skew and reducible quasi-periodic Hopf bifurcations; Nonlinearity **13**, p. 377–396 (2000)

274. *J.C. Tatjer:* Some bifurcations related to homoclinic tangencies for 1-parameter families of symplectic diffeomorphisms; p. 595–599 in Hamiltonian Systems with Three or More Degrees of Freedom, S'Agaró 1995 (ed. C. Simó) NATO Adv. Sci. Inst. C **533**, Kluwer (1999)

275. *D.V. Treshchëv:* The mechanism of destruction of resonant tori of Hamiltonian systems; Math. USSR-Sb. **68**, p. 181–203 (1991)

276. *D.V. Treshchëv:* Hyperbolic tori and asymptotic surfaces in Hamiltonian systems; Russ. J. Math. Phys. **2**(1), p. 93–110 (1994)

277. *A. Vanderbauwhede:* Centre Manifolds, Normal Forms and Elementary Bifurcations; Dynamics Rep. **2**, p. 89–169 (1989)

278. F. *Verhulst:* Asymptotic analysis of Hamiltonian systems; p. 137–183 in *Asymptotic Analysis II – Survey and New Trends* (ed. F. Verhulst) LNM **985,** Springer (1983)

279. J. *Villanueva:* *Normal Forms around Lower Dimensional Tori of Hamiltonian Systems*; Ph.D. thesis, Universitat Politècnica de Catalunya (1997)

280. H. *Waalkens, A. Junge and H.R. Dullin:* Quantum monodromy in the two-centre problem; *J. Phys. A: Math. Gen.* **36,** p. L307–L314 (2003)

281. T. *Wagenknecht:* Bifurcation of a reversible Hamiltonian system from a fixed point with fourfold eigenvalue zero; *Dynamical Systems* **17**(1), p. 29–44 (2002)

282. F.O.O. *Wagener:* On the skew Hopf bifurcation; Ph.D. thesis, Rijksuniversiteit Groningen (1998)

283. F.O.O. *Wagener:* A note on Gevrey regular KAM theory and the inverse approximation lemma; *Dynamical Systems* **18**(2), p. 159–163 (2003)

284. F.O.O. *Wagener:* On the quasi-periodic d-fold degenerate bifurcation; *J. Diff. Eq.* **216**, p. 261–281 (2005)

285. F.O.O. *Wagener:* A parametrised version of Moser's modifying terms theorem – *in preparation*

286. C.T.C. *Wall:* Geometric properties of generic differentiable manifolds; p. 707–774 in: *Geometry and Topology, Rio de Janeiro 1976* (eds. J. Palis and M. do Carmo) LNM **597**, Springer (1977)

287. G. *Wassermann:* *Stability of unfoldings*; LNM **393**, Springer (1974)

288. G. *Wassermann:* Stability of unfoldings in space and time; *Acta math.* **135**, p. 57–128 (1975)

289. G. *Wassermann:* *Classification of singularities with compact Abelian symmetry*; Regensburger Math. Schriften **1** (1977)

290. A. *Weinstein:* Normal Modes for Nonlinear Hamiltonian Systems; *Inv. math.* **20**, p. 47–57 (1973)

291. A. *Weinstein:* Bifurcations and Hamilton's Principle; *Math. Z.* **159**, p. 235–248 (1978)

292. A. *Weinstein:* The local structure of Poisson manifolds; *J. Diff. Geom.* **18**, p. 523–557 (1983)

293. K. *Wirthmüller:* *Universell topologisch triviale Deformationen*; Ph.D. thesis, Universität Regensburg (1979)

294. J. *Xu and J. You:* Persistence of lower-dimensional tori under the first Melnikov's non-resonance condition; *J. Math. Pures Appl.* **80**(10), p. 1045–1067 (2001)

295. P. *Yanguas:* *Integrability, Normalization and Symmetries of Hamiltonian Systems in 1-1-1 Resonance*; Ph.D. thesis, Universidad Pública de Navarra (1998)

296. J. *You:* A KAM Theorem for Hyperbolic-Type Degenerate Lower Dimensional Tori in Hamiltonian Systems; *Comm. Math. Phys.* **192**(1), p. 145–168 (1998)

297. J. *You:* Lower dimensional tori of Reversible Hamiltonian Systems in the resonant zone; p. 301–314 in *Dynamical Systems, Beijing 1998*, (eds. J. Yunping and W. Lan) World Scientific (1999)

298. E. *Zehnder:* Generalized Implicit Function Theorems with Applications to Some Small Divisor Problems I; *Comm. Pure Appl. Math.* **28**, p. 91–140 (1975)

299. E. *Zehnder:* Generalized Implicit Function Theorems With Applications To Some Small Divisor Problems II; *Comm. Pure Appl. Math.* **29**, p. 49–111 (1976)

Index

Lecture Notes in Mathematics

For information about earlier volumes
please contact your bookseller or Springer
LNM Online archive: springerlink.com

Vol. 1805: F. Cao, Geometric Curve Evolution and Image Processing (2003)

Vol. 1806: H. Broer, I. Hoveijn. G. Lunther, G. Vegter, Bifurcations in Hamiltonian Systems. Computing Singularities by Gröbner Bases (2003)

Vol. 1807: V. D. Milman, G. Schechtman (Eds.), Geometric Aspects of Functional Analysis. Israel Seminar 2000-2002 (2003)

Vol. 1808: W. Schindler, Measures with Symmetry Properties (2003)

Vol. 1809: O. Steinbach, Stability Estimates for Hybrid Coupled Domain Decomposition Methods (2003)

Vol. 1810: J. Wengenroth, Derived Functors in Functional Analysis (2003)

Vol. 1811: J. Stevens, Deformations of Singularities (2003)

Vol. 1812: L. Ambrosio, K. Deckelnick, G. Dziuk, M. Mimura, V. A. Solonnikov, H. M. Soner, Mathematical Aspects of Evolving Interfaces. Madeira, Funchal, Portugal 2000. Editors: P. Colli, J. F. Rodrigues (2003)

Vol. 1813: L. Ambrosio, L. A. Caffarelli, Y. Brenier, G. Buttazzo, C. Villani, Optimal Transportation and its Applications. Martina Franca, Italy 2001. Editors: L. A. Caffarelli, S. Salsa (2003)

Vol. 1814: P. Bank, F. Baudoin, H. Föllmer, L.C.G. Rogers, M. Soner, N. Touzi, Paris-Princeton Lectures on Mathematical Finance 2002 (2003)

Vol. 1815: A. M. Vershik (Ed.), Asymptotic Combinatorics with Applications to Mathematical Physics. St. Petersburg, Russia 2001 (2003)

Vol. 1816: S. Albeverio, W. Schachermayer, M. Talagrand, Lectures on Probability Theory and Statistics. Ecole d'Eté de Probabilités de Saint-Flour XXX-2000. Editor: P. Bernard (2003)

Vol. 1817: E. Koelink, W. Van Assche(Eds.), Orthogonal Polynomials and Special Functions. Leuven 2002 (2003)

Vol. 1818: M. Bildhauer, Convex Variational Problems with Linear, nearly Linear and/or Anisotropic Growth Conditions (2003)

Vol. 1819: D. Masser, Yu. V. Nesterenko, H. P. Schlickewei, W. M. Schmidt, M. Waldschmidt, Diophantine Approximation. Cetraro, Italy 2000. Editors: F. Amoroso, U. Zannier (2003)

Vol. 1820: F. Hiai, H. Kosaki, Means of Hilbert Space Operators (2003)

Vol. 1821: S. Teufel, Adiabatic Perturbation Theory in Quantum Dynamics (2003)

Vol. 1822: S.-N. Chow, R. Conti, R. Johnson, J. Mallet-Paret, R. Nussbaum, Dynamical Systems. Cetraro, Italy 2000. Editors: J. W. Macki, P. Zecca (2003)

Vol. 1823: A. M. Anile, W. Allegretto, C. Ringhofer, Mathematical Problems in Semiconductor Physics. Cetraro, Italy 1998. Editor: A. M. Anile (2003)

Vol. 1824: J. A. Navarro González, J. B. Sancho de Salas, $\mathscr{C}^\infty$ – Differentiable Spaces (2003)

Vol. 1825: J. H. Bramble, A. Cohen, W. Dahmen, Multiscale Problems and Methods in Numerical Simulations, Martina Franca, Italy 2001. Editor: C. Canuto (2003)

Vol. 1826: K. Dohmen, Improved Bonferroni Inequalities via Abstract Tubes. Inequalities and Identities of Inclusion-Exclusion Type. VIII, 113 p, 2003.

Vol. 1827: K. M. Pilgrim, Combinations of Complex Dynamical Systems. IX, 118 p, 2003.

Vol. 1828: D. J. Green, Gröbner Bases and the Computation of Group Cohomology. XII, 138 p, 2003.

Vol. 1829: E. Altman, B. Gaujal, A. Hordijk, Discrete-Event Control of Stochastic Networks: Multimodularity and Regularity. XIV, 313 p, 2003.

Vol. 1830: M. I. Gil', Operator Functions and Localization of Spectra. XIV, 256 p, 2003.

Vol. 1831: A. Connes, J. Cuntz, E. Guentner, N. Higson, J. E. Kaminker, Noncommutative Geometry, Martina Franca, Italy 2002. Editors: S. Doplicher, L. Longo (2004)

Vol. 1832: J. Azéma, M. Émery, M. Ledoux, M. Yor (Eds.), Séminaire de Probabilités XXXVII (2003)

Vol. 1833: D.-Q. Jiang, M. Qian, M.-P. Qian, Mathematical Theory of Nonequilibrium Steady States. On the Frontier of Probability and Dynamical Systems. IX, 280 p, 2004.

Vol. 1834: Yo. Yomdin, G. Comte, Tame Geometry with Application in Smooth Analysis. VIII, 186 p, 2004.

Vol. 1835: O.T. Izhboldin, B. Kahn, N.A. Karpenko, A. Vishik, Geometric Methods in the Algebraic Theory of Quadratic Forms. Summer School, Lens, 2000. Editor: J.-P. Tignol (2004)

Vol. 1836: C. Năstăsescu, F. Van Oystaeyen, Methods of Graded Rings. XIII, 304 p, 2004.

Vol. 1837: S. Tavaré, O. Zeitouni, Lectures on Probability Theory and Statistics. Ecole d'Eté de Probabilités de Saint-Flour XXXI-2001. Editor: J. Picard (2004)

Vol. 1838: A.J. Ganesh, N.W. O'Connell, D.J. Wischik, Big Queues. XII, 254 p, 2004.

Vol. 1839: R. Gohm, Noncommutative Stationary Processes. VIII, 170 p, 2004.

Vol. 1840: B. Tsirelson, W. Werner, Lectures on Probability Theory and Statistics. Ecole d'Eté de Probabilités de Saint-Flour XXXII-2002. Editor: J. Picard (2004)

Vol. 1841: W. Reichel, Uniqueness Theorems for Variational Problems by the Method of Transformation Groups (2004)

Vol. 1842: T. Johnsen, A.L. Knutsen, K3 Projective Models in Scrolls (2004)

Vol. 1843: B. Jefferies, Spectral Properties of Noncommuting Operators (2004)

Vol. 1844: K.F. Siburg, The Principle of Least Action in Geometry and Dynamics (2004)

Vol. 1845: Min Ho Lee, Mixed Automorphic Forms, Torus Bundles, and Jacobi Forms (2004)

Vol. 1846: H. Ammari, H. Kang, Reconstruction of Small Inhomogeneities from Boundary Measurements (2004)

Vol. 1847: T.R. Bielecki, T. Björk, M. Jeanblanc, M. Rutkowski, J.A. Scheinkman, W. Xiong, Paris-Princeton Lectures on Mathematical Finance 2003 (2004)

Vol. 1848: M. Abate, J. E. Fornaess, X. Huang, J. P. Rosay, A. Tumanov, Real Methods in Complex and CR Geometry, Martina Franca, Italy 2002. Editors: D. Zaitsev, G. Zampieri (2004)

Vol. 1849: Martin L. Brown, Heegner Modules and Elliptic Curves (2004)

Vol. 1850: V. D. Milman, G. Schechtman (Eds.), Geometric Aspects of Functional Analysis. Israel Seminar 2002-2003 (2004)

Vol. 1851: O. Catoni, Statistical Learning Theory and Stochastic Optimization (2004)

Vol. 1852: A.S. Kechris, B.D. Miller, Topics in Orbit Equivalence (2004)

Vol. 1853: Ch. Favre, M. Jonsson, The Valuative Tree (2004)

Vol. 1854: O. Saeki, Topology of Singular Fibers of Differential Maps (2004)

Vol. 1855: G. Da Prato, P.C. Kunstmann, I. Lasiecka, A. Lunardi, R. Schnaubelt, L. Weis, Functional Analytic

Recent Reprints and New Editions

4. Careful preparation of the manuscripts will help keep production time short besides ensuring satisfactory appearance of the finished book in print and online. After acceptance of the manuscript authors will be asked to prepare the final LaTeX source files (and also the corresponding dvi-, pdf- or zipped ps-file) together with the final printout made from these files. The LaTeX source files are essential for producing the full-text online version of the book (see http://www.springerlink.com/openurl.asp?genre=journal&issn=0075-8434 for the existing online volumes of LNM).

   The actual production of a Lecture Notes volume takes approximately 8 weeks.

5. Authors receive a total of 50 free copies of their volume, but no royalties. They are entitled to a discount of 33.3 % on the price of Springer books purchased for their personal use, if ordering directly from Springer.

6. Commitment to publish is made by letter of intent rather than by signing a formal contract. Springer-Verlag secures the copyright for each volume. Authors are free to reuse material contained in their LNM volumes in later publications: A brief written (or e-mail) request for formal permission is sufficient.

Addresses:

Professor J.-M. Morel, CMLA,
École Normale Supérieure de Cachan,
61 Avenue du Président Wilson, 94235 Cachan Cedex, France
E-mail: Jean-Michel.Morel@cmla.ens-cachan.fr

Professor F. Takens, Mathematisch Instituut,
Rijksuniversiteit Groningen, Postbus 800,
9700 AV Groningen, The Netherlands
E-mail: F.Takens@math.rug.nl

Professor B. Teissier, Institut Mathématique de Jussieu,
UMR 7586 du CNRS, Équipe "Géométrie et Dynamique",
175 rue du Chevaleret
75013 Paris, France
E-mail: teissier@math.jussieu.fr

For the "Mathematical Biosciences Subseries" of LNM:

Professor P. K. Maini, Center for Mathematical Biology,
Mathematical Institute, 24-29 St Giles,
Oxford OX1 3LP, UK
E-mail : maini@maths.ox.ac.uk

Springer, Mathematics Editorial, Tiergartenstr. 17,
69121 Heidelberg, Germany,
Tel.: +49 (6221) 487-8410
Fax: +49 (6221) 487-8355
E-mail: lnm@springer-sbm.com